RNA Genetics

Volume I
RNA-Directed
Virus Replication

Editors

Esteban Domingo
Scientist
Instituto de Biologia Molecular
Facultad de Ciencias
Universidad Autonoma de Madrid
Canto Blanco, Madrid, Spain

John J. Holland
Professor
Department of Biology
University of California, San Diego
La Jolla, California

Paul Ahlquist
Associate Professor
Institute for Molecular Virology
and
Department of Plant Pathology
University of Wisconsin-Madison
Madison, Wisconsin

CRC Press
Taylor & Francis Group
Boca Raton London New York

CRC Press is an imprint of the
Taylor & Francis Group, an **informa** business

First published 1988 by CRC Press
Taylor & Francis Group
6000 Broken Sound Parkway NW, Suite 300
Boca Raton, FL 33487-2742

Reissued 2018 by CRC Press

Library of Congress Cataloging-in-Publication Data

RNA genetics.

   Includes biblioigraphies and index.
   Contents: v. 1. RNA-directed virus replication --
v. 2. Retroviruses, viroids, and RNA recombination --
v. 3. Variability of RNA genomes.
   1. Viruses, RNA. 2. Viral genetics. I. Domingo,
Esteban. II. Holland, John J. III. Ahlquist, Paul.
[DNLM: 1. RNA--genetics. QU 58 R6273]
QR395.R57   1988   574.87'3283   84-22432
ISBN 0-8493-6666-6 (vol. 1)
ISBN 0-8493-6667-4 (vol. 2)
ISBN 0-8493-6668-2 (vol. 3)

A Library of Congress record exists under LC control number: 87022432

Publisher's Note
The publisher has gone to great lengths to ensure the quality of this reprint but points out that some imperfections in the original copies may be apparent.

Disclaimer
The publisher has made every effort to trace copyright holders and welcomes correspondence from those they have been unable to contact.

ISBN 13: 978-1-315-89732-5 (hbk)
ISBN 13: 978-1-351-07642-5 (ebk)

Visit the Taylor & Francis Web site at http://www.taylorandfrancis.com and the
CRC Press Web site at http://www.crcpress.com

THE EDITORS

Esteban Domingo was born in Barcelona, Spain (1943), where he attended the University of the city, receiving a B.Sc. in Chemistry (1985) and a Ph.D. in Biochemistry (1969). Subsequently he worked as postdoctoral fellow with Robert C. Warner at the University of California, Irvine (1970 to 1973) and with Charles Weissmann at the University of Zürich (1974 to 1976). Work with Weissmann on phage Qβ led to the measurements of mutation rates and the demostration of extensive genetic heterogeneity in phage populations.

He is presently staff scientist of the Consejo Superior de Investigaciones Científicas at the Virology Unit of the Centro de Biología Molecular and Professor of Molecular Biology at the Universidad Autónoma in Madrid. For the last 10 years his group has contributed to research on the genetics of foot-and-mouth diseases virus (FMDV). They have characterized the extreme genetic heterogeneity of this virus and have established cell lines persistently infected with FMDV. Dr. Domingo has been adivser to international organizations on FMDV and problems of research in biology and medicine. His current interests include the relevance of FMDV variability to viral pathogenesis and the design of new vaccines, subjects on which his group collaborates with several laboratories in Europe and America.

John J. Holland, Ph.D., is Professor of Biology at the University of California at San Diego, La Jolla, California. He received his Ph.D. in Microbiology at U.C.L.A., did postdoctoral work, and was Assistant Professor in the Department of Microbiology, University of Minnesota. He was Assistant Professor and Associate Professor of Microbiology at the University of Washington in Seattle and Professor and Chairman of the Department of Molecular Biology and Biochemistry at the University of California at Irvine before moving to the University of California at San Diego, and spent a sabbatical year as a visiting scientist at the University of Geneva, Switzerland. He has published numerous papers in the field of virology, and presently, he and his colleagues do research on rapid evolution of RNA viruses, virus-immunocyte interactions, defective interfering particles, persistent infections, and related areas.

Paul Ahlquist, Ph.D., is Associate Professor of Molecular Virology and Plant Pathology at the University of Wisconsin, Madison. He is a member of the American Society for Virology and currently serves on the executive committee of the International Committee on Taxonomy of Viruses. His research interests include virus genome structure, organization and evolution, RNA replication and gene expression mechanisms, and viral gene functions.

CONTRIBUTORS

Amiya K. Banerjee, Ph.D.
Chairman
Department of Molecular Biology
Research Institute
The Cleveland Clinic Foundation
Cleveland, Ohio

Christof K. Biebricher, Ph.D.
Research Biochemist
Max-Planck Institut für
Biophysikalische Chemie
Göttingen, West Germany

Gerry W. Both, Ph.D.
Principal Research Scientist
Division of Molecular Biology
CSIRO
North Ryde, New South Wales
Australia

Jeremy Bruenn, Ph.D.
Associate Professor
Department of Biological Sciences
State University of New York at Buffalo
Buffalo, New York

John M. Coffin, Ph.D.
Professor
Cancer Research Center
Tufts University School of Medicine
Boston, Massachusetts

Theo W. Dreher, Ph.D.
Department of Agricultural Chemistry
Oregon State University
Corvallis, Oregon

Rik I. L. Eggen, Ph.D.
Research Assistant
Department of Molecular Biology
Agricultural University
Wageningen, Netherlands

Manfred Eigen, Ph.D.
Professor and Director
Max-Planck Institut für
Biophysikalische Chemie
Göttingen, West Germany

Timothy C. Hall, Ph.D.
Professor
Department of Biology
Texas A&M University
College Station, Texas

Robert M. Krug, Ph.D.
Molecular Biology Program
Memorial Sloan-Kettering Cancer
 Center
New York, New York,

Richard J. Kuhn, B.S.
Graduate Student
Department of Microbiology
School of Medicine
State Unitversity of New York at Stony
 Brook
Stony Brook, New York

Michael M. C. Lai, M.D., Ph.D.
Professor
Department of Microbiology
University of Southern California
 School of Medicine
Los Angeles, California

Paul S. Masters, Ph.D.
Research Scientist
Laboratories for Virology
Wadsworth Center for Laboratories and
 Research
New York State Department of Health
Albany, New York

Bert L. Semler, Ph.D.
Assistant Professor
Department of Microbiology and
 Molecular Genetics
College of Medicine
University of California at Irvine
Irvine, California

Ellen G. Strauss, Ph.D.
Senior Research Associate
Division of Biology
California Institute of Technology
Pasadena, California

James H. Strauss, Ph.D.
Professor
Division of Biology
California Institute of Technology
Pasadena, California

Ab van Kammen, Ph.D.
Professor
Department of Molecular Biology
Agricultural University
Wageningen, Netherlands

Eckard Wimmer, Ph.D.
Professor and Chairman
Department of Microbiology
School of Medicine
State University of New York at Stony
 Brook
Stony Brook, New York

RNA GENETICS

Volume I
RNA-DIRECTED VIRUS REPLICATION

RNA Replication of Positive Strand RNA Viruses
Kinetics of RNA Replication by Qβ Replicase
Replication of the Poliovirus Genome
RNA Replication in Comoviruses
Replication of the RNAs of Alphaviruses and
Flaviviruses
RNA Replication of Brome Mosaic Virus and
Related Viruses
Replication of Coronavirus RNA

RNA Replication of Negative Strand RNA Viruses
Replication of Nonsegmented Negative
Strand RNA Viruses
Influenza Viral RNA Transcription and
Replication

RNA Replication of Double-Stranded RNA Viruses
Replication of the Reoviridae: Information Derived
from Gene Cloning and Expression
Replication of the dsRNA Mycoviruses

Volume II
RETROVIRUSES, VIROIDS, AND RNA RECOMBINATION

Reverse Transcribing Viruses and Retrotransposons
Replication of Retrovirus Genomes
Reverse Transcription in the Plant Virus,
Cauliflower Mosaic Virus
Hepatitis B Virus Replication
Retrotransposons

Replication of Viroids and Satellites
Structure and Function Relationships in
Plant Viroid RNAs
Replication of Small Satellite RNAs and
Viroids: Possible Participation of
Nonenzymic Reactions

Recombination in RNA Genomes
Genetic Recombination in Positive Strand
RNA Viruses
The Generation and Amplification of Defective Interfering RNAs
Deletion Mutants of Double-Stranded RNA Genetic
Elements Found in Plants and Fungi
Evolution of RNA Viruses

Volume III
VARIABILITY OF RNA GENOMES

Genetic Heterogeneity of RNA Genomes
High Error Rates, Population Equilibrium,
and Evolution of RNA Replication Systems
Variability, Mutant Selection, and Mutant
Stability in Plant RNA Viruses
Molecular Genetic Approaches to Replication and
Gene Expression in Brome Mosaic and
Other RNA Viruses
Sequence Variability in Plant Viroid RNAs

**Gene Reassortment and Evolution in
Segmented RNA Viruses**
Genetic Diversity of Mammalian Reoviruses
Influenza Viruses: High Rate of Mutation
and Evolution

Role of Genome Variation in Disease
Antigenic Variation in Influenza Virus Hemagglutinins
Variation of the HIV Genome: Implications
for the Pathogenesis and Prevention of AIDS
Biological and Genomic Variability Among Arenaviruses
Modulation of Viral Plant Diseases by
Secondary RNA Agents
Modulation of Viral Disease Processes by
Defective Interfering Particles

Role of Genome Variation in Virus Evolution
Sequence Space and Quasispecies Distribution

TABLE OF CONTENTS

RNA REPLICATION OF POSITIVE STRAND RNA VIRUSES

Chapter 1
Kinetics of RNA Replication by Qβ Replicase 1
Christof K. Biebricher and Manfred Eigen

Chapter 2
Replication of the Poliovirus Genome 23
Bert L. Semler, Richard J. Kuhn, and Eckard Wimmer

Chapter 3
RNA Replication in Comoviruses 49
Rik Eggen and Ab van Kammen

Chapter 4
Replication of the RNAs of Alphaviruses and Flaviviruses 71
James H. Strauss and Ellen G. Strauss

Chapter 5
RNA Replication of Brome Mosaic Virus and Related Viruses 91
T. W. Dreher and T. C. Hall

Chapter 6
Replication of Coronavirus RNA 115
Michael M. C. Lai

RNA REPLICATION OF NEGATIVE STRAND RNA VIRUSES

Chapter 7
Replication of Nonsegmented Negative Strand RNA Viruses 137
Paul S. Masters and Amiya K. Banerjee

Chapter 8
Influenza Viral RNA Transcription and Replication 159
Robert M. Krug

RNA REPLICATION OF DOUBLE-STRANDED RNA VIRUSES

Chapter 9
Replication of the Reoviridae: Information Derived from Gene Cloning and
Expression 171
G. W. Both

Chapter 10
Replication of dsRNA Mycoviruses 195
Jeremy Bruenn

INDEX 211

RNA Replication of Positive Strand RNA Viruses

Chapter 1

KINETICS OF RNA REPLICATION BY Qβ REPLICASE

Christof K. Biebricher and Manfred Eigen

TABLE OF CONTENTS

I.   Introduction.. 2

II.   The Infection Cycle of RNA Coliphages.. 3

III.   Characterization of the Replication Apparatus 4

IV.   Replication of Viral RNA... 5

V.   Mechanism of RNA Replication 7

VI.   Kinetics of Nucleotide Incorporation... 8

VII.   Interaction of Complementary Strands 11

VIII.   Competition Among RNA Species... 13

IX.   Noninstructed RNA Synthesis 14

X.   Conclusions.. 16

Acknowledgments... 16

Appendix.. 17

References.. 18

I. INTRODUCTION

RNA viruses are unique in having their information encoded in RNA. Their diversity, their widespread occurrence, and host range — prokaryotes, plants, and animals — reflect their evolutionary success. Even though RNA is generally believed to be the older form of information carrier, RNA genomes are extremely rare in cellular organisms — RNA plasmids[1] generally being regarded as inherited relics of viral infections — and so RNA replication does not occur in the normal processing of the genetic information. As a consequence, replication of viral RNA has several unique features.

Besides their extrinsic information — the message encoded by the nucleotide sequence — nucleic acids also possess intrinsic information, the specific folding of the sequence chain due to molecular interactions.[2] The intrinsic information is responsible for the recognition of nucleic acids by the expression machinery; together with recognition boxes, it affects the formatting of the extrinsic information. While the intrinsic information of double-stranded DNA is rather limited, there is an enormous potential for utilizing the intrinsic information of single-stranded RNA up to the sophistication level of enzymic catalysis.[3,4] It seems likely that the replication process makes extensive use of such intrinsic information. Therefore, there is selection pressure for preserving or improving the intrinsic information, as well as the extrinsic information, and both types of information must be reconciled with one another during evolution.

Normally, RNA synthesis in the cell occurs by transcription from the DNA genome. The reading accuracy of nucleotides as dictated by their chemical nature suffices, since errors are neither propagated to other RNA molecules nor transmitted to offspring of the cell. There is therefore no selection pressure favoring error-correcting mechanisms as there is in DNA replication. When RNA itself is replicated, however, error propagation has serious consequences. RNA viruses have to live with this limited replication accuracy; the amount of information they can transmit and thus their chain lengths are constrained by this fact.[5] Other consequences of the limited replication accuracy are rapid genetic drift and high sequence variability of RNA viral genomes[6,7] (RNA variability and its biological implications are reviewed in other areas of this book).

In contrast to cellular genomes, viral genomes are not confined to different compartments after each duplication step. Late in an infection cycle, many copies of the RNA genome are present together with their expression products. In this case, selection pressure acts on the whole genome distribution, since some mutants unable to launch a successful infection on their own can still be amplified in a cell. Such mutants can survive several infection cycles if cells are infected with high multiplicities. Recloning and selecting for the full information of the individual genome is only observed when a single virus invades a new cell. This "swamping" effect also favors genetic variability and allows "hitchhiking" of foreign defective RNA, e.g., of satellite RNA. It is thus no surprise that only a small fraction of intact RNA viruses are infectious.

The RNA in positive-strand viruses has several roles: it acts as messenger for expression of the viral proteins, as template in RNA replication and as core in the assembly of virion particles. At least the first two roles are difficult to reconcile with one another because of the different polarity of the reactions involved: while protein synthesis proceeds in the 5′ to 3′ direction, RNA replication begins at the 3′ end of the template RNA and proceeds towards its 5′ end. The reactions must be regulated in vivo so as to avoid clashes.

The two complementary viral strands found in infected cells may form double-stranded structures by Watson-Crick base pairing. In the first replication round, single-stranded viral RNA has to be transcribed by a viral RNA-dependent RNA polymerase. It is unlikely that one and the same enzyme catalyzes transcription from both single- and double-stranded RNA. Thus, if no other RNA-dependent RNA polymerase activity is present in the cell,

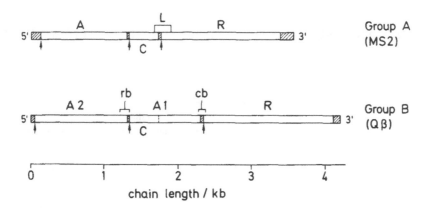

FIGURE 1. Gene maps of leviviruses. Upper: group A (MS2), lower group B (Qβ). Ribosome binding sites are indicated by arrows. rb = replicase binding site (interfering with ribosome binding at C initiation site); cb = coat protein binding site (interfering with ribosome binding at R initiation site). Analogous sites (not shown) are probably also found at group A viruses. The cistrons are listed in Table 1. Note that there are 4 genes, but only 3 ribosome binding sites for group A and group B phages. The lysis gene L of group A phages is translated by a frameshift error in translation of the C gene. C gene translation is terminated by a termination signal and L synthesis begins at the AUG start codon following the termination signal. The A1 gene of group B phages (coding for a phage capsid constituent possibly also involved in host lysis) is translated by occasionally reading through the stop codon of the C gene. Both L and A1 production probably utilize the limited accuracy of the translation machinery; in principle, accidental mutation events in viral RNA replication — e.g., a nucleotide deletion in the C gene or a base substitution in the stop codon — would lead to the same phenomenon. Information compiled from and based on References 21, 23, and 24.

viral RNA replication is unlikely to involve a double-stranded replicative form, except for a short stretch bound to the enzyme at the replication site.

II. THE INFECTION CYCLE OF RNA COLIPHAGES

RNA coliphages are abundant ($> 10^4$ infectious particles/mℓ) in sewage.[8] The discovery of the first RNA phage, f2,[9] aroused great interest, and soon similar phages were detected elsewhere.[10-14] The spread of coliphages over the world and their classification have been extensively investigated. The RNA coliphages have been classified into 2 groups (A and B) and 4 subgroups by serological and biochemical criteria.[15-18]

For about a decade RNA coliphages were a popular and highly successful system for the study of gene expression. The results obtained in that time period are compiled in two books[19,20] and a short overview.[21]

Coliphages, belonging to the family Leviviridae,[22] are small, icosahedral viruses with diameters of about 23 nm containing 1 molecule of single-stranded RNA with a nucleotide chain length of 3500 (group A) or 4200 (group B) coding for 4 genes that suffice for all necessary viral functions. The sequences of phages MS2[23] and Qβ[24] have been determined; they suggest extensive intramolecular base-pairing leading to a flower-like structure.[23] The organization of the phage genome is somewhat different for group A and B (Figure 1). The phage "head" is composed of 180 identical coat proteins (C), some of which are replaced for group B phages by a coat protein (A1) extended at its amino terminus by additional amino acid residues due to reading through a stop signal. The functions of the virion tail (receptor recognition) and of the assembly of mature phages are supplied by the adsorption or maturation protein (A or A2). After entering the cell, the viral RNA is recognized as a messenger by the ribosomes and translated. The product of the replicase gene R performs

Table 1

COMPONENTS AND PRODUCTS OF LEVIVIRUSES FROM *ESCHERICHIA COLI*

	Group					
	A		B			
Subgroup	I	II	III	IV	Number	
Example	MS2	GA	Qβ	SP	per virion	Remarks
RNA						
M_m/Mdalton	1.21	1.20	1.39	1.49	1	Strong secondary structure
L_c	3569 b		4220 b			
Adsorption protein (gene A or A2)						
M_m/kdalton	44.8	44.5	45.0	48.0	1	Necessary for adsorption and maturation
L_c	392 aa		419 aa			
Position	130-1308		62-1321			
Coat protein (gene C)						
M_m/kdalton	14.0	12.9	16.9	17.3	≈180	Hydrophobic, forms icosa-hedral shell
L_c	129 aa		132 aa			
Position	1335-1724		1345-1743			
Readthrough protein (gene A1)						
M_m/kdalton	—	—	38.5	39.0	≈7	Produced by erroneous reading through a stop codon
L_c			328 aa			
Position			1345-2334			
Replicase subunit (gene R)						
M_m/kdalton	60.0	57.0	65.0	?	0	Complexes with S1 and EF TuTs from host
L_c	544 aa		589 aa			
Position	1761-3395		2353—4119			
Lysis protein (gene L)						
L_c	74 aa		—		0	Produced by frame shift
Position	1678-1902					

Note: L_c = chain length (b = bases, aa = amino acids), M_m = molecular mass. Data compiled mainly from References 17 and 24.

in collaboration with a few host proteins all processes necessary in replication. Late in the infection cycle, translation must compete with replication and packaging of the RNA. The gene product of gene L, which is read in a different frame, is necessary at group A phages for lysis of the bacteria.[25,26] In group B phages, the maturation protein is responsible for triggering cell lysis.[27] The time-dependent regulation of synthesis of the different gene products is highly sophisticated and effective.[19,26] It makes extensive use of the secondary and tertiary structure of the RNA and its dynamic changes during translation and replication (Table 1, Figure 2).

III. CHARACTERIZATION OF THE REPLICATION APPARATUS

Extracts of RNA phage-infected *Escherichia coli* cells were found to contain an RNA-dependent RNA polymerase activity absent in uninfected cells.[28] Phage replicases containing the gene product of the R gene and 2 to 3 subunits from the host were subsequently isolated in high purity (Table 2).[29-31] In contrast to most other replicases,[32] the replicase of phage Qβ is remarkably stable and has been thus investigated most extensively. The subunits contributed by the host are the elongation factors of protein biosynthesis (EF)Tu and EFTs[33] and the ribosomal protein S1.[34] Replicase without the latter subunit (core enzyme) is unable to replicate viral plus strand, but performs all other reactions normally.[35] Replication of the viral strand also requires the participation of an additional host factor.[36] Both S1[37] and host factor[38] are RNA-binding proteins; their biological role is probably to avoid binding of viral RNA engaged in protein biosynthesis (Figure 2). The replication apparatus of other RNA coliphages seem to be analogous,[32,39] but host factors may differ for different phage groups.[40] Replicase (and host factor) amplify in vitro infectious viral RNA autocatalytically.[41] The role of the different subunits is not fully understood;[42] the viral subunit appears to provide the active center for phosphodiester formation, while EFTu and EFTs are involved in the initiation reaction.

IV. REPLICATION OF VIRAL RNA

After phenol extraction of virus-infected cells, three forms of viral RNA are detected: single-stranded plus strands, a double-stranded "Hofschneider-structure",[43] also called "replicative form (RF)", and a partially double-stranded "Franklin-structure",[44] called "replicative intermediate (RI)". However, when proteins were not removed, intracellular RNA was found to be RNase sensitive,[45] suggesting that the double strands were isolation artifacts. RF and RI are only infectious after melting into single strands. Replication thus requires a single-stranded template, and yields a complementary replica and the template as single strands.[46,47]

The mechanism of viral RNA replication is complicated in vivo by the fact that the viral RNA strand must also be expressed to provide replicase and other gene products, and is eventually packaged. The interference by processes competing with replication for the RNA can be avoided by investigating the mechanism of RNA replication in vitro. The system requires the nucleoside triphosphates ATP, CTP, GTP, UTP, Mg^{2+} ions, single-stranded template RNA, the enzyme Qβ replicase, and the host factor. The first replication round produces a "negative" replica strand with the complementary sequence. The negative strand is used by the replicase as template for the production of positive strand, and thus two cross-catalytical cycles are observed (Figures 2 to 4).[21]

Qβ replicase is quite specific in its choice of templates: in vivo, it only accepts Qβ RNA itself and its complementary strand.[48,49] In vitro, other templates have also been found to direct synthesis of their complementary strands (transcription) by Qβ replicase; however, they can not replicate autocatalytically (for reviews see References 42, 49, and 50). There is, however, a class of short-chain, self-replicating RNAs which can be autocatalytically amplified, the so-called midi-,[51] mini-,[52,53] micro-,[54] and nano-variants.[55] The name "variant" is misleading; since their sequences are not sufficiently related to each other or to Qβ RNA, itself, to suggest descendence from a common ancestor, they constitute different RNA species. They originate in so-called "template-free" incorporations by a *de novo* synthesis process not yet fully understood (Section IX).[56-60] For their effective replication, core replicase suffices. They resemble a heterogenous, self-replicating RNA class sedimenting with 6 S found in Qβ-infected cells late in the infection cycle. These short RNA species are especially suitable for studying the principles of RNA replication (Table 3).[50]

The nucleotide sequences of self-replicating RNA species have a C-cluster at the 3' end (and consequently a G-cluster at the 5' end) in common;[55] its presence is necessary, but not sufficient for replication. The extrinsic information, however, is not sufficient for providing the ability to replicate. In one species the folding of the sequence produced by replication was metastable and could refold into a more stable structure; only the metastable form was found to replicate.[61] Comparison of the electro-optic properties of replicating species also suggested structural similarities.[61]

The tertiary structure of the RNA (the intrinsic information) is thus crucial for replication

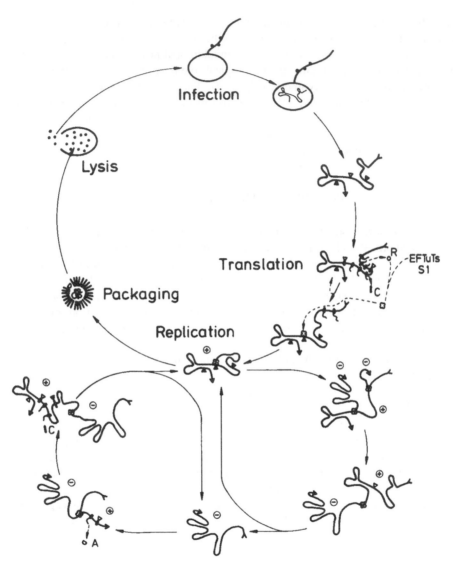

FIGURE 2.   Schematic model of a levivirus infection cycle. 🟐 Ribosomes, ⬭ replicase, ← 5′ end, ⤙ 3′ end, ∇ ribosome recognition site available, ▼ unavailable. Free viruses adsorb to sex pili with the help of the A or A2 cistron. The RNA enters the bacterial cytoplasm together with the A (A2) protein, while the empty shell remains outside. After proteolytic degradation of the adsorption protein, ribosomes bind to the C ribosome recognition site to initiate C gene translation. During translation of the C gene, the structure of RNA is altered to render the ribosome recognition site of the R gene available, which is then translated. The R gene product complexes with EF TuTs and ribosomal protein S1 from the host to form replicase. Replicase can not act as long as translation is going on. If sufficient replicase is present to compete with ribosomes, it binds to the rb site, blocking the ribosome recognition site of the C gene. Ribosomes already engaged in synthesis complete their job and fall off. The free 3′ end is then able to thread into the active site of the replicase and minus strand synthesis is initiated. (A more detailed scheme of the replication cycle itself is shown in Figure 3). More than a single replicase molecule may act on one template RNA strand, resulting in a Franklin structure (RI) after deproteination. After completion of the replica strand, the replicase has reached the 5′ end of the template. Deproteination of this complex results in a double-stranded Hofschneider structure (RF). The replication round is completed with the liberation of the replica minus strand, and then the template plus strand from the replicase. Replication of the minus strand occurs analogously. In newly synthesized plus strands the ribosome binding site for the A gene is available and A gene (with progressing replica chain length also C and R gene) translation occurs. Increasing amounts of C protein inhibit R gene translation by binding to the cb site. C protein and A protein binding is the start also for the spontaneous assembly of the phage capsid. Lysis of the cell requires in group A phages translation of the L gene product which is a side reaction of C translation (see Figure 1). Lysing bacteria liberate some 1000 phage particles each, only 5 to 20% of which are infectious. The model shown follows proposals originally made by Weissmann.[21]

Table 2
POLYPEPTIDES OF Qβ REPLICATION APPARATUS[42]

Polypeptide	M_m/kdalton	Origin	Possible role
Beta	65	Virus	Phosphodiester formation, specificity
EFTu	45	Host	Initiation, binding of 5′ terminal GTP
EFTs	35	Host	Must be complexed to EFTu
S1	70	Host	Binding of special site of plus strand
Host factor	6×15	Host	RNA binding protein, role unclear

Note: The first polypeptides complex to form core replicase. The complex between core replicase and ribosomal protein S1 is called holo replicase. Core replicase is sufficient for all reactions except replication of viral plus strand RNA for which holo replicase and host factor are required.

to occur. Therefore, the template can not be considered merely as a substrate: it shares the catalytic functions in the active replication complex with the replicase. (RNA is known to participate in, or even to effect, other enzymic reactions, most often those involved in its own processing,[3,4,62] in a few cases, also the processing of other RNA.[2,3])

V. MECHANISM OF RNA REPLICATION

Kinetic studies showed that the template specificity of Qβ replicase is controlled kinetically by the lifetimes of the initiation complexes.[63] In a first step, replicase is bound to template, apparently at an interior site.[64-68] At least 2 GTP molecules then must be bound at the 3′ end of the template. One of the GTP molecules is bound by the active site of the viral subunit of the replicase; the 5-terminal GTP is perhaps bound to the EFTu subunit.[42,69] Priming of the replication process then occurs by phosphodiester formation and pyrophosphate release. Chain elongation is performed by consecutive association of the complementary bases followed by phosphodiester formation and pyrophosphate release. Since Watson-Crick base-pairing is the basis of copying, a double-helical structure is involved at the replication fork. As replication proceeds, the strands are released from the enzyme in single-stranded form (Figures 3 and 4). The considerable energy for melting the transient double helix must be provided by the replication process itself. Reformation of a double helix between replica and template tails during replication is suppressed by formation of rigid intramolecular RNA structures. All replicating RNA species investigated exhibit remarkable structural stability;[61] however, introduction of an unstructured oligo (A) cluster into the sequences of Qβ RNA[70] or MDV-1[71] by RNA recombination techniques is apparently tolerated for replication. Forced removal of the enzyme from the replication fork (deproteination) converts the partially single-stranded structure immediately to the fully paired structure.[21] The elongation rate differs strongly from position to position, presumably because of folding and refolding of the single-stranded template and replica. Chain elongation pauses at specific positions where the normal elongation rates (approximately 100 bases per second) drops about two orders of magnitude.[72] The completed replica chain is adenylated and then released;[50,73,74] the resulting inactive enzyme-template complex eventually dissociates slowly to recycle template and enzyme (Figures 3 and 4).[58,75]

VI. KINETICS OF NUCLEOTIDE INCORPORATION

Kinetic profiles can be determined conveniently by measuring the incorporation rate of radioactive triphosphates into RNA.[58] The profiles (Figure 5) reveal two main replication phases:

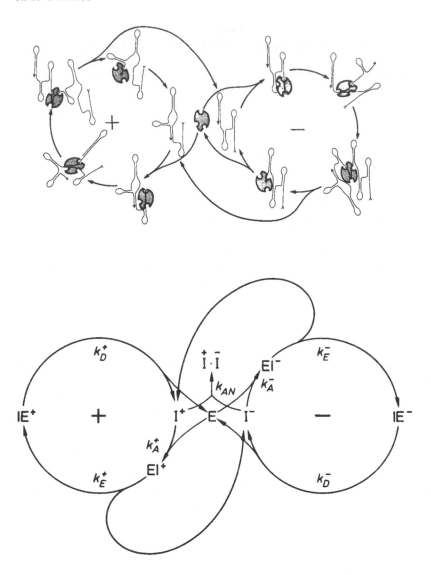

FIGURE 3. Schematic model of RNA replication. Enzyme having one catalytic site for phosphodiester formation and binding sites for the template and replica strands binds to the 3′ end of a template. A replica chain is initiated by phosphodiester formation between two complexed GTP molecules. The replica is elongated and the growing replica and the already-read tail of the template are released from different sites as single strands. After the replica chain is completed it is adenylated and released. The resulting inactive template-enzyme complex dissociates. Above: artists view of the replication mechanism. Below: simplified replication mechanism.

1. An exponential growth phase where enzyme is in excess and both replica and template can start further replication rounds.
2. A linear growth phase where enzyme is saturated with template and a further increase of the replication rate by newly synthesized RNA is not possible. Finally, the incorporation levels off, mainly due to product inhibition.

Determination of overall replication rates in both growth phases showed about a threefold higher rate constant value in the exponential phase for all species examined. This cannot be attributed to multiple replication sites on one template strand, because short-chained RNA

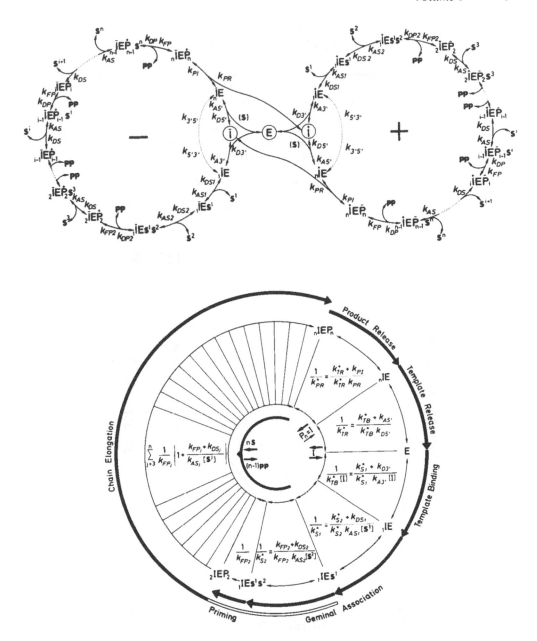

FIGURE 4. Minimal complete replication mechanism. Above: intermediates and rates. Below: steady-state calculation giving retention times for the elementary steps, valid for the linear growth phase. Consult Reference 75 for explanation of symbols and standard rate values. (From Biebricher, C. K., Eigen, M., and Gardiner, W. C., *Biochemistry*, 22, 2544, 1983. With permission. Copyright 1983 American Chemical Society.)

species cannot be occupied by more than one replicase molecule at one time, in contrast to the much longer viral RNA. Instead, this observation can only be accounted for as a consequence of the growth rate in the linear phase, being mainly controlled by the low rate of enzyme recycling after completion and release of the replica, while in the exponential growth phase, the released replica can start a new replication cycle immediately with excess enzyme.

A minimal replication mechanism has been composed from the kinetically separable steps described in Section V (Figure 4). It consists of two crosscatalytic cycles, one each for the

Table 3
TEMPLATES FOR Qβ REPLICASE[46]

RNA species	Replication	Remarks	Origin
Qβ RNA	Autocatalytic	Needs host factor to accept plus strand	Virus
Poly(C)	Transcription	Large excess of template necessary	Synthetic
Poly(NC)	Transcription	Large excess of template necessary	Synthetic
mRNA, rRNA	Transcription(?)	Mn^{2+}, high GTP, primer required; inefficient	Cellular
6 S RNA	Autocatalytic	Found in infected cells only, heterogeneous	*De novo*(?)
"Variants"	Autocatalytic	Produced in vitro by template-free synthesis	*De novo*

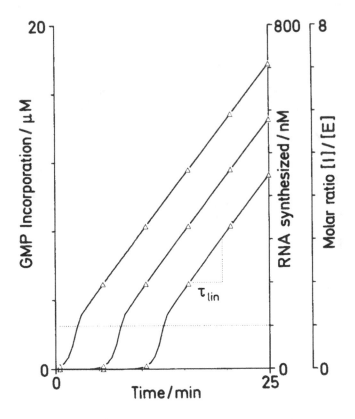

FIGURE 5. Determination of replication rates from incorporation profiles measured at different dilutions of the template input. The rate in the linear phase is calculated from the slope as moles of RNA synthesized per mole of enzyme. The overall incorporation rate in the exponential phase is determined from the time lag caused by the dilution of the template. (From Biebricher, C. K., *Chem. Scr.,* 26B, 57, 1986. With permission.)

plus and minus strands.[75] The rate constant values for some elementary steps can be measured, and for the others, reasonable estimates can be made. It is thus possible to simulate the kinetic behavior of RNA replication by numerical integration of the system of differential equations on a computer. The computed profiles show excellent agreement with the experimentally determined incorporation profiles, and the mechanism was also compatible otherwise with all experimental results. Computed profiles give useful insights into the influence of critical parameters for replication whose variation by experimental methods is impossible.[75]

In the exponential growth phase, all replication intermediates grow coherently with the same growth rate; in the linear growth phase, their concentrations reach steady-state values. On that basis, the rate equations of the replication mechanism can also be solved analytically (Figure 4). The rate equations derived describe the replication behavior in both phases correctly. For many purposes, simplified mechanisms (Figure 3) suffice to provide correct results and give compact equations (Appendix).[63,76,77]

The experimental incorporation rates in both phases show a Michaelis-Menten dependence upon substrate concentration;[58] the analytical solutions give also the same mathematical form, although the Michaelis constant and the turnover rate are complicated functions of the elementary reaction rates.[75] The overall replication rate in the linear phase was found experimentally to be independent of enzyme concentration in the range from 50 to 500 nM; above 100 nM, the exponential growth rate was also independent of enzyme concentration. Binding of enzyme to efficiently replicating RNA templates thus must occur rapidly and does not influence the replication rates at high enough enzyme concentrations.

The concentration of the intermediates is directly proportional to the retention times for the pertaining step, i.e., inversely proportional to the rate values for the elementary steps. In the linear growth phase, the majority of the enzyme is idling, awaiting the slow dissociation of the replicase-template complex. In the exponential growth phase, the proportion of template-bound enzyme actively engaged in replication is higher (Figure 6).

VII. INTERACTION OF COMPLEMENTARY STRANDS

In the exponential growth phase, almost all of the RNA is complexed to enzyme, and synthesized RNA is predominantly represented as enzyme-bound, single-stranded template. In the linear growth phase, nucleotide incorporation produces mainly free, single-stranded template strands.[63] However, free complementary strands may combine to form double strands, and their template activity is irreversibly lost.[61] The loss remains undetected in incorporation measurements, but is readily noticed by product analysis. Electrophoresis of replicating RNA species reveals bands for double- and single-stranded RNA; their proportion varies considerably from species to species (Figure 7). The exceptionally high proportion of single strands in MDV-1 RNA is remarkable. In contrast, with species with chain lengths less than 100, few single strands are detected. For each species a definite steady-state template concentration is established where production of single strands is balanced by double-strand formation.[77]

The kinetic model can be expanded simply by the addition of the double-strand formation step. Again, simple equations can be found quantitatively describing the steady-state concentrations. For the species we investigated, including those producing predominantly double-stranded RNA, the rate of double-strand formation between template and replica during replication itself was found to be negligible: pulse labeling showed single-strand synthesis in the steady-state of double strand, and mutant spectra of RNA species produced hybrid double strands.[78] If the experimentally found reversible binding of enzyme to double-stranded RNA is also taken into account, the simulated incorporation profiles confirm a marked product inhibition influence upon the experimentally determined profiles.

Double-strand formation causes nucleotide incorporation rates into plus and minus strands to be almost identical, even though the specific growth rates may be different in the plus and minus cycle. Asymmetries in the rate constant values of crucial replication steps result in unequal occupation of the enzyme, with plus and minus template intermediate concentrations adjusting themselves so as to balance slower replication rates with higher shares of enzyme. The overall growth rate of an RNA species is a modified harmonic mean of the rates of the complementary cycles; thus, there is a strong selection pressure against extreme

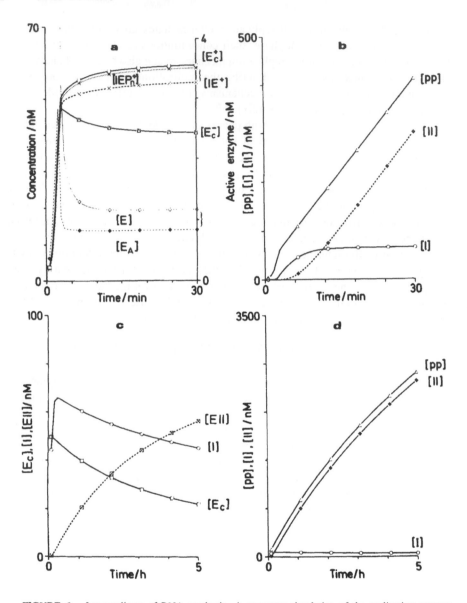

FIGURE 6.   Intermediates of RNA synthesis. A computer simulation of the replication process with standard rate constant values is shown, except that an asymmetry was introduced by chosing $k^+_{DS'} = 4 \times 10^{-3}$ s^{-1}, $k^-_{DS'} = 6 \times 10^{-3}$ s^{-1}. (a) Replication intermediates: note the coherent growth of the intermediates in the exponential phase (<4 min). In the linear phase, the intermediates reach steady state where the majority of the template-bound enzyme awaits template release (IE$^+$) or replica release (IEP$_n^+$) and is thus idling. Only a small part of the enzyme is actively engaged in synthesis (right scale, E$_A$) or free (E). Note the resulting asymmetry in E$_c^+$ and E$_c^-$. (b) Replication products: in the exponential phase, incorporation (pp) results mainly in the production of enzyme-bound RNA, while in the early linear phase (4 to 7 min), free template (I) is the main product. Later, single-stranded template reaches steady state and double strand (II) is the only product. (c) and (d) Product inhibition: after prolonged synthesis, the large concentration of double strand (II) competes with template (I) for the enzyme (EII, E$_c$), (c) resulting in a drop of enzyme-bound (E$_c$) and free (I) template and causing product inhibition of the incorporation profiles, seen in (d) as curvature. Other product inhibition terms[75] are not shown here. (From Biebricher, C. K., *Chem. Scr.*, 26B, 57, 1986. With permission.)

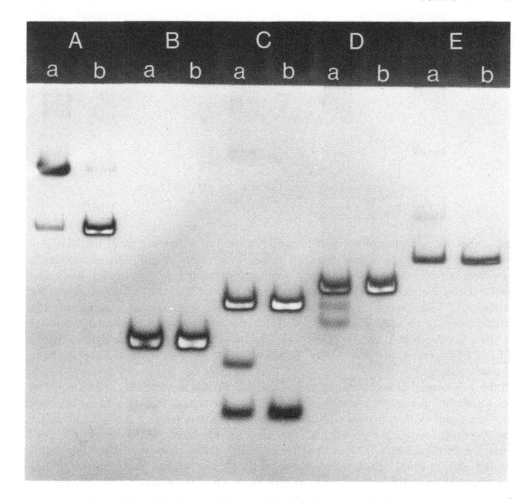

FIGURE 7.   Single- and double-stranded RNA formed by replication. Incorporation mixtures of different species (A—E) were applied to an electrophoresis gel after replication (a) directly, (b) after heating for 1 min at 100°C. The RNA species were A: MDV-1; B: MNV-11; C: SV-11; D: SV-5; E: SV-7. The gel was stained with acridin orange giving a bright green fluorescence with double-stranded (bright bands) and red fluorescence (dark bands) with single-stranded RNA. Note the large amount of single-stranded MDV-1 (upper band) while MNV-11 shows only small amounts of plus and minus strands (lower bands). The sample of SV-11 contains two single-stranded RNAs, a metastable RNA active as template (medium band), and a stable RNA inactive as template (lower band). (Data obtained from Biebricher, C. K., Diekmann, S., and Luce, R., *J. Mol. Biol.*, 154, 629, 1982.)

asymmetries.[77] Asymmetry and double-strand formation in viral replication has not yet been studied in detail. Viral RNA is also highly structured;[23] furthermore, the high sequence complexity (chain length) leads one to expect a low rate of double-strand formation.[79] In vivo, synthesis of viral strand is strongly favored over minus strand production, partly because a large fraction of the plus strands are occupied by ribosomes or packaged, thus being unavailable as templates, and partly because both in vivo and in vitro, host factor is often present in suboptimal amounts.[80]

VIII. COMPETITION AMONG RNA SPECIES

The RNA replication system of Qβ offers an opportunity for the study of evolution in the test tube, based on mutation and selection.[81,82] This aspect will be discussed in the context of the RNA quasispecies in Chapter 12 of Volume III; here, we shall restrict our analysis

Table 4
REPLICATION RATES AND RETENTION
TIMES IN THE EXPONENTIAL AND LINEAR
GROWTH PHASE[76]

	κ $/10^{-2}s^{-1}$	ρ $/10^{-3}s^{-1}$	τ_{exp} $/s$	τ_{lin} $/s$	τ_D $/s$	τ_E $/s$
MNV-11	1.92	6.37	52	157	143	14
MDV-1	1.42	3.60	70	278	263	15

Note: Conditions: template as indicated, 500 μM [NTP] (each),
120 nM holo Qβ replicase. The τ-values are retention times
(e.g., $\tau_E = 1/k_E$).

to the competition kinetics of given different RNA species. Different species replicating in the same test tube do not interfere directly with each other, since no diffusible gene products are formed, but they do share the same environment, including replicase, triphosphates, and growth factors. For computer simulations and analytical treatments of growth kinetics, the assumption of independent growth cycles sharing enzyme and substrate is sufficient. The selective values of classical population genetics do not apply; instead, a reasonable description for selective value of a species was found to be its relative population change $(1/[^1l_o])d[^1l_o]/dt$; species with the highest selective values are enriched in the population in all growth phases.[76,63]

In the exponential phase, where substrate and enzyme are present in large excess, RNA species grow independently, just as they would in separate compartments, and their selection values are equal to their overall replication rates. The fastest growing species thus enrich themselves relatively in the population. This situation has been often investigated experimentally and theoretically.[5,76,82] In the linear growth phase, however, when enzyme is saturated with template, all species present must compete for the small amount of free enzyme and cannot grow independently of each other. Overall replication rate and selection values are not correlated. Indeed, experiments have shown that often the more slowly growing species is selected (Table 4). Crucial for selection in the linear phase are the rates of template binding and double-strand formation.[50,76] Finally, a stable coexistence is reached where both species reach steady-state concentrations (Figure 8, Appendix).

IX. NONINSTRUCTED RNA SYNTHESIS

When high concentrations of highly purified Qβ replicase are added to incorporation mixtures in the absence of template, after long lag periods a sudden outgrowth of self-replicating RNA is observed.[57,58,60] The lag times scatter and are much longer than required for amplifying a single template strand to macroscopic appearance. The resulting RNA products differ from experiment to experiment in sequence, chain length, and growth rates,[53,58] and cannot have been present *ab initio* in the reaction mixture, e.g., as impurities in the enzyme preparation (*de novo* synthesis).[59,60] At low enzyme or substrate concentrations, or at higher ionic strengths or low substrate concentrations, no *de novo* synthesis takes place and RNA molecules can be cloned with Qβ replicase. The kinetic behavior of template-instructed and *de novo* synthesis is clearly different (Figure 9).[58]

Other DNA and RNA polymerases are also capable of *de novo* synthesis.[83] Their mechanism is probably analogous to that of Qβ-replicase. Two phases of *de novo* synthesis can be distinguished. In the first one, triphosphates are condensed slowly to more or less random oligonucleotides.[60,83] Once a self-replicating molecule is formed by chance, it is rapidly amplified, undergoing continuous evolution, to macroscopic levels. The formation of a

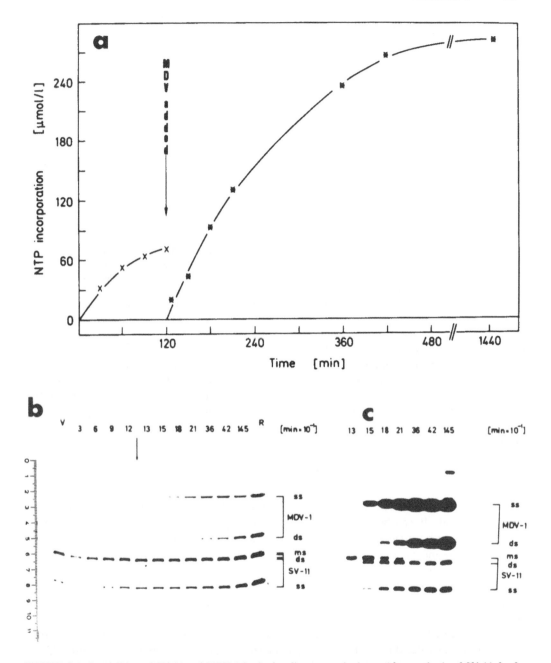

FIGURE 8. Competition of SV-11 and MDV-1 in the late linear growth phase. After synthesis of SV-11 for 2 hr in the steady state of double strand formation, MDV-1 RNA strands equivalent to 1/100,000 of the input enzyme molecules and [^{32}P]-labeled GTP were added and the incubation continued. The incorporation profile was determined (a) and the competition of the species was analyzed by separating the products on gel electrophoresis (b,c). The RNA bands were located by staining (b) and autoradiography (c). 10 min after addition of MDV-1, the active template of SV-11 (ms) is still the only labeled band, while still 10 min later, MDV-1 already dominates. After longer incorporation, the active MDV-1 template (ss) is abundant while the active SV-11 template is no longer detectable. (From Biebricher, C. K., in *Evolutionary Biology*, Vol. 16, Hecht, M. K., Wallace, B., and Prance, C. T., Eds., Plenum Press, New York, 1983, 1.)

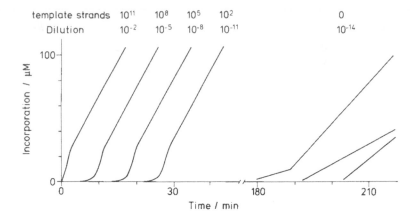

FIGURE 9.   Incorporation profiles obtained by serial dilution of RNA templates. MNV-11 (10[11] strands per microliter) was diluted serially in 1000-fold steps. The four first dilutions still contain template strands; the incorporation profiles and products are reproducible. The two last dilutions no longer contain template strands: RNA is synthesized only after long lag times; the incorporation profiles and products are irreproducible. Alterations of RNA products and replication rates are also observed during growth. (Note the kinks found in the profiles).

molecule which can be replicated is a very rare and nonreproducible event. The early products are probably ineffective templates; rapid optimization during amplification by mutation and selection is observed.[53,58] The RNA species obtained are adapted to the specific conditions chosen, which may be quite exotic ones.[57] Noninstructed nucleotide incorporation proceeds at a rate of about five orders of magnitude lower than template-instructed incorporation;[60] it is thus probably a side reaction of the replicase enzyme with an unknown — if any — biological role.

X. CONCLUSIONS

The replication mechanism of short-chained self-replicating RNA by RNA replicase is well understood. It can be quantitatively described by compact mathematical equations. There is no reason for suspecting essential differences in the replication mechanism of viral RNA itself; the mechanism is somewhat more complicated by the involvement of the host factor, the function of which is still unclear, and the possibility of multiple replication forks on one template strand. In vivo, the multiple role of the plus strands in replication, translation, and packaging complicates the reactions to a level of complexity which is not yet accessible to kinetic studies.

The competition of several self-replicating RNA species is also well understood. Selection values can be calculated from the kinetic parameters of each of the competing species. The determination of selection values can also be extended to cases where the different species are competing mutants of a quasispecies distribution. Thus, studies of self-replicating RNA species are expected to contribute to understanding the genetic variability of RNA viruses, discussed in several articles in this book.

ACKNOWLEDGMENT

We are indebted to Dr. W. C. Gardiner, Jr. for many suggestions and critical reading of the manuscript.

APPENDIX

Analytical Treatment of Replication Kinetics

Definitions:

$[E_o]$ = total concentration of enzyme (complexed and free)

$[I_o]$ = total concentration of RNA (complexed and free)

$[I_i]$ = any replication intermediate

Equilibrated exponential growth phase: $[I_o] \ll [E_o]$

Coherent exponential growth of all intermediates $(d[I_i]/[I_i]dt = \kappa)$

Equilibrated linear growth phase: $[I_o] > [E_o]$

Steady state for replication intermediates $(d[I_i]/dt = 0)$

Simplified Palindromic Three-Step Mechanism (Figure 4)

Definitions:

$$[E_c] = [EI] + [IE]$$

$$[I_o] = [I] + [E_c]$$

$$[E_o] = [E] + [E_c]$$

Differential equations:

$$d[I]/dt = -k_A[E][I] + k_D[IE] + k_E[EI]$$

$$d[EI]/dt = k_A[E][I] - k_E[EI]$$

$$d[IE]/dt = k_E[EI] - k_D[IE]$$

Experimentally found for $[E_o] > 100$ nM: $k_A[E] > k_E > k_D$ $(500, 60, 6 \times 10^{-3} \text{ s}^{-1})$

Linear growth phase:

$$\rho = v_{max}/[E_0] = k_E k_D/(k_E + k_D)$$

Exponential growth phase:

$$\kappa^3 + (k_A[E] + k_E + k_D)\kappa^2 + k_D(k_A[E] + k_E)\kappa - k_A[E]k_E k_D = 0$$

$$\kappa = k_D/2\{[1 + 4k_E/k_D]^{1/2} - 1\} \approx (k_E k_D)^{1/2}$$

Double Strand Formation

In the linear phase, the loss term for template by double strand formation $d[II]/dt = k_{ds}[I^+][I^-]$ $= 1/4k_{ds}[I]^2$ (for the palindromic case) must be considered, where $[II]$ is the concentration of double strand. A steady state for the free template is reached where

$$[I] = (2v/k_{ds})^{1/2}$$

$$\text{where} \quad v = \rho[E_c]$$

Only for double stranded RNA is a net increase observed with

$$d[II]/dt = v/2$$

Competition Between Different Species

Case A $[^1I], [^2I] \ll [E_o]$. Double strand formation can be neglected.

$$\frac{[^1I]}{[^2I]^t} = \frac{[^1I]}{[^2I]^{t=0}} \, e^{(^1\kappa - ^2\kappa)t}$$

Case B: $[^1I] \ll [E_o] < [^2I]$, $k_{hy} \ll k_{ds}$. Species 1 grows exponentially with the rate 1_κ as it would in the absence of the other species at the remaining steady-state free enzyme concentration. Hybridization considered: exponential growth with the rate $(1_\kappa - k_{hy}[^2I])$.

Case C: $[^1I], [^2I] > [E_o]$
Steady state conditions in the linear growth phase:

$$\frac{[^1I]}{[^2I]} = \frac{^1k_A^2 k_{ds} - ^2k_A k_{hy}}{^2k_A^1 k_{ds} - ^1k_A k_{hy}}$$

Both, numerator and denominator positive: both species coexist.
Numerator negative, denominator positive: species 1 dies out.
Numerator positive, denominator negative: species 2 dies out.
Numerator negative, denominator negative: one of the species dies out, depending on initial conditions.
$k_{hy} \ll k_{ds}$ favors coexistence, $k_{hy} \cong k_{ds}$ favors extinction.

REFERENCES

1. **Wickner, R. B.**, Double-stranded RNA replication in yeast: the killer system, *Annu. Rev. Biochem.*, 55, 373, 1986.
2. **Dickerson, R. F., Kopka, M. L., and Pjura, P.**, Pathways of information readout in DNA, *Chem. Scr.*, 26B, 139, 1986.
3. **Guerrier-Takada, C., McClain, W. H., and Altman, S.**, Cleavage of tRNA precursors by the RNA subunit of E. coli ribonuclease P (M1 RNA) is influenced by 3'-proximal CCA in the substrates, *Cell*, 38, 219, 1984.
4. **Cech, T. R. and Bass, B. L.**, Biological catalysis by RNA, *Annu. Rev. Biochem.*, 55, 599, 1986.
5. **Eigen, M. and Schuster, P.**, The hypercycle — a principle of natural self-organization. Part A. Emergence of the hypercycle, *Naturwissenschaften*, 64, 541, 1977.
6. **Holland, J., Spindler, K., Horodyski, F., Grabau, E., Nichol, S., and Van de Pol, S.**, Rapid evolution of RNA genomes, *Science*, 215, 1577, 1982.
7. **Domingo, E., Martínez-Salas, E., Sobrino, F., de la Torre, J. C., Portela, A., Ortín, J., López-Galindez, C., Pérez-Breña, Villanueva, N., Nájera, R., Van de Pol, S., Steinhauer, D., DePolo, N., and Holland, J.**, The quasispecies (extremely heterogeneous) nature of viral genome populations: biological relevance — a review, *Gene*, 40, 1, 1985.
8. **Havelaar, A. H., Hogeboom, W. M., and Pot, R. F.**, Specific RNA bacteriophages in sewage: methodology and occurrence, *Water Sci. Technol.*, 17, 645, 1984.
9. **Loeb, T. and Zinder, N. D.**, A bacteriophage containing RNA, *Proc. Natl. Acad. Sci. U.S.A.*, 47, 282, 1961.
10. **Davis, J. E., Strauss, J. H., and Sinsheimer, R. L.**, Bacteriophage MS2: another RNA phage, *Science*, 134, 1427, 1961.
11. **Paranchych, W. and Graham, A. F.**, Isolation and properties of an RNA-containing bacteriophage, *J. Cell. Comp. Physiol.*, 60, 199, 1962.

12. **Davern, C. I.**, The isolation and characterization of an RNA bacteriophage, *Aust. J. Biol. Sci.*, 17, 719, 1964.

13. **Hofschneider, P. H.**, Untersuchungen über kleine *E. coli* K12 Bakteriophagen, *Z. Naturforsch.*, 18b, 203, 1963.

14. **Marvin, D. A. and Hoffmann-Berling, H.**, Physical and chemical properties of two new small bacteriophages, *Nature (London)*, 197, 517, 1963.

15. **Watanabe, I., Miyake, T., Sakurai, T., Shiba, T., and Ohno, T.**, Isolation and grouping of RNA phages, *Proc. Jpn. Acad.*, 43, 204, 1967.

16. **Miyake, T., Haruna, I., Shiba, T., Itoh, Y. H., Yamane, K., and Watanabe, I.**, Grouping of RNA phages based on the template specificity of their RNA replicases, *Proc. Natl. Acad. Sci. U.S.A.*, 68, 2022, 1971.

17. **Furuse, K., Hirashima, A., Harigai, H., Ando, K., Watanabe, K., Kurusawa, K., Inokuchi, Y., and Watanabe, I.**, Grouping of RNA coliphages based on analysis of the sizes of their RNAs and proteins, *Virology*, 97, 328, 1979.

18. **Yonesaki, T., Furuse, K., Haruna, I., and Watanabe, I.**, Relationships among four groups of RNA coliphages based on the template specificity of GA replicase, *Virology*, 116, 379, 1982.

19. **Zinder, N. D., Ed.**, *RNA Phages*, Cold Spring Harbor Laboratory, Cold Spring Harbor, New York, 1975.

20. **Gren, E. A.**, *Regulatory Mechanisms of Replication of the RNA Bacteriophages*, Academy of Sciences, Latvian S.S.R., Riga, U.S.S.R., 1974 (in Russian).

21. **Weissmann, C.**, The making of a phage, *FEBS Lett. (Suppl.)*, 40, S10, 1974.

22. **Mattheus, R. E. F.**, Leviviridae, *Intervirology*, 17, 136, 1982.

23. **Fiers, W., Contreras, R., Duerinck, F., Haegemann, G., Iserentant, D., Merregaert, J., Min Jou, W., Molemans, F., Raeymakers, A., Van den Berghe, A., Volckaert, G., and Isebaert, M.**, Complete nucleotide sequence of bacteriophage MS2 RNA: primary and secondary structure of the replicase gene, *Nature (London)*, 260, 500, 1976.

24. **Mekler, P.**, Determination of Nucleotide Sequences of the Bacteriophage Qβ Genome: Organization and Evolution of an RNA Virus, Ph.D. thesis, University of Zurich, Switzerland, 1983.

25. **Beremand, M. N. and Blumenthal, T.**, Overlapping genes in RNA phage: a new protein implicated in lysis, *Cell*, 18, 257, 1979.

26. **Kastelein, R. A., Remaut, E., Fiers, W., and van Duin, J.**, Lysis gene expression of RNA phage MS2 depends on a frameshift during translation of the overlapping coat protein gene, *Nature (London)*, 295, 35, 1982.

27. **Karnik, S. and Billeter, M.**, The lysis function of RNA bacteriophage Qβ is mediated by the maturation (A_2) protein, *EMBO J.*, 2, 1521, 1983.

28. **Haruna, I., Nozu, K., Ohtaka, Y., and Spiegelman, S.**, An RNA replicase induced by and selective for a viral RNA: isolation and properties, *Proc. Natl. Acad. Sci. U.S.A.*, 50, 905, 1963.

29. **Kamen, R.**, Characterization of the subunits of Qβ replicase, *Nature (London)*, 228, 527, 1970.

30. **Kondo, M., Gallerani, R., and Weissmann, C.**, Subunit structure of Qβ replicase, *Nature (London)*, 228, 525, 1970.

31. **Kamen, R.**, A new method for the purification of Qβ RNA-dependent RNA polymerase, *Biochim. Biophys. Acta*, 262, 88, 1972.

32. **Fedoroff, N. and Zinder, N. D.**, Structure of the poly(G) polymerase component of bacteriophage f2 replicase, *Proc. Natl. Acad. Sci. U.S.A.*, 68, 1838, 1971.

33. **Blumenthal, T., Landers, T. A., and Weber, K.**, Bacteriophage Qβ replicase contains the protein biosynthesis elongation factors EF Tu and EF Ts, *Proc. Natl. Acad. Sci. U.S.A.*, 69, 1313, 1972.

34. **Wahba, A. J., Miller, M. J., Niveleau, A., Landers, T. A., Carmichael, G. G., Weber, K., Hawley, D. A., and Slobin, L. I.**, Subunit I or Qβ replicase and 30 S ribosomal protein S1 of *Escherichia coli*. Evidence for the identity of the two proteins, *J. Biol. Chem.*, 249, 3314, 1974.

35. **Kamen, R., Kondo, M., Romer, W., and Weissmann, C.**, Reconstitution of Qβ replicase lacking subunit α with protein-synthesis-interference factor i, *Eur. J. Biochem.*, 31, 44, 1972.

36. **Franze de Fernandez, M. T., Eoyang, L., and August, J. T.**, Factor fraction required for the synthesis of bacteriophage Qβ RNA, *Nature (London)*, 219, 588, 1968.

37. **Guerrier-Takada, C., Subramanian, A.-R., and Cole, P. E.**, The activity of discrete fragments of ribosomal protein S1 in Qβ replicase function, *J. Biol. Chem.*, 22, 13649, 1983.

38. **de Haseth, P. L. and Uhlenbeck, O. C.**, Interaction of *Escherichia coli* host factor protein with Qβ ribonucleic acid, *Biochemistry*, 19, 6146, 1980.

39. **Yonesaki, T. and Haruna, I.**, *In vitro* replication of bacteriophage GA RNA. Subunit structure and catalytic properties of GA replicase, *J. Biochem. (Tokyo)*, 89, 741, 1981.

40. **Yonesaki, T. and Aoyama, A.**, *In vitro* replication of bacteriophage GA RNA. Involvement of host factor(s) in GA RNA synthesis, *J. Biochem.*, 89, 303, 1981.

41. **Haruna, I. and Spiegelman, S.**, Autocatalytic synthesis of a viral RNA *in vitro*, *Science*, 150, 884, 1965.

42. **Blumenthal, T. and Carmichael, G. G.,** RNA replication: function and structure of Qβ replicase, *Annu. Rev. Biochem.,* 48, 525, 1979.
43. **Ammann, J., Delius, H., and Hofschneider, P. H.,** Isolation and properties of an intact phage-specific replicase from RNA phage M12, *J. Mol. Biol.,* 10, 557, 1964.
44. **Franklin, R. M.,** Purification and properties of the replicative intermediate of the RNA bacteriophage R17, *Proc. Natl. Acad. Sci. U.S.A.,* 55, 1504, 1966.
45. **Cramer, J. H. and Sinsheimer, R. L.,** Replication of bacteriophage MS2. Phage-specific ribonucleoprotein particles found in MS2-infected, *E. coli, J. Mol. Biol.,* 62, 189, 1971.
46. **Weissmann, C., Feix, G., and Slor, H.,** In vitro synthesis of phage RNA: the nature of the intermediates. *Cold Spring Harbor Symp. Quant. Biol.,* 33, 83, 1968.
47. **Eikhom, T. S.,** Isolation of free minus strands from Qβ-infected *Escherichia coli, J. Mol. Biol.,* 93, 99, 1975.
48. **Haruna, I. and Spiegelman, S.,** Specific template requirements of RNA replicases, *Proc. Natl. Acad. Sci. U.S.A.,* 54, 579, 1965.
49. **Spiegelman, S.,** An approach to the experimental analysis of precellular evolution, *Q. Rev. Biophys.,* 4, 213, 1971.
50. **Biebricher, C. K.,** Darwinian selection of RNA molecules *in vitro,* in *Evolutionary Biology,* Vol. 16, Hecht, M. K., Wallace, B., and Prance, C. T., Eds., Plenum Press, New York, 1983, 1.
51. **Mills, D. R., Kramer, F. R., and Spiegelman, S.,** Complete nucleotide sequence of a replicating RNA molecule, *Science,* 180, 916, 1973.
52. **Kacian, D. L., Mills, D. R., and Spiegelman, S.,** The mechanism of Qβ replication: sequence at the 5'-terminus of a S RNA template, *Biochim. Biophys. Acta,* 238, 212, 1971.
53. **Biebricher, C. K., Eigen, M., and Luce, R.,** Product analysis of RNA generated *de novo* by Qβ replicase, *J. Mol. Biol.,* 148, 369, 1981.
54. **Mills, D. R., Kramer, F. R., Dobkin, C., Nishihara, T., and Spiegelman, S.,** Nucleotide sequence of microvariant RNA: Another small replicating molecule, *Proc. Natl. Acad. Sci. U.S.A.,* 72, 4252, 1975.
55. **Schaffner, W., Ruegg, K. J., and Weissmann, C.,** Nanovariant RNAs: nucleotide sequence and interaction with bacteriophage Qβ replicase, *J. Mol. Biol.,* 117, 877, 1977.
56. **Banerjee, A. K., Rensing, U., and August, J. T.,** Replication of RNA viruses. Replication of a natural 6 S RNA by the Qβ RNA polymerase, *J. Mol. Biol.,* 45, 181, 1969.
57. **Sumper, M. and Luce, R.,** Evidence for *de novo* production of self-replicating and environmentally adapted RNA structures by bacteriophage Qβ replicase, *Proc. Natl. Acad. Sci. U.S.A.,* 72, 162-166, 1975.
58. **Biebricher, C. K., Eigen, M., and Luce, R.,** Kinetic analysis of template-instructed and *de novo* RNA synthesis by Qβ replicase, *J. Mol. Biol.,* 148, 391, 1981.
59. **Hill, D., and Blumenthal, T.,** Does Qβ replicase synthesize RNA in the absence of template?, *Nature (London),* 301, 350, 1983.
60. **Biebricher, C. K., Eigen, M., and Luce, R.,** Template-free RNA synthesis by Qβ replicase, *Nature (London),* 321, 89, 1986.
61. **Biebricher, C. K., Diekmann, S., and Luce, R.,** Structural analysis of self-replicating RNA synthesized by Qβ replicase, *J. Mol. Biol.,* 154, 629, 1982.
62. **Cech, T. R.,** The generality of self-splicing RNA: relationship to nuclear mRNA splicing, *Cell,* 44, 207, 1986.
63. **Biebricher, C. K.,** Darwinian evolution of self-replicating RNA molecules, *Chem. Scr.,* 26B, 57, 1986.
64. **Weber, H., Billeter, M. A., Kahane, S., Weissmann, C., Hindley, J., and Porter, A.,** Molecular basis for repressor activity of Qβ replicase, *Nature (London) New Biol.,* 237, 166, 1972.
65. **Vollenweider, H. J., Koller, T., Weber, H., and Weissmann, C.,** Physical mapping of Qβ replicase binding sites on Qβ RNA, *J. Mol. Biol.,* 101, 367, 1976.
66. **Meyer, F., Weber, H., and Weissmann, C.,** Interactions of Qβ replicase with Qβ RNA, *J. Mol. Biol.,* 153, 631, 1981.
67. **Mills, D. R., Nishihara, T., Dobkin, C., Kramer, F. R., Cole, P. E., and Spiegelman, S.,** The role of template structure in the recognition mechanism of Qβ replicase, in *Nucleic Acid — Protein Recognition,* Vogel, H. J., Ed., Academic Press, New York, 1977, 533.
68. **Nishihara, T., Mills, D. R., and Kramer, F. R.,** Localization of the Qβ replicase recognition site in MDV-1 RNA, *J. Biochem. (Tokyo),* 93, 669, 1983.
69. **Landers, T. A., Blumenthal, T., and Weber, K.,** Function and structure in RNA phage Qβ replicase: the roles of the different subunits in transcription of synthetic templates, *J. Biol. Chem.,* 249, 5801, 1974.
70. **Shen, T. J. and Jiang, M. Y.,** RNA recombination. Introducing poly(A) into Qβ phage, *Sci. Sin.,* 25, 485, 1982.
71. **Miele, E. A., Mills, D. R., and Kramer, F. R.,** Autocatalytic replication of a recombinant RNA, *J. Mol. Biol.,* 171, 281, 1983.
72. **Mills, D. R., Dobkin, C., and Kramer, F. R.,** Template-determined, variable rate of RNA chain elongation by Qβ replicase, *Cell,* 15, 541, 1978.

73. **Bausch, J. N., Kramer, F. R., Miele, E. A., Dobkin, C., and Mills, D. R.,** Terminal adenylation in the synthesis of RNA by Qβ replicase, *J. Biol. Chem.,* 258, 1978, 1983.

74. **Dobkin, C., Mills, D. R., Kramer, F. R., and Spiegelman, S.,** RNA replication: required intermediates and the dissociation of template, product, and Qβ replicase, *Biochemistry,* 18, 2038, 1979.

75. **Biebricher, C. K., Eigen, M., and Gardiner, W. C.,** Kinetics of RNA replication, *Biochemistry,* 22, 2544, 1983.

76. **Biebricher, C. K., Eigen, M., and Gardiner, W. C.,** Kinetics of RNA replication: competition and selection among self-replicating RNA species, *Biochemistry,* 24, 6550, 1985.

77. **Biebricher, C. K., Eigen, M., and Gardiner, W. C.,** Kinetics of RNA replication: plus-minus asymmetry and double-strand formation, *Biochemistry,* 23, 3186, 1984.

78. **Biebricher, C. K.,** Replication and selection kinetics of short-chained self-replicating RNA molecules, in *Positive Strand RNA viruses,* Vol. 54, U.C.L.A. Symp. Molecular and Cellular Biology Brinton, M. A. and Rueckert, R., Eds., Alan R. Liss, New York, 1986, 9.

79. **Wetmur, J. G. and Davidson, N.,** Kinetics of renaturation of DNA, *J. Mol. Biol.,* 31, 349, 1968.

80. **Carmichael, G. G., Weber, K., Niveleau, A., and Wahba, A. J.,** The host factor required for RNA phage Qβ RNA replication *in vitro, J. Biol. Chem.,* 250, 3607, 1975.

81. **Mills, D. R., Peterson, R. L., and Spiegelman, S.,** An extracellular Darwinian experiment with a self-duplicating nucleic acid molecule, *Proc. Natl. Acad. Sci. U.S.A.,* 58, 217, 1967.

82. **Kramer, F. R., Mills, D. R., Cole, P. E., Nishihara, T., and Spiegelman, S.,** Evolution *in vitro:* sequence and phenotype of a mutant RNA resistant to ethidium bromide, *J. Mol. Biol.,* 89, 719, 1974.

83. **Kornberg, A., Bertsch, L. L., Jackson, J. F., and Khorana, H. G.,** Enzymatic synthesis of DNA. XVI. Oligonucleotides as templates and the mechanism of their replication, *Proc. Natl. Acad. Sci. U.S.A.,* 51, 315, 1964.

Chapter 2

REPLICATION OF THE POLIOVIRUS GENOME

Bert L. Semler, Richard J. Kuhn, and Eckard Wimmer

TABLE OF CONTENTS

I.   Introduction 24

II.   Poliovirus Genomic RNA and the Generation of Functional
Virus Polypeptides 24
    A.   Genome Structure and Encoded Proteins 24
    B.   Survey of the Poliovirus Infection Cycle 27
    C.   Selected Differences Between Poliovirus RNA and the Genomic
        RNAs of Other Picornaviruses 29

III.   Experimental Observations of In Vivo Poliovirus RNA Replication 30
    A.   Identification of Virus-Specific Replication Complexes 30
    B.   Poliovirus Proteins Associated with the RNA Replication Complex 30
    C.   Host-Cell Proteins Involved in Poliovirus RNA Replication 32
    D.   Viral RNA Structures Found in Infected Cells 33

IV.   In Vitro Replication of Poliovirus RNA 34
    A.   Soluble, Purified Replication Systems 34
    B.   Membrane-Associated Replication Systems 36

V.   Models for RNA Synthesis and Initiation 37
    A.   Initiation 37
        1.   Initiation of RNA Synthesis by Protein Priming 37
        2.   The Role of "Host Factor" 39
        3.   The Hairpin Model 39
        4.   The Problem of Template Recognition and of Initiation
            at Different Termini 41
    B.   Elongation 41

VI.   Summary and Future Directions 41

Acknowledgments 42

References 42

I. INTRODUCTION

The picornaviruses represent an important group of eukaryotic pathogens, the intracellular life cycles of which are characterized by several unique macromolecular/biosynthetic events. Poliovirus is a member of the enterovirus group of the Picornaviridae and perhaps best exemplifies the unique aspects of the picornavirus group. For example, poliovirus is known to cause rapid shut-off of host-cell protein synthesis following infection of susceptible cells. Polioviruses (and all picornaviruses) also employ a unique strategy for regulating the expression of its viral genes by synthesizing precursor polyproteins that are then cleaved by viral-encoded proteinases. Finally, poliovirus has evolved a unique mechanism for replication of its genomic RNA by utilizing a viral RNA polymerase that requires protein/nucleic acid components (as yet undefined) to generate a 5′ protein-nucleotidyl moiety during the initiation step. This review will examine the accumulated evidence that is responsible for the current hypotheses describing the mechanisms(s) of poliovirus RNA replication. Although the primary focus of this literature survey will be on poliovirus RNA synthesis, work on other picornaviruses will be discussed in order to broaden the perspective on selected topics. Readers are encouraged to consult other review articles and book chapters for a somewhat different scope and emphasis on picornavirus replication.[1-4]

The present review will attempt to summarize the experimental results obtained over the last 2 decades that have uncovered many important aspects of poliovirus RNA replication. Because the data to be covered requires a fairly in-depth knowledge of picornavirus molecular biology, a description of viral gene organization and polypeptide expression will be presented first. In addition, an overview of the virus life cycle will be included along with specific examples of how poliovirus differs from some of the other picornaviruses. The data obtained from in vivo experiments with poliovirus-infected cells will then be presented in Section III. This section will analyze the nature of polio replication complexes with respect to subcellular fractionation, responses to detergent treatments, and the viral/cellular proteins involved. The discussion of results from in vivo observations will be extended to an examination of the different poliovirus-specific RNA species produced during an infection. The topic of in vitro synthesis of polio RNA will then be presented in some detail. Specific attention will be given to the viral and cellular components required for in vitro RNA synthesis, to the comparison of results obtained from purified vs. crude systems, and to the distinction between the initiation events and RNA chain elongation activities that have been reported to occur in vitro. The different templates used for in vitro RNA synthesis by the polio RNA polymerase will be discussed along with the various product RNAs arising from polio-specific RNA synthesis in vitro. Finally, models for both initiation and elongation of RNA chains by the poliovirus replicase will be presented. These models will have different variations on the proposed mechanisms for initiation reactions that include protein priming, host-factor or terminal uridylyl transferase-mediated priming, and plus vs. minus strand priming.

II. POLIOVIRUS GENOMIC RNA AND THE GENERATION OF FUNCTIONAL VIRUS POLYPEPTIDES

A. Genome Structure and Encoded Proteins

The genomic RNA of poliovirus is of positive polarity (i.e., message sense) and was shown to be the only viral component necessary to initiate a complete virus life cycle once inside of a mammalian cell.[5,6] The RNA packaged into the nonenveloped, icosahedral virion particles contains a 5′ genome-linked protein (VPg) and 3′ polyadenylate tract.[7-10] The 5′ terminal VPg is covalently attached to the 5′ uridylate moiety of polio RNA via an O^4-(5′-uridylyl)-tyrosine linkage.[11,12] The function of VPg at the 5′ end of polio RNA rather than the usual 7mG (7-methylguanosine) cap structure found on almost all eukaryotic RNAs of plus-strand polarity is unknown. There have been proposals and some data that suggest VPg may have a role in RNA replication (see below). However, it is known that this 22-amino acid peptide is not required for infectivity because treatment of virion RNA with proteinase K prior to DEAE-dextran-facilited uptake by HeLa cells does not reduce the specific infectivity of poliovirus RNA.[13] VPg is not present on polio mRNA that is found associated with cellular polyribosomes. Rather, the 5' end of mRNA is terminated with pUp,[14-16] and this lack of VPg is the only difference between polio virion RNA and polio mRNA.[13]

The genomic RNAs for all three serotypes of poliovirus (including the Sabin vaccine derivatives) have been completely sequenced.[17-22] For the Mahoney strain of type 1, the viral RNA is 7441 nucleotides in length,[17-19] exclusive of the 3' poly(A) tract. Included in the viral RNA is a 5' noncoding region of 742 nucleotides (see Reference 23), a long open-reading frame beginning at nucleotide 743 and extending 6627 nucleotides and a 3' noncoding region of 71 nucleotides. The functions of the noncoding sequences at both ends of the viral genome are unknown. Presumably these terminal sequences carry signals for replicase recognition/binding to initiate RNA synthesis at the 3' ends of both plus and minus strands. In addition, the 5' noncoding region of polio RNA may carry sequences required for ribosome binding prior to initiation of translation of the precursor polyprotein. Such ribosome-binding signals are probably important in directing ribosomes to the initiator AUG codon at nucleotide 743 since there are 8 AUG codons that precede this codon, and these AUGs are apparently not used as initiation codons.[23]

The coding region of the polio genome contains a single, long, open-reading frame that encodes 2209 consecutive amino acids that would give rise to a polyprotein having a molecular weight of 247,000 daltons. Such a giant polypeptide is not actually generated during a poliovirus infection because at least two viral proteinases are known to cleave precursor polypeptides during translation[24,25] to yield the functional capsid and noncapsid proteins. The mature, viral polypeptides synthesized during a poliovirus infection have been grouped according to where they are encoded in the viral genome (using the nomenclature of Rueckert and Wimmer).[26] As shown in Figure 1, the proteins derived from the left end of the viral RNA are cleaved from the P1 precursor. These proteins include the capsid proteins VPO, VP4, VP2, VP3, and VP1. The middle region of the genome encodes the P2 precursor that can be cleaved to yield 2A, 2BC, and 2C, and in the final cleavage also 2B.[27] The proteins derived from the P2 precursor are all nonstructural polypeptides. The functions of 2B and 2C are largely unknown. There was a previous suggestion that 2C is a viral proteinase.[28] However, if 2C is a proteinase, it does not carry the activity that cleaves the majority of polio polypeptides at either glutamine-glycine (Q-G) or tyrosine-glycine (Y-G) bonds[24,25] (see below). Experiments with guanidine-resistant mutants of poliovirus and host-range mutants of human rhinovirus type 2 suggest the protein 2C may have a role in RNA synthesis.[29,30] The only P2 cleavage product with a known function is the 2A polypeptide that was recently shown to be the enzyme responsible for Y-G cleavage of the polio polyprotein[25] (see below).

The polypeptides derived from the right end of the viral RNA are processed from the P3 precursor polypeptide. The proteins cleaved from the P3 precursor include 3D^{pol} (the viral RNA polymerase), 3C^{pro} (a viral proteinase), VPg (the genome-linked protein), and 3AB (a membrane-associated precursor to VPg). There are two other precursor polypeptides derived from P3: (1) 3CD, a 72-kdalton protein that contains the amino acid sequences of 3C^{pro} and 3D^{pol} and (2) a protein originally designated as 4a, which may be an unstable precursor to the functional RNA polymerase.[31,32] Note that in Figure 1, polypeptide 4a is not shown. It is a protein that contains all of the 3D^{pol} sequences as well as 60 amino acid residues derived from the carboxy terminus of 3C^{pro}.[32] In addition, there are two polypeptides (3C' and 3D') that are alternate cleavage products of precursor 3CD. Except for 3C^{pro} and

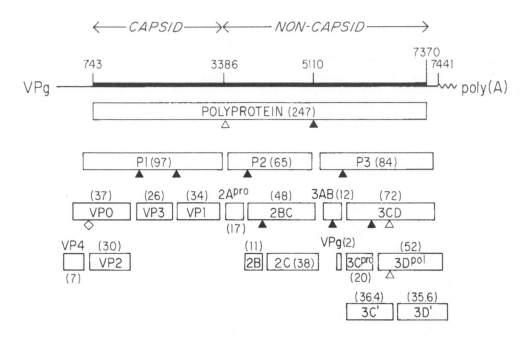

FIGURE 1. Protein processing map of the polypeptides encoded by the poliovirus genomic RNA. The horizontal line with the nucleotide numbers above it depicts the viral RNA. The thickened portion of the RNA denotes the polyprotein coding region. The boxes beneath the genomic RNA represent polio-specific polypeptides. The filled triangles (▲) represent Q-G cleavage sites. The open triangles (△) represent Y-G cleavage sites. The open diamond (◇) depicts an N-S cleavage site in VPO. The numbers in parentheses are the polypeptide molecular weights ($\times 10^{-3}$ daltons).

$3D^{pol}$, the functions of the P3-derived proteins are unknown. Both 3CD and 3C' contain the amino acid sequence of proteinase $3C^{pro}$, but it is not known whether either of these proteins can process polio precursors. It has been suggested[33,34] that protein 3AB may be involved in initiation of RNA synthesis within a membrane-bound replication complex, but no biochemical evidence for such a role has been presented. Thus, like most of the P2-derived polypeptides, the majority of the proteins generated by cleavage of the P3 precursor have not been functionally defined as to their biochemical roles during a poliovirus infection.

The assignment of the map positions of the polio proteins (described above) to specific regions on the viral genome is a result of both pactamycin[35,36] and tryptic peptide mapping experiments.[27,37,38] A more precise cleavage map was obtained when data from amino acid sequence analysis was compared to the amino acid sequence predicted by the complete nucleotide sequence of the viral RNA.[17,39-42] These data revealed the exact amino acid pairs that are cleaved in the polyprotein precursor to generate more than 17 viral-encoded proteins. Most of the viral proteins are cleaved from precursors between Gln–Gly (Q-G) amino acid pairs. Thus, Gln becomes the carboxy-terminal amino acid of one of the cleavage products, and Gly becomes the amino-terminal residue of the other cleavage product. In addition to the Q-G pairs, two other amino acid pairs in the polyprotein sequence were shown to be utilized as cleavage sites. Cleavage at Tyr–Gly (Y-G) sites was shown to occur at the VP1-P2 junction and at the 3C'-3D' border.[24,40,42] A cleavage between Asn–Ser (N-S) in VPO to generate VP4 and VP2 has also been demonstrated.

Cleavage of the poliovirus polyprotein at defined sites requires a specific interaction between the viral proteinases and their precursor substrates. Since there are three classes of cleavage sites in polio proteins (i.e., Q-G, Y-G, and N-S), one would predict that more than 1 proteinase is responsible for complete polyprotein processing. Indeed, there have been 2 viral-encoded proteinases identified as the enzymes responsible for Q-G and Y-G cleavages in the polio polyprotein.[24,25] In two separate studies employing similar experimental approaches, Hanecak et al.[24,43] and Toyoda et al.[25] used cleavage inhibition with monospecific antibodies to proteins 3C and 2A, respectively, to demonstrate that 3C carried out Q-G cleavages and that 2A was responsible for Y-G cleavages. The antibodies were also used in subsequent experiments to immunoprecipitate polio-specific polypeptides generated from cDNA cloned into bacterial expression vectors. In these experiments, gene segments containing the sequences of 3C and 2A (plus additional polio sequences corresponding to part of the precursor polypeptides from which these two proteins are derived) were cloned downstream from inducible prokaryotic promoters in bacterial expression plasmids. Upon induction, bacteria harboring the plasmids synthesized precursor polypeptides that were subsequently cleaved by either the 2A or 3C proteinase. Demonstration that the cleavage activities resided in the 2A or 3C proteins was provided by linker insertion mutagenesis or deletion analysis. Interestingly, the results from the 3C expression experiments strongly suggest that generation of the 3C polypeptide itself occurs by an intramolecular cleavage mechanism.[43] Such a mechanism was originally proposed for the generation of the encephalomyocarditis (EMC) virus 3C proteinase.[44]

The functional diversity of structural and nonstructural polypeptides required for successful completion of the polio infectious cycle cannot occur at the level of viral RNA replication or selective translation (since there is only single known species of viral mRNA and only a single known start site for in vivo protein synthesis). Instead, poliovirus controls the relative amounts of viral proteins in the cell during the course of an infection by protein processing. The control of protein processing by the 3C and 2A proteinases must certainly involve recognition of the Q-G and Y-G amino acid pairs, respectively, that will ultimately be cleaved. However, processing determinants other than the amino acid pairs that are cleaved must exist because 4 of 13 Q-G pairs and 8 of 10 Y-G pairs predicted to occur in the amino acid sequence of the polio polyprotein are apparently not used as cleavage sites.[45] Secondary recognition signals may also exist that "guide" the proteinases to the correct Q-G or Y-G cleavage sites in precursor polypeptides. Alternatively, the availability of Q-G and Y-G cleavage sites may be based entirely upon the correct folding of the precursor substrates that places a Q-G or Y-G site in a position (or context) that is accessible for recognition and cleavage by 3C or 2A. If the folding of precursor polypeptides is important in the presentation of cleavage sites to the viral proteinases, then the order in which specific processing events occur could be a major factor in controlling the relative amounts of viral-specific proteins in the cell during the course of an infection. It is the regulation of protein production that must determine the sequence of events leading to successful completion of the poliovirus life cycle. An overview of this life cycle is presented in the next section.

B. Survey of the Poliovirus Infection Cycle

The major landmarks of molecular events that occur during a poliovirus infection of a primate cell in culture are shown schematically in Figure 2. The viral infection is initiated by attachment of virion particles to the cell surface via a specific receptor. For poliovirus, these receptors are only normally present on cells of primate origin.[46] Following adsorption and uncoating of the virion particles, the viral RNA released into the cytoplasm associates with the protein synthesis machinery of the cell to engage in viral-specific translation. Two events that occur early during a poliovirus infection are important to note at this point. First, polio mRNA that is associated with polyribosomes does not contain VPg at its 5′ terminus. Either as a prerequisite for association with ribosomes or as a consequence of that association, VPg is removed from the viral RNA by a so-called unlinking enzyme.[47] This enzyme is present in uninfected HeLa cells as well as in several other different mammalian cell types. The contribution of the unlinking activity to the expression of polio genetic information is unknown. The presence of VPg at the 5′ ends of viral RNAs does not interfere with ribosome

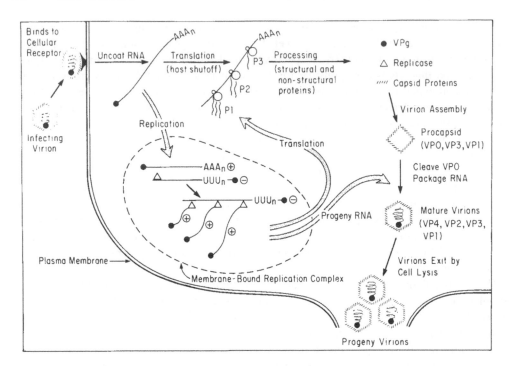

FIGURE 2.   Diagrammatic overview of the poliovirus life cycle. Taken in part from References 1 and 3.

binding in vitro, nor does VPg interfere with in vitro translation of polio RNA.[48] Perhaps polio uses the unlinking activity to regulate which progeny RNAs are destined to be mRNA and which RNAs are to be packaged into virion particles.

A second important event that occurs early during a poliovirus infection is the shut-off of host-cell protein synthesis. The inhibition is quite rapid, and within 1 hr postinfection cell protein synthesis is severely impaired. This inhibition leads to the preferential translation of polio-specific RNAs over the capped, cellular mRNAs. The defect in host-cell protein synthesis is thought to involve the inactivation or proteolytic cleavage of polypeptides associated with a cap-binding protein complex.[49-54] Although an intact genomic RNA appears to be a requirement for polio shut-off of cellular translation,[55] it is not known which polio-specific polypeptide(s) is the mediator of this inhibitory action. Recent data employing partial purification of an activity from polio-infected cells that cleaves a large polypeptide component (p220) of the cap-binding protein complex suggest that the two known viral proteinases ($3C^{pro}$ and $2A^{pro}$) are not directly involved in the shut off of cap-dependent, cellular translation.[56-58]

Among the protein products synthesized from the genomic RNA of infecting virions are viral replicase complexes of unknown structure. The replicase complex is known, however, to contain the virus-encoded core RNA polymerase $3D^{pol}$ [59,60] (formerly called NCVP4, p63, p3-4b) that is most likely associated with other viral[31,61,62] and host[63,64] proteins when it is active in the cell. The replicase complex must first copy some of the infecting plus strand viral RNAs into complementary strand intermediates (minus strands) that will subsequently serve as templates for the synthesis of new plus strand RNAs. As will be discussed below, the RNA synthesis occurs in membrane-bound replication complexes. The progeny plus strands of RNA (generated in the replication complexes) that are synthesized early in infection then engage in further virus-specific protein synthesis, resulting in a large amplification of polio gene products in the infected cell. After several rounds of protein synthesis and viral RNA replication, infected cells accumulate a pool of capsid protein precursors that self-assemble into procapsids.[65] The procapsids then associate with the progeny plus strand viral RNAs to form mature virions. The association between progeny RNAs and the procapsid structures must be specific since it is only after this interaction that the cleavage of VPO (one of the proteins in the procapsid, refer to Figure 2) to VP4 and VP2 occurs. The origin of the enzyme that makes this so-called "morphogenetic cleavage" at an N-S amino acid pair in VPO is presently unknown. However, recent data from crystallographic studies on the three-dimensional structure of human rhinovirus and poliovirus suggest that autocatalytic cleavage of VPO may occur as a result of viral RNA insertion into the procapsids.[66,67] The particles produced by packaging progeny RNA and cleaving VPO are mature, infectious virions. These virions eventually cause lysis of infected cells and then initiate infection of other cells.

C. Selected Differences Between Poliovirus RNA and the Genomic RNAs of Other Picornaviruses

The overall genome structures and polyprotein organization of all picornaviruses are roughly equivalent. The viral RNAs uniformly possess a 5'-linked VPg protein and a 3' polyadenylate tract. The picornaviral genomes all contain a long 5' noncoding region (500 to 1300 nucleotides) followed by a long open reading frame of 6500 to 7000 nucleotides. (Refer to References 17 to 22 for poliovirus; 68 to 70 for foot-and-mouth disease virus [FMDV]; 71 for EMC virus; 72 to 74 for rhinovirus; 75 to 77 for hepatitis A virus; and 78 for coxsackie virus). However, for the genomic RNAs of aphthoviruses (FMDV) and the cardioviruses (EMC virus, mengo virus), there is a poly(C) tract of approximately 50 to 200 nucleotides that is located within 150 to 500 nucleotides of the 5' end of the genome. The function of the poly(C) tract is presently unknown, but it is known that both the poly(C) tract and the 5' noncoding sequences upstream from it are probably required for FMDV infectivity.[79]

A second difference between poliovirus RNA and that of FMDV and EMC virus is the encoding of leader polypeptides at the amino terminus of the polyprotein coding region in the latter two genomes. Such leader polypeptides are apparently not encoded in the enterovirus (polio, coxsackie, hepatitus A) and rhinovirus genomes. The role of the leader polypeptides during a virus infection has not been established. However, recent experiments employing both prokaryotic and eukaryotic expression/mutagenesis of cDNA clones of the FMDV genome corresponding to the L/L'-P1 region suggest that the L leader polypeptide is a viral protease that catalyzes its own release from the viral polyprotein.[80]

A further distinction between the genome of poliovirus and that of FMDV comes from differences in the proteins encoded in the polyprotein. For example, the FMDV genome does not encode a 2A polypeptide that is normally encoded by the middle region of polio RNA and is derived from the P2 precursor by proteolytic cleavage.[40] The FMDV genome also codes for 3 VPg polypeptides, the coding sequences of which are arranged tandemly within the P3 region of the viral polyprotein.[81] All other picornaviruses that have been sequenced appear to code for a single VPg polypeptide.

Finally, there are numerous differences between poliovirus and the other picornaviruses in terms of the sites of proteolytic processing that are specified by the viral RNA.[45] As mentioned above, the poliovirus polyprotein is processed primarily at Q-G and Y-G amino acid pairs. The sequence data for EMC virus suggest that cleavage of the viral polyprotein occurs at glutamine-serine (Q-S), Q-G, tyrosine-proline (Y-P), and alanine-aspartic acid (A-D) amino acid pairs. The accumulated data for FMDV suggests that glutamic acid-glycine (E-G), glutamine-leucine (Q-L), glutamine-isoleucine (Q-I), L-N, A-D, glutamine-threonine (Q-T), and possibly other sites are used in protein processing. Sequence data for rhinovirus, hepatitus A virus, and coxsackievirus RNAs also indicate a great degree of variation in the putative signals used in the cleavage of viral precursor polypeptides. Moreover, as mentioned above, for the L protein of FMDV and as suggested for complete processing of the EMC virus polyprotein,[82] the genetic map location and actual number of viral-encoded proteinases may vary considerably among the picornaviruses. Such differences in the nature of the viral-specific cleavage activities would not be unexpected given the large numbers of amino acid pairs that are utilized as cleavage sites.

III. EXPERIMENTAL OBSERVATIONS OF IN VIVO POLIOVIRUS RNA REPLICATION

The first evidence for a virus-specific RNA polymerase activity in poliovirus-infected cells was presented by Baltimore et al.[83] In that study, a microsomal fraction from virus-infected HeLa cells was shown to contain an RNA polymerase activity that was not present in uninfected HeLa cells. The appearance of the RNA polymerase activity during the time course of a polio infection correlated well with the appearance of infectious virus particles, with the polymerase activity peaking at 3 to 4 hr postinfection. These data marked the beginning of numerous biochemical and genetic studies aimed at understanding the mechanism of RNA replication by poliovirus. The following section will summarize some of the data arising from analysis of viral replication complexes and polio-specific RNA structures found in viral-infected cells.

A. Identification of Virus-Specific Replication Complexes

The initial experiments aimed at identifying the polio replicase complex involved disruption of virus-infected cells at various times post-infection and then subjecting the cell-free extracts to a variety of centrifugation and detergent/salt treatments. Crude fractionation of infected cells followed by sucrose gradient sedimentation showed that polio RNA polymerase activity was associated with a large, heterogeneous replication complex that is membrane-bound.[84-86] Treatment of infected-cell extracts with detergent (deoxycholate) released the replication complex from cytoplasmic membranes. This released replication complex was shown to sediment in sucrose gradients with a sedimentation value of 250 S and was apparently free of cellular ribosomes.[86] Similar structures associated with viral RNA polymerase activity were also observed in FMDV-infected cells[87] and mengovirus-infected cells.[88] The precise nature of the interaction of the polio replication complex with cytoplasmic membranes is unknown. However, it has been demonstrated by isopycnic banding of cytoplasmic extracts from polio-infected HeLa cells that the majority of viral RNA polymerase activity is found in the smooth membrane fraction.[34,89,90] In addition, the majority of labeled, virus-specific RNA is found in the smooth membrane fraction of HeLa cells following short pulse-labeling with ^{3}H-uridine.[89] Further evidence for the importance of the association of the polio replication complex with smooth membranes comes from the demonstration of a significant increase in smooth membrane content of infected cells compared to that of uninfected cells.[91]

B. Poliovirus Proteins Associated with the RNA Replication Complex

The composition of the ''crude'' replication complex isolated after deoxycholate treatment of infected cell extracts is quite heterogeneous. A large number of viral[92] and cellular polypeptides are associated with the crude replication complex along with polio-specific RNA species.[84,86,93] Treatment of the replication complex with either SDS or pronase releases viral RNAs that sediment between 30 S and 70 S in sucrose gradients.[86] Further steps in the isolation of the polio replication complex were taken in order to obtain a soluble, stable preparation of the viral RNA polymerase activity. These steps included the solubilization of the particulate fraction of an infected-cell extract using a combination of deoxycholate and Nonidet P40 (NP-40). This treatment resulted in the release of a 70 S RNA polymerase complex that was active in the synthesis of polio-specific RNA.[94] A subsequent step that was later added to the purification of the viral RNA polymerase complex was the precipitation of the complex with 2 *M* LiCl followed by sucrose gradient centrifugation.[59] Although there were numerous host-cell polypeptides associated with the above RNA polymerase complex, the primary viral protein found in the complex was 3D^{Pol}. Another report employing essentially the same purification scheme suggested that other viral proteins (P1, 3CD, 2C, VP1 and others) were also present in purified preparations of the polio RNA polymerase activity.[95] Nevertheless, the data of Lundquist et al.[59] strongly suggested that viral protein 3D^{Pol} was the major polio protein involved in RNA polymerization.

The experiments reviewed so far in this section employed assays for polio RNA polymerase activity that did not respond to the addition of exogenous viral RNA. That is, the replication complexes could only catalyze the copying of endogenous viral RNA already associated with the complex. A major advance in studies on polio RNA replication was made by Flanegan and Baltimore,[96] who reported the isolation of a soluble, polio-specific RNA polymerase activity that would copy a poly(A) template hybridized to an oligo(U) primer. The requirement for a poly(U) polymerase activity associated with the poliovirus replication complex had been previously demonstrated in studies describing the biosynthesis of the 3' poly(A) tract of polio RNA.[90,97] Subsequently, it was shown that the same poly(U) polymerase activity could also catalyze the in vitro copying of exogenously added poliovirion RNA in the presence of an oligo(U) primer.[60] A highly purified preparation of the primer-dependent polio RNA polymerase activity contained only the viral protein 3D^{Pol}.[98] The purified polio RNA polymerase (3D^{Pol}), therefore, represents a template- and primer-dependent enzyme capable of efficiently catalyzing the elongation reaction of polio RNA replication.

The isolation of a different form of the poliovirus RNA polymerase that was able to initiate the copying of exogenously added virion RNA in vitro in the absence of an oligo(U) primer was described by Dasgupta et al.[99] This so-called "7S" form of the polio polymerase had been purified by phosphocellulose and poly(U)-Sepharose® chromatography. It contained viral proteins 3D^{Pol} and 3CD as well as host proteins thought to be required for initiation of RNA synthesis (Reference 64; see below). Although viral protein 3CD did not, by itself, carry out any in vitro synthesis of polio RNA,[98] its presence suggested that perhaps viral proteins other than 3D^{Pol} were required for a polio replicase activity capable of *de novo* initiation of RNA synthesis. Data that have been interpreted to show involvement of protein 3CD in the polio replicase activity were published by Bowles and Tershak[61] in their study on the breakdown of 3CD at the nonpermissive temperature for a type 3 strain of poliovirus. This proteolytic breakdown correlated well with a concommitant reduction in the in vivo level of RNA synthesis. In addition, Korant has suggested that stabilization of protein 3CD by the addition of protease inhibitor correlated with a lengthened period of polio RNA synthesis in vivo compared to infected cells that were not treated with the protease inhibitor.[100]

There is also evidence for viral polypeptides other than 3D^{Pol} and 3CD being involved in the in vivo replication of polio RNA. Viral polypeptide 2C has been implicated as having a role in polio RNA synthesis by several lines of evidence. First, as noted above, protein 2C has been detected in partially-purified preparations of the poliovirus replication complex.[95] Second, the polio RNA replication complex that is associated with cellular[34] or artificial phospholipid membranes[156] contains polypeptide 2C. Finally, the most convincing evidence for the involvement of protein 2C in polio RNA synthesis comes from studies with guanidine, an inhibitor of the growth of picornaviruses.[101] Guanidine (at millimolar levels in infected cells) has been shown to interfere with polio RNA synthesis, possibly at the initiation step in RNA replication.[101,102] Poliovirus infections of primate cells in the presence of guanidine give rise to mutant viruses that are either resistant to the presence of guanidine in the growth medium or dependent upon its presence for growth.[103] The most precise experiments that determined the site of action of guanidine on picornavirus proteins were carried out using genetic recombination,[104-106] mapping of electrophoretic variants of viral-specific polypeptides,[107,108] and RNA sequence analysis of the genomes of the guanidine-resistant and guanidine-dependent mutants.[109] Collectively, these studies demonstrate that poliovirus polypeptide 2C (and the equivalent protein in FMDV) is the target site for the inhibitory action of guanidine. Thus, the accumulated biochemical and genetic evidence suggests a role for protein 2C in the in vivo replication of poliovirus RNA.

The genome-linked protein, VPg, has been implicated as having a role in polio RNA replication because it is found at the 5' ends of all nascent viral RNAs in poliovirus-infected cells,[110,111] an observation suggesting that it may be a primer for initiation of RNA synthesis. Several polypeptide precursors to poliovirus VPg have been detected in extracts of infected cells using immunoprecipitation with antibodies directed against synthetic peptides.[33,34,112,113] Three of these precursor polypeptides (3AB, P2-3AB, 2C-3AB; formerly P3-9, 3b/9, X/9, respectively) contain a 22-amino acid hydrophobic domain that may serve as a membrane-anchoring region.[33,34] The same three proteins have been shown to be enriched in membrane fractions from poliovirus-infected HeLa cells that are active in in vitro RNA synthesis.[34] It is possible that VPg is attached to the 5' ends of nascent viral RNAs at the moment of initiation (or shortly thereafter) by cleavage from one of the above precursor polypeptides. Interestingly, VPg-producing cleavages occur at Q-G sites (on both its amino and carboxy terminus) that are catalyzed by the viral proteinase 3C. Since the 3C^{pro} polypeptide has been found in the membrane-bound replication complex,[34,92] it is tempting to speculate that initiation of RNA synthesis in vivo may be controlled by the proteolytic processing of VPg-precursor polypeptides. Another unstable precursor polypeptide (4a) has been found in the polio replication complex,[31] and this protein is also found tightly associated with cytoplasmic membranes.[114]

C. Host-Cell Proteins Involved in Poliovirus RNA Replication

As mentioned above, the crude and partially purified RNA replication complexes from extracts of virus-infected HeLa cells are associated with numerous host-cell proteins. The first demonstration of a role for any of these proteins in polio RNA replication came when Dasgupta et al.[63] reported that a "host factor" present in uninfected HeLa cells was required for initiation of in vitro RNA synthesis in response to exogenously added polio RNA in the absence of an oligo(U) primer. Subsequent purification of the host factor showed that it is a 67,000-dalton protein and is located primarily in the soluble fraction of disrupted cells.[115-117] Additional evidence for the role of host factor in RNA replication comes from the ability of antibodies against purified host factor to immunoprecipitate the poliovirus RNA polymerase (3D^{pol}) from extracts of viral-infected cells.[117] The results from the above data suggest that host factor and 3D^{pol} are physically associated in infected cells. This conclusion is further supported by the demonstration that host factor can selectively bind to an affinity column containing the viral RNA polymerase bound to Sepharose®.[115] A note of caution should be added to the interpretation of data from the host factor-stimulated polio RNA polymerase experiments. In the data mentioned above,[63,115] there was no demonstration of a clear-cut template specificity for poliovirus RNA over heterologous RNAs tested in the RNA polymerase reactions carried out in the presence of host factor. Thus, the requirement of this host factor for in vivo RNA replication has not been demonstrated.

More recent experiments with host factor have shown that the normal function of this polypeptide in uninfected HeLa cells is that of a protein kinase that phosphorylates eukaryotic initiation factor-2 (eIF-2) as well as itself.[118] How such an activity could be involved in poliovirus RNA replication is, at present, unclear. However, there are additional experiments that suggest that host factor may not be a protein kinase. Rather, Andrews and Baltimore[64] have purified a 68,000-dalton "host factor" from the soluble portion of HeLa cell extracts and have demonstrated that this protein contains terminal uridylyl transferase activity. The purified enzyme is capable of adding uridine residues to the 3' poly(A) end of viral RNA.

It was proposed[64] that the additional uridine residues can anneal back onto the poly(A) and generate a hairpin primer for the polio RNA polymerase. As will be discussed below in detail, the data and proposals regarding host factor as a terminal uridylyl transferase are consistent with (but not proof of) a model for poliovirus replication that includes an intermediate RNA molecule that is twice the size of poliovirus genomic RNA.[119] It should be be noted, however, that other investigators have presented data suggesting that the synthesis of double-length RNA molecules by the polio RNA polymerase in the presence of host factor may be the result of nonspecific endonucleolytic cleavage of template RNA to generate an internal 3'-OH group that serves as a primer for initiation.[120,121] It remains to be determined whether the two different activities (phosphorylation and terminal uridylylation) can actually be ascribed to the same protein or "host factor" and what the role of such activities is during the in vivo replication of polio RNA.

D. Viral RNA Structures Found in Infected Cells

There are three different forms of polio-specific viral RNAs found in infected cells. These include (1) single-stranded RNA (ssRNA) that can be either virion RNA or mRNA;[122] (2) replicative intermediates (RI) which are heterogenous RNA molecules that are partially double stranded and partially single stranded;[86,93] and (3) double-stranded RNA or replicative form (RF).[123] During a polio infection, newly synthesized RNA consists of a single major species of RNA that sediments at 35 S in sucrose gradients and is identical to the RNA isolated from purified virions.[122] The 35 S RNA is single stranded, has messenger polarity, and is also identical to the viral-specific RNA associated with polyribosomes in infected cells.[124-126] As mentioned above, the only known difference between the ssRNA isolated from virions compared to that isolated from polyribosomes of infected cells is the absence of VPg from the 5' end of ribosome-associated polio RNA (i.e., mRNA).

The replicative intermediate forms of polio RNA are found in small quantities in virus-infected cells and are usually detected by pulse labeling with a radioactive precursor. The RIs are partially RNAse-resistant and have been shown to sediment heterogeneously in sucrose gradients at 20 to 70 S.[93] The RI structure consists of a full-length minus strand template with four to eight nascent daughter strands of plus strand polarity.[127,128] A structure containing a full-length plus strand template with nascent daughter strands of minus strand polarity has not been detected.[129] As mentioned above, all nascent plus strands of RNA on replicative intermediates contain VPg at their 5' ends as do their template minus strands found in these structures.[110,111]

The replicative form (RF) of polio-specific RNA detected in infected cells is a complete copy of the viral plus strand RNA that is hydrogen bonded to a full-length minus strand copy of genomic RNA.[123] One end of the RF molecule contains a heteropolymeric duplex, while the other end contains a poly(A)/poly(U) homopolymeric duplex.[130,131] Both ends of the RF molecule contain a VPg covalently attached to the 5' uridylate moiety.[132] One unexpected structural aspect of RF is the observation that the 3'-terminal poly(A) tract is longer than the 5'-terminal poly(U) by up to 100 nucleotides.[130] This finding is in contrast to the perfect duplex structure reported for the heteropolymeric end of RF,[130] although another study found that a minority fraction of RF molecules carried several additional adenylate residues on the 3' ends of their minus strands.[133] In addition, there appear to be some forms of RF that are missing a VPg molecule at one end.[134] Interestingly, it has been reported[119] that a small population of RF molecules are covalently linked at one end through a hairpin structure, producing an RNA molecule that migrates in denaturing gels as a single-stranded molecule twice the length of the viral genome. Such structures may be intermediates in a proposed pathway for RNA replication based on a snap-back molecule that leads to self-generation of a primer for RNA synthesis (see below). Finally, it is not known how RF molecules fit into the scheme of RNA synthesis in vivo. Perhaps these structures are responsible for the generation of newly synthesized minus strands as a result of inefficient, "single run" synthesis.[135]

IV. IN VITRO REPLICATION OF POLIOVIRUS RNA

The experiments which will be described below have contributed greatly toward the development of an in vitro replication system for poliovirus and for picornaviruses in general. However, progress in this area has not been as forthcoming as one would have predicted following the purification of the viral polymerase 3D^{pol}. The prototype in vitro replication system was developed many years before by Spiegelman and colleagues. Working with the bacteriophage Qβ, they demonstrated that incubation of the phage RNA along with a purified viral protein resulted in the *de novo* synthesis of authentic viral RNA.[136] It turned out, however, that the purified phage protein was actually a complex of the phage-encoded polymerase along with three host-encoded proteins which normally functioned in protein translation[137] (see Chapter 1). An in vitro replication system, as its name implies, is strictly defined as a system which has the ability to synthesize a replica of the input template RNA. That is, plus strands are copied into minus strands which, in turn, serve as templates for the synthesis of new plus strands. To date, there is no in vitro replication system of picornavirus RNA which meets this criterion. In order to understand this inability to develop a system which replicates authentic picornavirus RNA, it is necessary to examine the strategies that have been used to study poliovirus replication. The strategies can be clearly divided into two general approaches: the first utilizes purified polypeptides and an exogenous source of RNA, while the second utilizes a crude membrane mixture from infected cells and is dependent on endogenous RNA.

A. Soluble, Purified Replication Systems

Although the bulk of virus-specific polymerase activity can be found in membraneous replication complexes from infected cells,[84-86] the membranes themselves do not appear to be required for the activity. Treatment of these membrane complexes with detergents successfully solubilized the polymerase activity, while leaving it independent of added template.[86,87,94] Analysis of the products synthesized in this solubilized system indicated that both plus and minus strand RNA were produced,[157] but initiation did not appear to occur in these in vitro systems. As described above, the assay developed by Flanegan and Baltimore was instrumental in the isolation of a soluble, template-dependent poly(U) polymerase.[96] This activity, along with the copurifying RNA polymerase activity, was shown to be dependent upon an oligo(U) primer. Such a finding was quite interesting, since it suggested that 3D^{pol} was capable of catalyzing only RNA chain elongation. It was also apparent that oligo(U) did not function as the primer in vivo since it has never been found either free in the cytoplasm or bound to the poly(A) tails of messenger RNA. An in vitro system would therefore have to address both the mechanism of RNA initiation and also the linking of VPg to progeny strands.

One step in this direction was taken by Dasgupta et al.[63] Using the Qβ system as an example, these investigators looked in an uninfected HeLa cell ribosomal salt wash for an activity which would allow initiation of RNA synthesis in the absence of an oligo(U) primer. As mentioned above, they isolated a protein termed "host factor" which did not appear to correspond to any of the known initiation or elongation factors for protein synthesis, but released the polio RNA polymerase from its primer-dependent block. Antibodies made against purified host factor, isolated from uninfected cells, were able to inhibit this primer-independent activity, but had no effect on reactions in which oligo(U) was used as the primer.[138] This suggested that host factor was required for the initiation of minus strand RNA synthesis in vitro. It should be pointed out that the experiments carried out in the purified soluble system utilize vRNA (plus-sense) and therefore represent only minus-strand RNA synthesis. Moreover, this synthesis is characterized by a single-cycle transcription: the polymerase copies the template vRNA and when it reaches the end, it is released from template and product, unable to initiate another round of transcription by using the newly synthesized double-strand RNA as a template.

Unfortunately, the above finding failed to explain the mechanism by which VPg is linked to the nascent RNA and at what step in replication the linkage occurs. An answer to these questions appeared to come from in vitro experiments using purified 3D^{Pol}, host factor, and vRNA as template. Immunoprecipitation of the newly synthesized RNA products using anti-VPg antibodies revealed VPg-related polypeptides which were covalently bound to the RNA.[113,140,141] Since the only sources of VPg-related polypeptides were the purified 3D^{Pol} and the template vRNA, it was assumed that small amounts of VPg-containing polypeptides copurified with the polymerase and served as donor for VPg on newly synthesized RNA strands. It was further shown that this host factor-dependent in vitro transcription system required ATP for RNA initiation.[142] It was suggested that this ATP requirement was necessary for a protein kinase activity which copurified with host factor.[118] Whether such protein kinase activity is required for RNA initiation in vivo is unknown.

Experiments performed in another laboratory produced quite different results using the same set of purified components in an in vitro transcription system. Young et al.[119] showed that some of the product RNA synthesized in the presence of host factor was twice the size of the input template RNA. This was the result of a covalent linkage between the newly synthesized RNA and the template RNA. The product RNA had the structure of a double-stranded molecule resulting from snap-back at the 3′ end of the template strand and synthesis of a new complementary strand. Shortly thereafter, Andrews and Baltimore[64] reported a novel terminal uridylyl transferase activity (TUT) which copurified with the host factor they used in the in vitro experiments. Further work[143] showed that TUT could replace host factor (as determined by incorporation of a labeled nucleotide) in the purified in vitro transcription system, although the products of transcription of vRNA template were not presented. Based on these findings, these authors suggested that TUT added uridylate residues to the 3′ end of the template (that is, onto the poly(A) tract of the template RNA). When a sufficient number of uridylates are in place, the oligo(U) tail will hydrogen bond to the poly(A) and TUT will be released. The resulting molecule would supposedly be a suitable substrate for 3D^{Pol} elongation activity.

In light of the finding that the template RNA and the product RNA were covalently linked following the in vitro transcription reaction, it was suggested that the previously reported immunoprecipitation of *de novo* synthesized products with anti-VPg may be due to VPg present on the template RNA.[144,145] Treatment of the template RNA with proteinase K prior to the in vitro reaction or use of an RNA template lacking VPg prevented the immunoprecipitation of the product RNA by anti-VPg. The newly synthesized VPg-linked RNA strands may therefore be the result of end-labeling the fragmented, VPg-linked template RNA.[144,145] These results do not completely explain data by Morrow et al.[141] who reported that some immunoprecipitated RNAs were linked to a 45-kdalton VPg-related polypeptide.

Further analysis of product RNA synthesized in the presence of host factor, using single-strand polio cDNA as a probe, indicated that the product RNA was of both plus- and minus-strand polarity, and that a small fraction of the product could be larger than the template RNA.[146] In contrast, the product RNA primed with oligo(U) was exclusively of minus-strand polarity. This suggested that the generation of both plus- and minus-strand products was the result of the same mechanism, although, as it will be pointed out, the action of terminal uridylyl transferase can only explain the production of new minus strands.

Recent experiments from a number of different groups have been performed in an attempt to clarify the contrasting observations concerning the purified soluble replication system.

These studies have focused on the nature of the template RNA used in the in vitro reaction. Using RNA synthesized by an SP6 RNA polymerase from truncated cDNA clones representing both the 5' and 3' termini of plus- and minus-stranded RNA, Lubinski et al.[120] have shown that end-linked, double-stranded structures were formed in the reaction. It was suggested that the synthetic templates were degraded by endonuclease(s) such that snap-back structures could be formed that, in turn, served as primers for 3D^{pol}. In addition, neither the poly(A) tract of plus-strand RNA, nor heteropolymeric sequences at the 3' termini of either plus- or minus-strand RNA were required for the synthesis of dimer-length product.[120] Hey et al.[121] analyzed the end-linked products they obtained with host factor and 3D^{pol}, using virion RNA as template. None of the products contained poly(A)-poly(U) as was expected from the mechanism proposed by Flanegan and colleagues. Hey et al.[121] also suggested that the synthesis of end-linked structures is the result of endonucleolytic production of 3' termini suitable for priming the 3D^{pol} elongation reaction. This conclusion was further supported by the biochemical results of periodate oxidation of the 3' terminal adenylate residue of the template vRNA. If RNA synthesis is the result of addition of nucleotides to the poly(A) tract, then oxidation should prevent initiation on such a template. It was found, however, that the oxidized RNA remained an efficient template for 3D^{pol} when host factor was added during or prior to the reaction.[121] Indeed, it was found that host factor alone could remove poly(A) from the template RNA.[121] These data can be used to explain why, in reactions leading to end-linked structures, the products are extremely heterogeneous in length and why the end-linked poliovirus RF (twice the length of genome RNA) is only a minute fraction of these products.[119] It should be pointed out that no TUT activity was observed in the host factor preparations used by Lubinski et al.[120] or Hey et al.[121] Thus, it is not known whether host factor preparations used in different laboratories exhibit the same activity, a dilemma that should be solved by an exchange of reagents among the investigators.

B. Membrane-Associated Replication Systems

As was mentioned previously, poliovirus RNA synthesis occurs primarily in a membranous environment.[34,84-86,89] In vitro RNA synthesis assays using a crude membrane fraction from infected cells demonstrated that the membrane fraction is responsible for the majority of virus-specific RNA synthesis. This crude replication complex (CRC), as it has been called, can synthesize all three forms of viral-specific RNA under the appropriate in vitro conditions. Analysis of the RNA synthesized in vitro in the presence of [α-^{32}P]UTP by two-dimensional gel electrophoresis of RNase T1-generated oligonucleotides suggests that it is similar to authentic viral RNA.[147,148] Only very recently, however, has it been possible to synthesize VPg-linked RNA in vitro, that is, RNA resulting from faithful initiation of RNA synthesis.[149] The 5' VPg-linked RNase T1 oligonucleotide (VPg–pUUAAAACAGp) migrates slower than the other oligonucleotides during electrophoresis in the first dimension of the two-dimensional gel. If conditions are used which account for this aberrant migration, the 5' VPg-linked T1-oligonucleotide can be identified in a fingerprint of RNA synthesized by CRC in the presence of labeled [α-^{32}P]UTP.[149] Although plus-strand RNA is the predominant source of oligonucleotides in the fingerprints, oligonucleotides derived from minus strands can also be observed.[149] It thus appears that plus and minus RNA strands can be synthesized in CRC.

All known virus-specific polypeptides and their precursors can be found in the CRC. The affinity for membranes of most of these polypeptides is very high since neither high salt nor 4 M urea are sufficient to dissociate CRC.[34,114] Protease protection studies have shown that 3AB is partially protected by the presence of the membranes, an observation supporting the proposed insertion of 3AB into membranes due to its 22-amino acid hydrophobic region.[33,34] The addition of nonionic detergent prior to the addition of protease results in 3AB becoming protease-sensitive. Attempts to purify polypeptide 3AB have been hampered by tight association with other polypeptides (especially 2C) (our unpublished observations).

The interaction of 3AB with other viral polypeptides is probably an important feature of the viral replication complex.

The isolation of antibodies which recognized VPg and its precursors advanced the study of the CRC.[33,34,112] Using the anti-VPg antibodies, Takegami et al.[148] were able to develop an assay that detected the structures representing the 5' ends of both plus- and minus-strand RNA. A similar result was obtained also by Vartapetian et al.[150] who used a CRC obtained from *Ehrlich ascites* cells which had been infected with EMC virus. Small amounts of VPg-pUpU were also found in poliovirus-infected cells labeled in vivo by Crawford and Baltimore.[151] The assumed role of the membranes in initiation of polio RNA replication is supported by the observation that synthesis of VPg-pUpU in CRC was completely inhibited by the addition of nonionic detergent.[148,149]

Although the in vitro synthesis of VPg-pUpU could be reproducibly achieved by Takegami and colleagues,[148] the low yields of the nucleotidyl-peptide made it impossible to investigate its role in chain elongation. However, treatment of the CRC with DEAE-cellulose to remove endogenous nucleoside triphosphates and the addition of an ATP regenerating system greatly stimulated the yield of VPg-pUpU.[149] Conditions were then found which allowed [α-^{32}P]VPg-pUpU to be chased into longer products, as the chain elongation reaction was assayed by the detection of the 5' VPg-linked T1-oligonucleotide. Interestingly, conditions for efficient formation of VPg-pUpU are different from the optimal conditions of RNA chain elongation,[148,149] an observation suggesting two distinct processes. Unfortunately, the mechanism of VPg uridylylation has not been elucidated. In addition, it is not even known whether VPg-pUpU formation is dependent on a viral template. Attempts to remove endogenous template from CRC by micrococcal nuclease and "reactivate" the activities with added viral RNA have not met with success.

V. MODELS FOR RNA SYNTHESIS AND INITIATION

A. Initiation

The earliest analyses of virus-specific RNA structures found in infected cells strongly suggested that poliovirus RNA synthesis may proceed in steps unique among replication schemes of RNA viruses. First, the 5'-terminal nucleotide was found to be a pyrimidine (U) and not a purine, the preferred base for enzymes that initiate RNA synthesis *de novo*. Second, newly synthesized RNAs were all VPg-linked, even the nascent strands of RI, and no trace of pppN-termini could be found.[4] These observations were interpreted to suggest that VPg is involved in the initiation of RNA synthesis.[4] This notion was strongly supported by the finding that the poliovirus-specific RNA polymerase 3D^{pol} is a primer-dependent enzyme: it was considered possible that uridylylated VPg could serve as a primer for 3D^{pol} similar to the deoxycytidylylated terminal protein of adenovirus that serves as primer for the adenovirus-specific DNA polymerase (reviewed by Wimmer).[4] An alternative model was subsequently introduced by Flanegan and colleagues who proposed that RNA molecules allow "self-priming" on snap-back structures at their 3' termini such that covalently bound heteroduplex RNA molecules are formed.[119] VPg (or a precursor thereof) was then proposed to cleave the hairpin and subsequently link itself to the newly synthesized RNA strand. As we discussed above, this model has received considerable support through a number of publications. However, recent data have raised important questions about the validity of the hairpin priming mechanism for in vivo RNA synthesis.[120,121]

1. Initiation of RNA Synthesis by Protein Priming

If VPg is directly involved in the priming of RNA synthesis, one would have expected to find considerable amounts of the oligopeptide in infected cells. In fact, no free VPg was found in cells until Crawford and Baltimore[151] observed that the detection of unbound VPg

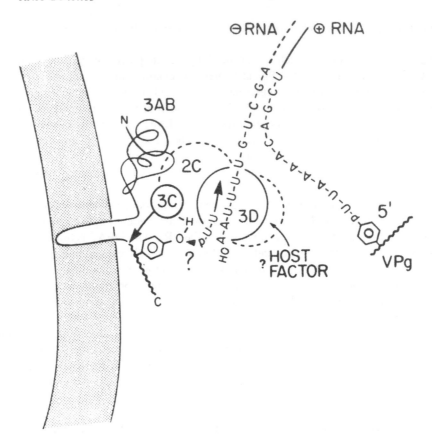

FIGURE 3.   Model of VPg-pUpU primed initiation of RNA synthesis. 3AB is the membrane-bound poliovirus protein, the COOH terminus (wavy line) of which is VPg, 3C is the virus-encoded proteinase (responsible for the cleavage between 3A and 3B), 3D, the primer-dependent RNA polymerase, and 2C, an auxiliary viral protein, mutations of which (g^r) lead to an altered phenotype of RNA synthesis in vivo. The possibility of the involvement of a "host factor" is indicated, although there is no evidence to support such factor in the membrane-bound replication complex. This model can account for the initiation of both plus and minus strand RNA. (Modified from Takegami, T., Kuhn, R. J., Anderson, C. W., and Wimmer, E., *Proc. Natl. Acad. Sci. U.S.A.*, 80, 7447, 1983.)

and uridylylated derivatives thereof is complicated by their physical properties. Takegami et al.,[34,148] on the other hand, argued that a precursor of VPg, most likely 3AB,[33,34,140] may participate in the initiation reaction. Since proteins larger than VPg have never been detected in a linkage with the nascent RNA strands, it was assumed that the precursor is rapidly cleaved by the proteinase 3C to yield VPg-RNA. 3AB is a membrane-bound polypeptide[33] and thus well suited to serve as donor for VPg in the membrane-associated machinery of poliovirus RNA replication. The model depicting the events leading to VPg-primed RNA synthesis is shown in Figure 3.[148] 3AB is uridylylated and subsequently cleaved by 3C^{pro}; at the same time, the uridylyl-VPg functions as primer for 3D^{pol}. Whether or not any host factor is involved in this process remains to be seen. Also, the role of polypeptide 2C in the events of initiation, if any, is unclear. It is possible that 2C functions as helicase for the release of the nascent strands, but this has not been proven.

Apart from studies of virus-specific proteins such as 3AB, this model is based upon experiments of RNA synthesis with the membraneous, crude replication complex isolated from infected HeLa cells. As mentioned in the previous section, VPg-pU and VPg-pUpU can be synthesized in vitro[148] and preformed VPg-pU can be chased into VPg-pUpU.[149]

Moreover, kinetic evidence suggests that the preformed VPg-pUpU can be chased into authentic elongation products.[149] All of these reactions are highly sensitive to the addition of nonionic detergents such as NP-40, an observation suggesting that an intact, membraneous environment must be maintained to uridylylate VPg and to release single stranded RNA from CRC.[149] The possibility cannot be excluded, however, that the uridylylation itself can proceed in the absence of membranes, but that the enzyme catalyzing this reaction is sensitive to detergent. A major question remains to be solved: what is the activity that catalyzes the uridylylation of the Tyr residue in VPg? Using mutants mapping in the 3' end of the viral RNA, Toyoda et al.[158] have obtained evidence that implicates $3D^{pol}$ in the formation of VPg-pUpU. Such an observation is not unreasonable since the adenovirus-specific DNA polymerase has also been shown to link dCMP onto the adenovirus protein.

A large panel of mutants in VPg are currently being constructed (using recombinant DNA techniques) by Kuhn et al.[159] with the objective of finding a phenotype that would shed light onto the events shown in Figure 3. It has become evident already that the NH_2-terminal sequence of VPg is highly sensitive to alteration. For example, if the VPg sequence H_2N–GAYTGL... (where the Y is the linker amino acid to the RNA) is changed to H_2N–GAYYGL... or H_2N–GATYGL..., the constructs are rendered noninfectious.[159] A change at amino acid Number 6, on the other hand, from L→M (H_2N–GAYTGM...) was found to yield infectious virus. This result is also of practical value, since VPg can now be labeled with ^{35}S-methionine which makes its detection easier. So far, no *ts* phenotype mapping in VPg has been found.

2. The Role of ''Host Factor''

As has been mentioned before, a host factor HF of 67,000 Mr was implicated in poliovirus RNA replication in an in vitro RNA synthesis system that did not require an oligo(U) primer. In the presence of HF, the product RNA appears to be full-length, that is, the product RNA represents complete, noncovalently linked complementary strands.[118] Indeed, if negative-stranded RNA (prepared from cDNA clones using SP6 polymerase) is used as template for $3D^{pol}$ and HF, infectious poliovirus RNA (containing nonviral sequences at either terminus) can be synthesized.[152] It was originally thought that RNA transcribed with $3D^{pol}$ and HF was linked to a protein containing VPg sequences and that the reaction was thus indicative of protein-primed initiation.[113,140,141] As was mentioned before, however, this conclusion has been disputed recently by two groups[144,145] who found that the majority (if not all) of the VPg-linked product RNA originates from the covalent linking of poliovirion template RNA to the transcripts. Currently, there are opposing views as to what ''host factor'' is and what role it plays in poliovirus replication.

3. The Hairpin Model

Flanegan and colleagues reported that in the in vitro transcription of poliovirion RNA, using purified $3D^{pol}$ and HF, molecules were produced in which template and product RNA are covalently linked at one end.[119] In contrast, when oligo(U) was added to the transcription mixture, none of the products were end-linked since initiation occurred on the oligonucleotide primer. It was concluded that, in the absence of oligo(U), HF serves to modify the 3' end of the template RNA topographically such that its 3'-terminal nucleoside can function as primer (Figure 4).

The hairpin model of initiation received support by Andrews et al.[64,143] who reported that HF is a terminal uridylyl transferase (TUT). Accordingly, these authors suggested that the 3' end of the template is uridylylated. The product then forms a snap-back structure, the 3' end of which serves as primer for $3D^{pol}$. The addition of U residues to 3' termini of RNA was found to occur most efficiently on poly(A) tails. These observations are credible for initiation of RNA synthesis at the 3'-terminal poly(A) of poliovirus plus strands, but make it difficult to explain initiation of plus strands at the 3' end of minus strands (see below).

TEMPLATE RNA AS PRIMER

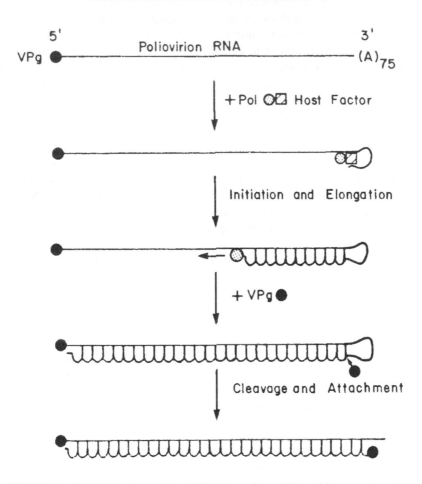

FIGURE 4.   Self-priming by template RNA. A host factor (either a kinase or a terminal uridylate transferase) produces a hairpin at the 3' end of the template RNA that serves as a primer for 3D^{pol}. After some elongation has taken place, VPg (or its precursor) will cleave the hairpin, thereby attaching itself to the 5' end of the nascent RNA strand. If nicking does not occur, an end-linked RF molecule is formed; as shown here, the homopolymeric segments of the RF would be covalently bound. We have named those structures "homo-linked" RF. Hairpin-mediated initiation and complete transcription of minus-stranded template would yield "hetero-linked" RF molecules. (This drawing was kindly provided by the courtesy of Dr. J. B. Flanegan.)

The self-priming hairpin model has recently suffered from new data that strongly suggests the involvement of endonuclease(s) in the in vitro reactions leading to random degradation of template with the fortuitous formation of snap-back structures.[120,121] Lubinski et al.,[120] in addition to the nicking mechanism, have proposed yet another mechanism that may function in the formation of end-linked products. This mechanism states that snap-back structures occur in certain full-length product RNA (a phenomenon resembling reverse transcription), but not in template RNA. Taken altogether, the transcription experiments with purified or semipurified polypeptides in completely soluble medium have never produced authentic poliovirus RNA. Instead, they have yielded rather conflicting results that have not yet uncovered how initiation of poliovirus RNA synthesis occurs in vivo. Moreover, even if end-linked molecules are intermediates in poliovirus replication, what protein supposedly cleaves the loops? VPg or a precursor to VPg?

4. The Problem of Template Recognition and of Initiation at Different Termini

Whatever the mechanism, the poliovirus RNA replication machinery must be capable of initiating RNA synthesis at two very different 3' termini: the homopolymeric segment [poly(A)] of plus strands, and the heteropolymeric terminus ($-CAGUUUUAA_{OH}$) of minus strands. The only feature common to these ends are two terminal $-AA_{OH}$ residues. Clearly VPg-pUpU, a structure that can be synthesized in a replication complex in vitro[148,149] and that has been detected in infected cells,[151] could recognize and bind to such a terminal dinucleotide prior to initiation. Similarly, both termini of plus or minus strands could be oligouridylylated by TUT to form snapback structures. If so, one would predict that the termini of minus strands isolated from infected cells would carry extra U residues that remained there after the assumed cutting by VPg. As mentioned above, no U residues have been found at the 3' termini of minus strands of RF.[130,133] Unfortunately, a similar study on minus strands of RI has not been performed.

Two A residues at the 3' terminus of an RNA can hardly be the sole determinant of template selection since nearly all cellular mRNAs are polyadenylylated. Thus, the poliovirus polypeptide must recognize sequences other than the ends of the RNA. Of course, numerous primary or secondary structures distant to the termini could serve as recognition signals. No such signals have been discovered as yet. Experiments to assay for these signals are difficult because (1) an in vitro replication system is not available, and (2) deletions in the genome of DI particles are restricted to the P1 capsid region (see below). These observations, which contrast with other RNA virus replication systems (e.g., vesicular stomatitis virus or togaviruses), do not allow an easy search for the minimum genome sequence still capable of replicating through complementation with wild-type virus. The in vitro construction of DI genomes, based upon the highly efficient phage T7 RNA polymerase transcription system,[153] has recently been achieved, but no "viable" deletion mutants have been found as yet (Bradley, Wimmer, Girard, and van der Werf, unpublished results).

B. Elongation

The RNA polymerase 3D^{pol} is probably the only activity produced by poliovirus capable of elongating the nascent RNA strand. It is unknown as yet whether the newly synthesized RNA is released from the template by strand displacement (in RI molecules with a double-stranded backbone) or prevented from hybridizing to the template RNA by polypeptides in the replication complex (in single-stranded RI molecules). If strand replacement is the mechanism, one would predict that an RNA-specific helicase would be involved in replication also.

When genomes of DI particles were sequenced, it was found that the deletions in the capsid region **always** occur in the P1 region and in frame of the polyprotein[154] (compare Volume II, Chapter 9). Kuge et al.[154] suggested that poliovirus genomes may need their own replication proteins (originating from the P2 and P3 regions) for RNA synthesis, that is, that some P2 or P3 proteins may function only *in cis*. This proposal finds support in a genetic analysis of Bernstein et al.[155] who found that specific mutants in the P2 and P3 region generated in vitro by manipulations of cDNA clones, cannot be complemented. More data, however, must be generated to define the specificity of viral proteins for heterologous or homologous template RNA. For example, if 3D^{pol} can indeed only act *in cis* (that is, on the RNA only from which it was translated) it is difficult to envision the events that lead to RI structures with multiple growing strands. It is possible that 3D^{pol} acts *in cis* only to produce minus strands but can funtion *in trans* to produce plus strands.

VI. SUMMARY AND FUTURE DIRECTIONS

The study of poliovirus RNA replication, now in its 24th year and pursued by many investigators, has proven to be exceptionally tedious and frustrating. As has been pointed out before[3] the search for the ultimate in vitro replication system followed the classical strategy by Spiegelman and contemporaries involving the purification of cellular and viral polypeptides and their subsequent reconstitution to a functional complex. So far, this strategy has been largely unsuccessful. The authors of this review are convinced that an essential component of the in vitro replication complex is a hydrophobic environment provided by cellular membranes where polypeptides can be anchored and processed to participate in RNA replication. Indeed, only a membraneous replication complex has yielded authentic virion RNA so far.[149] Clearly, the membrane may be replaceable by an artificial hydrophobic environment, but attempts to achieve this have so far not been successful.

The major problems that should be solved in the near future to elucidate the mechanism of poliovirus RNA replication are

1. An efficient assay for the uridylylation of VPg or its precursors in vitro. This would allow an elucidation of the mechanism of initiation.
2. The elucidation of essential components of the membraneous endogenous replication complex. This includes particularly the role of 2C and the question of the participation of any host cellular protein in RNA synthesis.
3. The problem of template selection, e.g., the search for recognition signals.
4. The explanation of why virus-specific protein synthesis and RNA replication are tightly linked.

ACKNOWLEDGMENTS

We thank our colleagues, particularly Naokazu Takeda, Chen-Fu Yang, Haruka Toyoda, Hiroomi Tada, Steven Pincus, and Martin Nicklin, for many stimulating discussions. We are indebted to Patricia Dewalt, Victoria Johnson, and Mary Frances Ypma-Wong for critical comments. We are grateful to Lynn Zawacki for preparation of the manuscript. Work described here was supported in part by NIH grants AI 15122, CA 28146 (E.W.), and AI 22693 and American Cancer Society grant MV-183 (B.L.S.). B.L.S. is supported by a Faculty Research Award from the American Cancer Society.

REFERENCES

1. **Rueckert, R. R.,** Picornaviruses and their replication, in *Virology,* Fields, B. N., Ed., Raven Press, New York, 1985, 705.
2. **Koch, F. and Koch, G.,** *The Molecular Biology of Poliovirus,* Springer-Verlag, Vienna, 1985.
3. **Kuhn, R. J. and Wimmer, E.,** The replication of picornaviruses, in *The Molecular Biology of Positive Strand RNA Viruses,* Rowlands, D. J., Mahy, B. W. J., and Mayo, M., Eds., Academic Press, New York, 1986, in press.
4. **Wimmer, E.,** Genome-linked proteins of viruses, *Cell,* 28, 199, 1982.
5. **Alexander, H. E., Koch, G., Morgan Mountain, I., and Van Damme, O.,** Infectivity of ribonucleic acid from poliovirus in human cell monolayers, *J. Exp. Med.,* 108, 493, 1958.
6. **Holland, J. J., McLaren, L. C., and Syverton, J. T.,** The mammalian cell-virus relationship. IV. Infection of naturally insusceptible cells with enterovirus ribonucleic acid, *J. Exp. Med.,* 110, 65, 1959.
7. **Lee, Y. F., Nomoto, A., Detjen, B. M., and Wimmer, E.,** A protein covalently linked to poliovirus genome RNA, *Proc. Natl. Acad. Sci. U.S.A.,* 74, 59, 1977.
8. **Flanegan, J. B., Pettersson, R. F., Ambros, V., Hewlett, M. J., and Baltimore, D.,** Covalent linkage of a protein to a defined nucleotide sequence at the 5′-terminus of virion and replicative intermediate RNAs of poliovirus, *Proc. Natl. Acad. Sci. U.S.A.,* 74, 961, 1977.

9. **Armstrong, J. A., Edmonds, M., Nakazato, H., Phillips, B. A., and Vaughan, M. H.,** Polyadenylic acid sequences in the virion RNA of poliovirus and Eastern Equine Encephalitis Virus, *Science,* 176, 526, 1972.

10. **Yogo, Y. and Wimmer, E.,** Polyadenylic acid at the 3′ terminus of poliovirus RNA, *Proc. Natl. Acad. Sci. U.S.A.,* 9, 1877, 1972.

11. **Rothberg, P. G., Harris, T. J. R., Nomoto, A., and Wimmer, E.,** O⁴-(5′-Uridylyl) tyrosine is the bond between the genome-linked protein and the RNA of poliovirus, *Proc. Natl. Acad. Sci. U.S.A.,* 75, 4868, 1978.

12. **Ambros, V. and Baltimore, D.,** Protein is linked to the 5′ end of poliovirus RNA by a phosphodiester linkage to tyrosine, *J. Biol. Chem.,* 253, 5263, 1978.

13. **Nomoto, A., Kitamura, N., Golini, F., and Wimmer, E.,** The 5′-terminal structures of poliovirion RNA and poliovirus mRNA differ only in the genome-linked protein VPg, *Proc. Natl. Acad. Sci. U.S.A.,* 74, 5345, 1977.

14. **Nomoto, A., Lee, Y. F., and Wimmer, E.,** The 5′ end of poliovirus mRNA is not capped with m⁷G(5′)ppp(5′)Np, *Proc. Natl. Acad. Sci. U.S.A.,* 73, 375, 1976.

15. **Hewlett, M. J., Rose, J. K., and Baltimore, D.,** 5′-terminal structure of poliovirus polyribosomal RNA is pUp, *Proc. Natl. Acad. Sci. U.S.A.,* 73, 327, 1976.

16. **Fernandez-Munoz, R. and Darnell, J. E.,** Structural difference between the 5′-termini of viral and cellular mRNA in the poliovirus-infected cells: possible basis for the inhibition of the host protein synthesis, *J. Virol.,* 18, 719, 1976.

17. **Kitamura, N., Semler, B. L., Rothberg, P. G., Larsen, G. R., Adler, C. J., Dorner, A. J., Emini, E. A., Hanecak, R., Lee, J. J., van der Werf, S., Anderson, C. W., and Wimmer, E.,** Primary structure, gene organization and polypeptide expression of poliovirus RNA, *Nature (London),* 291, 547, 1981.

18. **Racaniello, V. R. and Baltimore, D.,** Molecular cloning of poliovirus cDNA and determination of the complete nucleotide sequence of the viral genome, *Proc. Natl. Acad. Sci. U.S.A.,* 78, 4887, 1981.

19. **Nomoto, A., Omata, T., Toyoda, H., Kuge, S., Horie, H., Kataoka, Y., Genba, Y., Nakano, Y., and Imura, N.,** Complete nucleotide sequence of the attenuated poliovirus Sabin 1 strain genome, *Proc. Natl. Acad. Sci. U.S.A.,* 79, 5793, 1982.

20. **Stanway, G., Cann, A. J., Hauptmann, R., Hughes, P., Clarke, L. D., Mountford, R. C., Minor, P. D., Schild, G., and Almond, J. W.,** The nucleotide sequence of poliovirus type 3 leon 12 a₁b: comparison with poliovirus type 1, *Nucleic Acids Res.,* 11, 5629, 1983.

21. **Toyoda, H., Kohara, M., Kataok, Y., Suganuma, T., Omata, T., Imura, N., and Nomoto, A.,** Complete nucleotide sequences of all three poliovirus serotype genomes, *J. Mol. Biol.,* 174, 561, 1984.

22. **La Monica, N., Meriam, C., and Racaniello, V.,** Mapping of sequences required for mouse neurovirulence of poliovirus type 2 lansing, *J. Virol.,* 57, 515, 1986.

23. **Dorner, A. J., Dorner, L. F., Larsen, G. R., Wimmer, E., and Anderson, C. W.,** Identification of the initiation site of poliovirus polyprotein synthesis, *J. Virol.,* 42, 1017, 1982.

24. **Hanecak, R., Semler, B. L., Anderson, C. W., and Wimmer, E.,** Proteolytic processing of poliovirus polypeptides: antibodies to polypeptide P3-7C inhibit cleavage at glutamine-glycine pairs, *Proc. Natl. Acad. Sci. U.S.A.,* 79, 3973, 1982.

25. **Toyoda, H., Nicklin, M. J. H., Murray, M. G., Anderson, C. W., Dunn, J. J., Studier, F. W., and Wimmer, E.,** A second virus-encoded proteinase involved in proteolytic processing of poliovirus polyprotein, *Cell,* 45, 761, 1986.

26. **Rueckert, R. R. and Wimmer, E.,** Systematic nomenclature of picornavirus proteins, *J. Virol.,* 50, 957, 1984.

27. **Pallansch, M. A., Kew, O. M., Semler, B. L., Omilianowski, D. R., Anderson, C. W., Wimmer, E., and Reuckert, R. R.,** The protein processing map of poliovirus, *J. Virol.,* 49, 873, 1984.

28. **Korant, B., Chow, N., Lively, M., and Powers, J.,** Virus-specified protease in poliovirus-infected HeLa cells, *Proc. Natl. Acad. Sci. U.S.A.,* 76, 2992, 1979.

29. **Anderson-Sillman, K., Bartal, S., and Tershak, D. R.,** Guanidine-resistant poliovirus mutants produce modified 37-kilodalton proteins, *J. Virol.,* 50, 922, 1984.

30. **Yin, F. H. and Lomax, N. B.,** Host range mutants of human rhinovirus in which nonstructural proteins are altered, *J. Virol.,* 48, 410, 1983.

31. **Etchison, D. and Ehrenfeld, E.,** Viral polypeptides associated with the RNA replication complex in poliovirus-infected cells, *Virology,* 107, 135, 1980.

32. **Semler, B. L., Hanecak, R., Dorner, L. F., Anderson, C. W., and Wimmer, E.,** Poliovirus RNA synthesis *in vitro:* structural elements and antibody inhibition, *Virology,* 126, 624, 1983.

33. **Semler, B. L., Anderson, C. W., Hanecak, R., Dorner, L. F., and Wimmer, E.,** A membrane-associated precursor to poliovirus VPg identified by immunoprecipitation with antibodies directed against a synthetic heptapeptide, *Cell,* 28, 405, 1982.

34. **Takegami, T., Semler, B. L., Anderson, C. W., and Wimmer, E.**, Membrane fractions active in poliovirus RNA replication contain VPg precursor polypeptides, *Virology*, 128, 33, 1983.
35. **Summers, D. F., and Maizel, J. V.**, Determination of the gene sequence of poliovirus with pactamycin, *Proc. Natl. Acad. Sci. U.S.A.*, 68, 2852, 1971.
36. **Taber, R., Rekosh, D., and Baltimore, D.**, Effect of pactamycin on synthesis of poliovirus proteins: a method for genetic mapping, *J. Virol.*, 8, 395, 1971.
37. **Jacobson, M. F., Asso, J., and Baltimore, D.**, Further evidence on the formation of poliovirus proteins, *J. Mol. Biol.*, 49, 657, 1970.
38. **Rueckert, R. R., Matthews, T. J., Kew, O. M., Pallansch, M., McClean, C., and Omilianowski, D.**, Synthesis and processing of picornaviral protein, in *The Molecular Biology of Picornaviruses*, Perez-Bercoff, R., Ed., Plenum Press, New York, 1979, 113.
39. **Semler, B. L., Anderson, C. W., Kitamura, N., Rothberg, P. G., Wishart, W. L., and Wimmer, E.**, Poliovirus replication proteins: RNA sequence encoding P3-1b and the sites of proteolytic processing, *Proc. Natl. Acad. Sci. U.S.A.*, 78, 3464, 1981.
40. **Semler, B. L., Hanecak, R., Anderson, C. W., and Wimmer, E.**, Cleavage sites in the polypeptide precursors of poliovirus protein P2-X, *Virology*, 114, 589, 1981.
41. **Larsen, G. R., Anderson, C. W., Dorner, A. J., Semler, B. L., and Wimmer, E.**, Cleavage sites within the poliovirus capsid protein precursors, *J. Virol.*, 41, 340, 1982.
42. **Emini, E. A., Elzinga, M., and Wimmer, E.**, Carboxy-terminal analysis of poliovirus proteins: termination of poliovirus RNA translation and location of unique poliovirus polyprotein cleavage sites, *J. Virol.*, 42, 194, 1982.
43. **Hanecak, R., Semler, B. L., Ariga, H., Anderson, C. W., and Wimmer, E.**, Expression of a cloned gene segment of poliovirus in E. coli: evidence for autocatalytic production of the viral proteinase, *Cell*, 37, 1063, 1984.
44. **Palmenberg, A. C. and Rueckert, R. R.**, Evidence for intramolecular self-cleavage of picornaviral replicase precursors, *J. Virol.*, 41, 244, 1982.
45. **Nicklin, M. J. H., Toyoda, H., Murray, M. G. and Wimmer, E.**, Proteolytic processing in the replication of polio and related viruses, *Biotechnology*, 4, 33, 1986.
46. **Holland, J. J.**, Enterovirus entrance into specific host cells and subsequent alterations of cell protein and nucleic acid synthesis, *Bacteriol. Rev.*, 28, 3, 1964.
47. **Ambros, V., Pettersson, R. F., and Baltimore, D.**, An enzymatic activity in uninfected cells that cleaves the linkage between poliovirion RNA and the 5' terminal protein, *Cell*, 15, 1439, 1978.
48. **Golini, F., Semler, B. L., Dorner, A. J., and Wimmer, E.**, Protein-linked RNA of poliovirus is competent to form an initiation complex of translation *in vitro*, *Nature (London)*, 287, 600, 1980.
49. **Trachsel, H., Sonenberg, N., Shatkin, A. J., Rose, J. K., Leong, K., Bergmann, J. E., Gordon, J., and Baltimore, D.**, Purification of a factor that restores translation of vesicular stomatitis virus mRNA in extracts from poliovirus-infected HeLa cells, *Proc. Natl. Acad. Sci. U.S.A.*, 77, 770, 1980.
50. **Lee, K. A. W., and Sonenberg, N.**, Inactivation of cap-binding proteins accompanies the shut-off of host protein synthesis by poliovirus, *Proc. Natl. Acad. Sci. U.S.A.*, 79, 3447, 1982.
51. **Hansen, J., Etchison, D., Hershey, J. W. B., and Ehrenfeld, E.**, Association of cap-binding protein with eukaryotic initiation factor 3 in initiation factor preparations from uninfected and poliovirus-infected HeLa cells, *J. Virol.*, 42, 200, 1982.
52. **Etchison, D., Milburn, S. C., Edery, I., Sonenberg, N., and Hershey, J. W. B.**, Inhibition of HeLa cell protein synthesis following poliovirus infection correlates with the proteolysis of a 220,000-dalton polypeptide associated with eukaryotic initiation factor 3 and a cap binding protein complex, *J. Biol. Chem.*, 257, 14806, 1982.
53. **Etchison, D., Hansen, J., Ehrenfeld, E., Edery, I., Sonenberg, N., Milburn, S., and Hershey, J. W. B.**, Demonstration *in vitro* that eukaryotic initiation factor 3 is active but that a cap-binding protein complex is inactive in poliovirus-infected HeLa cells, *J. Virol.*, 51, 832, 1984.
54. **Lee, K. A. W., Edery, I., and Sonenberg, N.**, Isolation and structural characterization of cap-binding proteins from poliovirus-infected HeLa cells, *J. Virol.*, 54, 515, 1985.
55. **Helentijaris, T. and Ehrenfeld, E.**, Inhibition of host cell protein synthesis by UV-inactivated poliovirus, *J. Virol.*, 21, 259, 1977.
56. **Lloyd, R. E., Etchison, D., and Ehrenfeld, E.**, Poliovirus protease does not mediate cleavage of the 220,000-Da component of the cap-binding protein complex, *Proc. Natl. Acad. Sci. U.S.A.*, 82, 2723, 1985.
57. **Lee, K. A. W., Edery, I., Hanecak, R., Wimmer, E., and Sonenberg, N.**, Poliovirus protease 3C (P3-7C) does not cleave P220 of the eukaryotic mRNA cap-binding protein complex, *J. Virol.*, 55, 489, 1985.
58. **Lloyd, R. E., Toyoda, H., Etchison, D., Wimmer, E., and Ehrenfeld, E.**, Cleavage of the cap-binding protein complex polypeptide p220 is not effected by the second poliovirus protease 2A, *Virology*, 150, 229, 1986.
59. **Lundquist, R. E., Ehrenfeld, E., and Maizel, J. V.**, Isolation of a viral polypeptide associated with poliovirus RNA polymerase, *Proc. Natl. Acad. Sci. U.S.A.*, 71, 4773, 1974.

60. **Flanegan, J. B. and Baltimore, D.,** Poliovirus polyuridylic acid polymerase and RNA replicase have the same viral polypeptide, *J. Virol.,* 29, 352, 1979.
61. **Bowles, S. A. and Tershak, D. R.,** Proteolysis of noncapsid protein 2 of type 3 poliovirus at the restrictive temperature: breakdown of noncapsid protein 2 correlates with loss of RNA synthesis, *J. Virol.,* 27, 443, 1978.
62. **Wright, P. J. and Cooper, P. D.,** Poliovirus proteins associated with the replication complex in infected cells, *J. Gen. Virol.,* 30, 63, 1976.
63. **Dasgupta, A., Zabel, P., and Baltimore, D.,** Dependence of the activity of the poliovirus replicase on a host cell protein, *Cell,* 19, 423, 1980.
64. **Andrews, N. C. and Baltimore, D.,** Purification of a terminal uridylyl-transferase that acts as host factor in the *in vitro* poliovirus replicase reaction, *Proc. Natl. Acad. Sci. U.S.A.,* 83, 221, 1986.
65. **Rueckert, R. R.,** On the structure and morphogenesis of picornaviruses, in *Comprehensive Virology,* Vol. 6, Fraenkel-Conrat, H. and Wagner, R. R., Eds., Plenum Press, New York, 1976, 131.
66. **Rossmann, M. G., Arnold, E., Erickson, J. W., Frankenberger, E. A., Griffith, J. P., Hecht, H., Johnson, J. E., Kamer, G., Luo, M., Mosser, A. G., Rueckert, R. R., Sherry, B., and Vriend, G.,** Structure of a human common cold virus and functional relationship to other picornaviruses, *Nature (London),* 317, 145, 1985.
67. **Hogle, J. M., Chow, M., and Filman, D. J.,** Three-dimensional structure of poliovirus at 2.9A resolution, *Science,* 229, 1358, 1985.
68. **Carroll, A. R., Rowlands, D. J., and Clark, B. E.,** The complete nucleotide sequence of the RNA coding for the primary translation product of foot-and-mouth disease virus, *Nucleic Acids Res.,* 12, 2461, 1984.
69. **Forss, S., Strebel, K., Beck, E., and Schaller, H.,** Nucleotide sequence and genome organization of foot-and-mouth disease virus, *Nucleic Acids Res.,* 12, 6587, 1984.
70. **Robertson, B. H., Grubman, M. J., Weddell, G. N., Moore, D. M., Welsh, J. D., Fischer, T., Dowbenko, D. J., Yansura, D. G., Small, B., and Kleid, D. G.,** Nucleotide and amino acid sequence coding for polypeptides of foot-and-mouth disease virus type A12, *J. Virol.,* 54, 651, 1985.
71. **Palmenberg, A. C., Kirby, E. M., Janda, M. R., Drake, N. L., Duke, G. M., Potratz, K. F., and Collett, M. S.,** The nucleotide and deduced amino acid sequence of the encephalomyocarditis viral polyprotein coding region, *Nucleic Acids Res.,* 12, 2969, 1984.
72. **Stanway, G., Hughes, P., Mountford, R. C., Minor, P. D., and Almond, J. W.,** The complete nucleotide sequence of a common cold virus: rhinovirus 14, *Nucleic Acids Res.,* 12, 7859, 1984.
73. **Callahan, P. L., Mizutani, S., and Colonno, R. J.,** Molecular cloning and complete sequence determination of RNA genome of human rhinovirus type 14, *Proc. Natl. Acad. Sci. U.S.A.,* 82, 732, 1985.
74. **Skern, T., Sommergruber, W., Blaas, D., Gruendler, P., Fraundorfer, F., Pieler, C., Fogy, I., and Kuechler, E.,** Human rhinovirus 2: complete nucleotide sequence and proteolytic processing signals in the capsid protein region, *Nucleic Acids Res.,* 13, 2111, 1985.
75. **Baroudy, B. M., Ticehurst, J. R., Miele, T. A., Maizel, J. V., Purchell, R. H., and Feinstone, S. M.,** Sequence analysis of hepatitis A virus cDNA coding for capsid proteins and RNA polymerase, *Proc. Natl. Acad. Sci. U.S.A.,* 82, 2143, 1985.
76. **Najarian, R., Caput, D., Gee, W., Potter, S. J., Renard, A., Merryweather, J., Van Nest, G., and Dina, D.,** Primary structure and gene organization of human hepatitis A virus, *Proc. Natl. Acad. Sci. U.S.A.,* 82, 2627, 1985.
77. **Linemeyer, D. L., Menke, J. G., Martin-Gallardo, A., Hughes, J. V., Young, A., and Mitra, S. W.,** Molecular cloning and partial sequencing of hepatitis A viral cDNA, *J. Virol.,* 54, 247, 1985.
78. **Tracy, S., Liu, H.-L., and Chapman, N. M.,** Coxsackievirus B3: primary structure of the 5' non-coding and capsid protein coding regions of the genome, *Virus Res.,* 3, 263, 1985.
79. **Rowlands, D. J., Harris, T. J. R., and Brown, F.,** More precise location of the polycytidylic acid tract in foot-and-mouth disease virus RNA, *J. Virol.,* 26, 335, 1978.
80. **Strebel, K. and Beck, E.,** A second protease of foot-and-mouth disease virus, *J. Virol.,* 58, 893, 1986.
81. **Forss, S. and Schaller, H.,** A tandem repeat gene in a picornavirus, *Nucleic Acids Res.,* 10, 6441, 1982.
82. **Jackson, R. J.,** A detailed kinetic analysis of the *in vitro* synthesis and processing of encephalomyocarditis virus products, *Virology,* 149, 114, 1986.
83. **Baltimore, D., Franklin, R. M., Eggers, H. J., and Tamm, I.,** Poliovirus induced RNA polymerase and the effects of virus-specific inhibitors on its production, *Proc. Natl. Acad. Sci. U.S.A.,* 49, 843, 1963.
84. **Penman, S., Becker, Y., and Darnell, J. E.,** A cytoplasmic structure involved in the synthesis and assembly of poliovirus components, *J. Mol. Biol.,* 8, 541, 1964.
85. **Baltimore, D.,** *In vitro* synthesis of viral RNA by the poliovirus RNA polymerase, *Proc. Natl. Acad. Sci. U.S.A.,* 51, 450, 1964.
86. **Girard, M., Baltimore, D., and Darnell, J. E.,** The poliovirus replication complex: site for synthesis of poliovirus RNA, *J. Mol. Biol.,* 24, 59, 1967.

87. **Arlinghaus, R. B. and Polatnick, J.,** Detergent-solubilized RNA polymerase from cells infected with foot-and-mouth disease virus, *Science,* 158, 1320, 1967.
88. **Arlinghaus, R. B., Syrewicz, J. J., and Loesch, W. T., Jr.,** RNA polymerase complexes from mengovirus infected cells, *Arch. Gesch. Virusforsch.,* 38, 17, 1972.
89. **Caliguiri, L. A. and Tamm, I.,** The role of cytoplasmic membranes in poliovirus biosynthesis, *Virology,* 42, 100, 1970.
90. **Dorsch-Hasler, K., Yogo, Y., and Wimmer, E.,** Replication of picornaviruses. I. Evidence from *in vitro* RNA synthesis that poly(A) of the poliovirus genome is genetically coded, *J. Virol.,* 16, 1512, 1975.
91. **Mosser, A. G., Caliguiri, L. A., and Tamm, I.,** Incorporation of lipid precursors into cytoplasmic membranes of poliovirus-infected HeLa cells, *Virology,* 47, 39, 1972.
92. **Caliguiri, L. A. and Mosser, A. G.,** Proteins associated with the poliovirus RNA replication complex, *Virology,* 46, 375, 1971.
93. **Baltimore, D. and Girard, M.,** An intermediate in the synthesis of poliovirus RNA, *Proc. Natl. Acad. Sci. U.S.A.,* 56, 741, 1966.
94. **Ehrenfeld, E., Maizel, J. V., and Summers, D. F.,** Soluble RNA polymerase complex from poliovirus-infected HeLa cells, *Virology,* 40, 840, 1970.
95. **Roder, A. and Koschel, K.,** Virus-specific proteins associated with the replication complex of poliovirus RNA, *J. Gen. Virol.,* 28, 85, 1975.
96. **Flanegan, J. B. and Baltimore, D.,** Poliovirus-specific primer-dependent RNA polymerase able to copy poly(A), *Proc. Natl. Acad. Sci. U.S.A.,* 74, 3677, 1977.
97. **Spector, D. and Baltimore, D.,** Polyadenylic acid on poliovirus RNA. III. *In vitro* addition of polyadenylic acid to poliovirus RNAs, *J. Virol.,* 15, 1432, 1975.
98. **Van Dyke, T. A. and Flanegan, J. B.,** Identification of poliovirus polypeptide p63 as a soluble RNA-dependent RNA polymerase, *J. Virol.,* 35, 732, 1980.
99. **Dasgupta, A., Baron, M. H., and Baltimore, D.,** Poliovirus replicase: a soluble enzyme able to initiate copying of poliovirus RNA, *Proc. Natl. Acad. Sci. U.S.A.,* 76, 2679, 1979.
100. **Korant, B. D.,** Regulation of animal virus replication by protein cleavage, in *Proteases and Biological Control,* Reich, E., Rifkin, D. B., and Shaw, E., Eds., Cold Spring Harbor Laboratory, Cold Spring Harbor, New York, 1975, 621.
101. **Caliguiri, L. A. and Tamm, I.,** Guanidine and 2-(hydroxybenzyl)benzimadizole (HBB): selective inhibitors of picornavirus multiplication, in *Selective Inhibitors of Viral Functions,* Carter, W., Ed., CRC Press, Boca Raton, Florida, 1973, 257.
102. **Tershak, D. R.,** Inhibition of poliovirus polymerase by guanidine *in vitro, J. Virol.,* 41, 313, 1982.
103. **Cooper, P. D.,** The genetic analysis of poliovirus, in *The Biochemistry of Viruses,* Levy, H. B., Ed., Marcel Dekker, New York, 1969.
104. **McCahon, D., Slade, W. R., Priston, R. A. J., and Lake, J. R.,** An extended genetic recombination map for foot-and-mouth disease virus, *J. Gen. Virol.,* 35, 555, 1977.
105. **Tolskaya, E. A., Romanova, L. A., Kolesnikova, M. S., and Agol, V. I.,** Intertypic recombination in poliovirus: genetic and biochemical studies, *Virology,* 124, 121, 1983.
106. **Emini, E. A., Leibowitz, J., Diamond, D. C., Bonin, J., and Wimmer, E.,** Recombinants of Mahoney and Sabin strain poliovirus type 1: analysis of *in vitro* phenotypic markers and evidence that resistance to guanidine maps in the non-structural proteins, *Virology,* 137, 74, 1984.
107. **Anderson-Sillman, K., Bartal, S., and Tershak, D. R.,** Guanidine-resistant poliovirus mutants produce modified 37-kilodalton proteins, *J. Virol.,* 50, 922, 1984.
108. **Saunders, K., King, A. M. Q., McCahon, D., Newman, J. W. I., Slade, W. R., and Forss, S.,** Recombination and oligonucleotide analysis of guanidine-resistant foot-and-mouth disease virus mutants, *J. Virol.,* 56, 921, 1985.
109. **Pincus, S. E., Diamond, D. C., Emini, E. A., and Wimmer, E.,** Guanidine-selected mutants of poliovirus: mapping of point mutations to polypeptide 2C, *J. Virol.,* 57, 638, 1986.
110. **Nomoto, A., Detjen, B., Pozzatti, R., and Wimmer, E.,** The location of the polio genome protein in viral RNA and its implication for RNA synthesis, *Nature (London),* 268, 208, 1977.
111. **Pettersson, R. F., Ambros, V., and Baltimore, D.,** Identification of a protein linked to nascent poliovirus RNA and to the polyuridylic acid of negative strand RNA, *J. Virol.,* 27, 357, 1978.
112. **Baron, M. H. and Baltimore, D.,** Antibodies against the chemically synthesized genome-linked protein of poliovirus react with native virus-specific proteins, *Cell,* 28, 395, 1982.
113. **Morrow, C. D. and Dasgupta, A.,** Antibody to a synthetic nonapeptide to the NH_2 terminus of poliovirus genome-linked protein VPg reacts with native VPg and inhibits *in vitro* replication of poliovirus RNA, *J. Virol.,* 48, 429, 1983.
114. **Tershak, D. R.,** Association of poliovirus proteins with the endoplasmic reticulum, *J. Virol.,* 52, 777, 1984.
115. **Baron, M. H. and Baltimore, D.,** Purification and properties of a host cell protein required for poliovirus replication *in vitro, J. Biol. Chem.,* 257, 12351, 1982.

116. **Dasgupta, A.,** Purification of host factor required for *in vitro* transcription of poliovirus RNA, *Virology,* 127, 245, 1983.

117. **Dasgupta, A.,** Antibody to host factor precipitates poliovirus RNA polymerase from poliovirus-infected HeLa cells, *Virology,* 128, 252, 1983.

118. **Morrow, C. D., Gibbons, G. F., and Dasgupta, A.,** The host protein required for *in vitro* replication of poliovirus is a protein kinase that phosphorylates eukaryotic initiation factor-2, *Cell,* 40, 913, 1985.

119. **Young, D. C., Tuschall, D. M., and Flanegan, J. B.,** Poliovirus RNA-dependent RNA polymerase and host cell protein synthesis product RNA twice the size of poliovirus RNA *in vitro, J. Virol.,* 54, 256, 1985.

120. **Lubinski, J. M., Kaplan, G., Racaniello, V. R., and Dasgupta, A.,** Mechanism of *in vitro* synthesis of covalently linked dimeric RNA molecules by the poliovirus replicase, *J. Virol.,* 58, 459, 1986.

121. **Hey, T. D., Richards, O. C., and Ehrenfeld, E.,** Host factor-induced template modification during synthesis of poliovirus RNA *in vitro, J. Virol.,* 61, 802, 1987.

122. **Zimmermann, E. F., Heeter, M., and Darnell, J. E.,** RNA synthesis in poliovirus-infected cells, *Virology,* 19, 400, 1963.

123. **Baltimore, D.,** Purification and properties of poliovirus double-stranded ribonucleic acid, *J. Mol. Biol.,* 18, 421, 1966.

124. **Levintow, L.,** The reproduction of picornaviruses, in *Comprehensive Virology,* Vol. 2, Fraenkel-Conrat, H. and Wagner, R. R., Eds., Plenum Press, New York, 1974, 109.

125. **Rekosh, D., Lodish, H. F., and Baltimore, D.,** Translation of poliovirus RNA by an E. coli cell-free system, *Cold Spring Harbor Symp. Quant. Biol.,* 34, 747, 1969.

126. **Summers, D. F. and Levintow, L.,** Constitution and function of polyribosomes of poliovirus-infected HeLa cells, *Virology,* 27, 44, 1965.

127. **Girard, M.,** *In vitro* synthesis of poliovirus ribonucleic acid: role of the replicative intermediate, *J. Virol.,* 3, 376, 1969.

128. **Baltimore, D.,** The replication of picornaviruses, in *The Biochemistry of Viruses,* Levy, H. B., Ed., Marcel Dekker, New York, 1969, 103.

129. **Bishop, J. M., Koch, G., Evans, B., and Merriman, M.,** Poliovirus replicative intermediate: structural basis of infectivity, *J. Mol. Biol.,* 46, 235, 1969.

130. **Larsen, G. R., Dorner, A. J., Harris, T. J. R., and Wimmer, E.,** The structure of poliovirus replicative form, *Nucleic Acids Res.,* 8, 1217, 1980.

131. **Yogo, Y. and Wimmer, E.,** Sequence studies of poliovirus RNA. III. Polyuridylic acid and polyadenylic acid as components of the purified poliovirus replicative intermediate, *J. Mol. Biol.,* 92, 467, 1975.

132. **Wu, M., Davidson, N., and Wimmer, E.,** An electron microscope study of proteins attached to poliovirus RNA and its replicative form (RF), *Nucleic Acids Res.,* 5, 4711, 1978.

133. **Richards, O. C. and Ehrenfeld, E.,** Heterogeneity of the 3' end of minus strand RNA in poliovirus replicative form, *J. Virol.,* 36, 387, 1980.

134. **Richards, O. C., Ehrenfeld, E., and Manning, J.,** Strand-specific attachment of avidin spheres to double-stranded poliovirus RNA, *Proc. Natl. Acad. Sci. U.S.A.,* 76, 676, 1979.

135. **Perez-Bercoff, R.,** The mechanism of replication of picornavirus RNA, in *The Molecular Biology of Picornaviruses,* Perez-Bercoff, R., Ed., Plenum Press, New York, 1979, 293.

136. **Spiegelman, S. and Hayashi, M.,** The present status of the transfer of genetic information and its control, *Cold Spring Harbor Symp. Quant. Biol.,* 28, 161, 1963.

137. **Kamen, R. I.,** Structure and function of Qβ RNA replicase, in *RNA Phages,* Zinder, N. D., Ed., Cold Spring Harbor Laboratories, Cold Spring Harbor, New York, 1975, 203.

138. **Dasgupta, A., Hollingshead, P., and Baltimore, D.,** Antibody to a host protein prevents initiation by the poliovirus replicase, *J. Virol.,* 42, 1114, 1982.

139. **Baron, M. H. and Baltimore, D.,** *In vitro* copying of viral positive strand RNA by poliovirus replicase, *J. Biol. Chem.,* 257, 12359, 1982.

140. **Baron, M. H. and Baltimore, D.,** Anti-VPg antibody inhibition of the poliovirus replicase reaction and production of covalent complexes of VPg-related proteins and RNA, *Cell,* 30, 745, 1982.

141. **Morrow, C. D., Navab, M., Peterson, C., Hocko, J., and Dasgupta, A.,** Antibody to poliovirus genome-linked protein (VPg) precipitates *in vitro* synthesized RNA attached to VPg-precursor polypeptide(s), *Virus Res.,* 1, 89, 1984.

142. **Morrow, C. D., Hocko, J., Navab, M., and Dasgupta, A.,** ATP is required for initiation of poliovirus RNA synthesis *in vitro:* demonstration of tyrosine-phosphate linkage between *in vitro*-synthesized RNA and genome-linked protein, *J. Virol.,* 50, 515, 1984.

143. **Andrews, N. C., Levin, D., and Baltimore, D.,** Poliovirus replicase stimulation by terminal uridylyl transferase, *J. Biol. Chem.,* 260, 7628, 1985.

144. **Andrews, N. C. and Baltimore, D.,** Lack of evidence for VPg priming of poliovirus RNA synthesis in the host factor-dependent *in vitro* replicase reaction, *J. Virol.,* 58, 212, 1986.

145. **Young, D. C., Dunn, B. M., Tobin, G. S., and Flanegan, J. B.,** Anti-VPg antibody precipitation of product RNA synthesized *in vitro* by the poliovirus polymerase and host factor is mediated by VPg on the poliovirion RNA template, *J. Virol.,* 58, 715, 1986.

146. **Hey, T. D., Richards, O. C., and Ehrenfeld, E.,** Synthesis of plus- and minus-strand RNA from poliovirion RNA template *in vitro, J. Virol.,* 58, 790, 1986.

147. **Etchison, D. and Ehrenfeld, E.,** Comparison of replication complexes synthesizing poliovirus RNA, *Virology,* 111, 33, 1981.

148. **Takegami, T., Kuhn, R. J., Anderson, C. W., and Wimmer, E.,** Membrane-dependent uridylylation of the genome-linked protein VPg of poliovirus, *Proc. Natl. Acad. Sci. U.S.A.,* 80, 7447, 1983.

149. **Takeda, N., Kuhn, R. J., Yang, C.-F., Takegami, T., and Wimmer, E.,** Initiation of poliovirus plus-strand RNA synthesis in a membrane complex of infected HeLa cells, *J. Virol.,* 60, 43, 1986.

150. **Vartapetian, A. B., Koonin, E. V., Agol, V. I., and Bogdanov, A. A.,** Encephalomyocarditis virus RNA synthesis *in vitro* is protein-primed, *EMBO J.,* 3, 2583, 1984.

151. **Crawford, N. M. and Baltimore, D.,** Genome-linked protein VPg of poliovirus is present as free VPg and VPg-pUpU in poliovirus-infected cells, *Proc. Natl. Acad. Sci. U.S.A.,* 80, 7452, 1983.

152. **Kaplan, G., Lubinski, J., Dasgupta, A., and Racaniello, V. R.,** *In vitro* synthesis of infectious poliovirus RNA, *Proc. Natl. Acad. Sci. U.S.A.,* 82, 8424, 1985.

153. **Van der Werf, S., Bradley, J., Wimmer, E., Studier, F. W., and Dunn, J. J.,** Synthesis of infectious poliovirus RNA by purified T7 RNA polymerase, *Proc. Natl. Acad. Sci. U.S.A.,* 83, 2330, 1986.

154. **Kuge, S., Saito, I., and Nomoto, A.,** Primary structure of poliovirus defective interfering particle genomes and possible generation mechanisms of the particles, *J. Mol. Biol.,* 192, 473, 1986.

155. **Bernstein, H. D., Sarnow, P., and Baltimore, D.,** Genetic complementation among poliovirus mutants derived from an infectious cDNA clone, *J. Virol.,* 60, 1040, 1986.

156. **Butterworth, B. E., Shimshick, E. J., and Yin, F. H.,** Association of the poliviral RNA polymerase complex with phospholipid membranes, *J. Virol.,* 19, 457, 1976.

157. **Lundquist, R. E. and Mazel, J. V.,** Structural studies of the RNA component of the poliovirus replication complex, *Virology,* 85, 434, 1978.

158. **Toyoda, H., Yang, C.-F., Takeda, N., Nomoto, A., and Wimmer, E.,** Analysis of RNA synthesis of type 1 poliovirus by using an *in vitro* molecular genetic approach, *J. Virol.,* 61, 2816, 1987.

159. **Kuhn, R. J., Tada, H., Ypma-Wong, M. F., Dunn, J. J., Semler, B. L., and Wimmer, E.,** Construction of a mutagenesis cartridge for poliovirus VPg: isolation and characterization of viable and non-viable mutants, *Proc. Natl. Acad. Sci. U. S. A.,* 1988, in press.

Chapter 3

RNA REPLICATION IN COMOVIRUSES

Rik Eggen and Ab van Kammen

TABLE OF CONTENTS

I.   Introduction 50

II.   Cowpea Mosaic Virus is Type Member of the Comovirus Group 52

III.   Expression of Comovirus RNAs 53
        A.   Expression of CPMV-B-RNA 53
        B.   Expression of CPMV-M-RNA 54

IV.   Distinctive Features of the Structure of CPMV RNA and its
        Replicative Form 55

V.   Replication of CPMV RNA 57

VI.   Characterization of Purified CPMV RNA Replication Complex 58

VII.   Similarities Between CPMV and Picornaviruses 60

VIII.   CPMV RNA Replication Differs from the RNA Replication of TMV,
        BMV, AMV, and CMV 61

IX.   Initiation of CPMV RNA Replication 62

X.   Role of Protein Processing in Starting Viral RNA Replication 63

XI.   A Model for CPMV RNA Replication 63

XII.   Conclusion 65

Acknowledgment 65

References 66

I. INTRODUCTION

Among the various groups of plant RNA viruses with a positive strand RNA genome are viruses with a mono-, a bi-, and a tripartite genome (see Table 1).[1] Monopartite viruses, like tobamoviruses (e.g., tobacco mosaic virus, TMV), tymoviruses (e.g., turnip yellow mosaic virus, TYMV), and potyviruses (e.g., potato virus Y, PVY), contain the information for virus multiplication within a single RNA molecule, whereas in viruses with a bipartite genome, like comoviruses (e.g., cowpea mosaic virus, CPMV) and nepoviruses (e.g., tomato black ring virus, TBRV), this information is divided among two single-stranded RNA molecules and in viruses with a tripartite genome, like bromoviruses (e.g., brome mosaic virus, BMV) and alfalfa mosaic virus (AMV), among three RNA molecules. With bipartite and tripartite genome viruses the different segments of the genome are separately encapsidated so that infectious preparations of such viruses consist of mixtures of two or three nucleoprotein particles.

The different plant RNA virus groups are further characterized by the structures found at the 5' and 3' termini of their genomic RNAs. The predominant structures at the 5' ends are a cap ($= m^7$GpppG) similar to the structure found on eukaryotic messenger RNAs and a protein, VPg (viral protein, genome linked), linked to the 5'-terminal phosphate. Twelve different virus groups are known to carry a cap at the 5' ends of their genomic RNAs; among which are the bromoviruses, the tobamoviruses, the tymoviruses, and alfalfa mosaic virus, whereas for six groups it has been proven that their genomic RNAs are supplied with a VPg. To the latter viruses belong, among others, the comoviruses, the potyviruses, and the nepoviruses.

With respect to the 3' terminus three different structures are found. For six groups, among which the tobamoviruses, tymoviruses, and bromoviruses, it has been described that they have a tRNA-like structure which can be aminoacylated in vitro, whereas in six other plant virus groups, including the comoviruses, potyviruses, and nepoviruses, the genomic RNAs terminate with a poly(A) tail, and six more groups, including alfalfa mosaic virus and sobemoviruses (e.g., southern bean mosaic virus, SBMV) have neither a poly(A) tail nor a tRNA-like structure. The various RNA segments of a given bi- or tripartite genome virus always have the same structures at their respective 5' and 3' ends.

For the expression of their genetic information, three different main strategies are used by various groups.

1. With some groups of viruses the viral RNA is a monocistronic messenger resulting in the synthesis of a polyprotein from which functional smaller proteins are generated by proteolytic cleavages. This is found, for example, in comoviruses and potyviruses.
2. In other groups of viruses the expression of genomic RNA is limited to the 5' proximal gene and the virus produces subgenomic mRNAs to express other viral genes. This strategy is found, among others, in tobamoviruses, bromoviruses, and alfalfa mosaic virus.
3. Still other groups of plant viruses use a combination of the two previous strategies and express the information in their genomic RNAs partly by polyprotein synthesis and proteolytic processing and partly by synthesis of subgenomic mRNA for the expression of other viral genes. Tymoviruses and sobemoviruses are examples of plant viruses which express their genetic information by this strategy.

There is no clear correlation between the genomic 5'- and 3'-terminal structures and the mode of expression of the genomic RNAs with the exception of viruses which have a 5' VPg and a 3' poly(A) tail. This combination has so far been found with the monopartite potyviruses and the bipartite comoviruses and nepoviruses. All three groups happen to use

Table 1
ORGANIZATION AND STRATEGY OF THE POSITIVE-STRAND PLANT RNA VIRUSES

Virus group	Type member[a]	Molecular weight genomic RNAs	Terminal structures of genomic RNAs		Mode of expression
			5' end	3' end	
Monopartite Genome Viruses					
Tymoviruses	TYMV	2.0×10^6	cap	tRNA-like	Polyprotein processing subgenomic RNA
Tobamoviruses	TMV	2.0×10^6	cap	tRNA-like	Subgenomic RNA
Potexviruses	PVX	$2.0—2.5 \times 10^6$	cap	polyA	?
Sobemoviruses	SBMV	1.4×10^6	VPg	X_{OH}	Polyprotein processing subgenomic RNA
Luteoviruses	BYDV	2.0×10^6	VPg	X_{OH}	Polyprotein processing subgenomic RNA
Potyviruses	PVY	3.2×10^6	VPg	polyA	Polyprotein processing
Tobacco necrosis virus	TNV	1.4×10^6	ppA	X_{OH}	?
Tombus viruses	TBSV	1.5×10^6	cap	X_{OH}	?
Closteroviruses	SBYV	$2.2—6.5 \times 10^6$	?	?	?
Bipartite Genome Viruses					
Tobraviruses	TRV	2.4×10^6 $0.6—1.4 \times 10^6$	cap	X_{OH}	Readthrough, subgenomic RNA
Furoviruses	BNYVV	2.3×10^6 1.6×10^6	cap	polyA	Readthrough, subgenomic RNA ?
Comoviruses	CPMV	2.02×10^6 1.22×10^6	VPg	polyA	Polyprotein processing
Nepoviruses	TRSV	2.8×10^6 $1.3—2.4 \times 10^6$	VPg	polyA	Polyprotein processing
Pea enation mosaic virus	PEMV	1.8×10^6 1.4×10^6	VPg	?	?
Dianthoviruses	CRSV	1.5×10^6 0.5×10^6	?	?	?
Tripartite Genome Viruses					
Bromoviruses	BMV	1.1×10^6 1.0×10^6 0.7×10^6	cap	tRNA-like	Subgenomic RNA
Cucumoviruses	CMV	1.27×10^6 1.13×10^6 0.75×10^6	cap	tRNA-like	Subgenomic RNA
Hordeiviruses	BSMV	1.5×10^6 1.35×10^6 $1.05—1.35 \times 10^6$	cap	tRNA-like	Subgenomic RNA
Alfalfa mosaic virus	AMV	1.1×10^6 0.8×10^6 0.7×10^6	cap	X_{OH}	Subgenomic RNA

Table 1 (continued)
ORGANIZATION AND STRATEGY OF THE POSITIVE-STRAND PLANT RNA VIRUSES

Virus group	Type member[a]	Molecular weight genomic RNAs	Terminal structures of genomic RNAs		Mode of expression
			5' end	3' end	
Tripartite Genome Viruses					
Ilarviruses	TSV	1.1×10^6 0.9×10^6 0.7×10^6	cap	X_{OH}	Subgenomic RNA

[a] AMV: alfalfa mosaic virus, BMV: brome mosaic virus, BNYVV: beet necrotic yellow vein virus, BSMV: barley stripe mosaic virus, BYDV: barley yellow dwarf virus, CMV: cucumber mosaic virus, CPMV: cowpea mosaic virus, CRSV: carnation ringspot virus, PEMV: pea enation mosaic virus, PVX: potato virus X, PVY: potato virus Y, SBYV: sugar beet yellows virus, TBSV: tomato bushy stunt virus, TMV: tobacco mosaic virus, TNV: tobacco necrosis virus, TRSV: tobacco ringspot virus, TRV: tobacco rattle virus, TSV: tobacco streak virus, TSWV: tomato spotted wilt virus, TYMV: turnip yellow mosaic virus.

translation into a polyprotein with subsequent processing of the polyprotein in smaller viral proteins as expression strategy.

Expression of the viral genome is required for virus RNA replication, and it seems plausible that the variation in translation strategies, together with the different number of viral functions encoded by the genome of various groups of viruses, will also result in several mechanisms for viral RNA replication. Naturally there is a common basic pattern of viral RNA replication in the understanding that plant RNA viruses with a positive RNA genome replicate through the formation of a complementary negative strand that subsequently is used as template for the synthesis of progeny viral RNAs. Such a replication mechanism demands an RNA-dependent RNA polymerase (RNA replicase) for the transcription of both the positive and the negative viral RNA strands. In recent years it has become clear that probably most — if not all — plant RNA viruses encode their own specific RNA replicase just as RNA bacteriophages and animal RNA viruses (see Chapter 1). Since the transcription starts at the 3' end of the template molecule, the viral replicase must be able to specifically recognize the 3' termini of both positive and negative viral RNA strands, which are not similar. Besides the viral-encoded RNA polymerase, other virus-encoded and/or host cell proteins may be required for the specific selection of viral RNA molecules in the initiation of RNA replication.

Having explained to a certain extent the diversity of plant RNA viruses with their variation in genome structure, expression, and replication we shall exclusively engage further in this chapter with one virus group and discuss the present state of knowledge on the replication of comoviruses.

II. COWPEA MOSAIC VIRUS IS TYPE MEMBER OF THE COMOVIRUS GROUP

The comoviruses are a group of 14 different plant viruses that have the same structural organization of genomic RNAs and virus particles, and use the same mechanism for expression and replication of the viral RNAs (see Reference 2 for review). Cowpea mosaic virus (CPMV) is type member of the comoviruses and the only comovirus that has been thoroughly examined with respect to genome structure, translation, and replication strategy.[2,3] In the following, we shall therefore mainly discuss the results obtained from studies on CPMV. CPMV has acquired this exceptional position because in many respects it is an easy virus to work with. CPMV multiplies rapidly in its natural host *Vigna unguiculata* (L.) and as a result, purified CPMV can be obtained in gram amounts.[2,4] Therefore, the virus has lent itself very well for studies on the molecular properties of the virus particles and the genomic RNAs. Moreover, cowpea mesophyll protoplasts are in vitro efficiently and approximately synchronously infected with CPMV and this has greatly contributed to the understanding of the expression and replication mechanism of the virus.[5]

CPMV has a bipartite genome consisting of two positive strand RNA molecules that each are encapsidated in icosahedral particles with a diameter of 28 nm. The two nucleoprotein particles are denoted B and M components and have similar capsids composed of 60 copies of each of two different coat proteins, a large one with a molecular weight of 37 kdaltons, VP37, and a smaller one with a molecular weight of 23 kdaltons, VP23.[6-8] The B and M components differ in nucleic acid content, B containing a single RNA molecule (B-RNA) with a molecular weight of 2.04×10^6, and M, a RNA molecule (M-RNA) with a molecular weight of 1.22×10^6.[9,10] Both B- and M-RNA have a small protein, VPg, covalently linked to the 5' end and a poly(A) tail at the 3' end.[11-15] The RNAs are translated in vitro and in vivo into large polyproteins that are subsequently cleaved through a number of steps into several functional proteins.[2,3]

Both B and M components, or their RNAs, are required for virus multiplication in plants, but B-RNA is self supporting with respect to RNA replication in cowpea protoplasts.[16-19] B-RNA that is expressed and replicated in cowpea protoplasts in the absence of M-RNA is not assembled to virions.[18,19] Conversely, M-RNA is completely dependent on B-RNA expression for its replication. These findings demonstrate that B-RNA encodes functions required for replication whereas M-RNA carries information for the virus capsid proteins.

III. EXPRESSION OF COMOVIRUS RNAs

So far as different comoviruses have been examined, the genomic RNAs of comoviruses are in vitro translated as monocistronic messengers into large polyproteins corresponding to about 80% of their estimated coding capacity (see Reference 3 for review). The primary translation products are then processed into smaller viral proteins by specific proteolytic cleavages.[2,3]

Only for CPMV there is a rather complete picture of the expression mechanism of the viral RNAs, substantiated by knowledge of the nucleotide sequence of M- and B-RNA, a detailed analysis of the polyprotein processing and characterization of the viral proteins produced.[2,3,9,10] For CPMV the viral protein synthesis has been studied both in vitro and in vivo, resulting in the current model for the expression of the two CPMV RNAs as drawn in Figure 1 and briefly discussed in the following.

A. Expression of CPMV-B-RNA

The sequence of B-RNA, 5889 nucleotides excluding the poly(A) tail contains a single open reading frame of 5598 nucleotides, spanning from the AUG-codon at position 207 to the UAG-stopcodon at position 5805 (Figure 1).[10] In full agreement with the single long open reading frame, B-RNA is translated in vitro as well as in vivo into a 200-kdalton protein. The 200-kdalton protein is rapidly cleaved into 32- and 170-kdalton proteins, and this processing can even start before the chain of the 200-kdalton primary translation product is completed.[20-24] The 170-kdalton protein is then further processed via two alternative routes: either the 170-kdalton is cleaved into 60- and 110-kdalton proteins or, by another cleavage, into 84- and 87-kdalton proteins.[24,25] The 110- and 84-kdalton proteins can undergo an additional cleavage to give rise to the 87- and 60-kilodalton proteins, respectively, together with a 24-kdalton protein.[23-26] The 60-kdalton protein is the direct precursor of VPg.[27,28] Whereas the cleavage steps mentioned so far occur in vivo and in vitro, processing of the

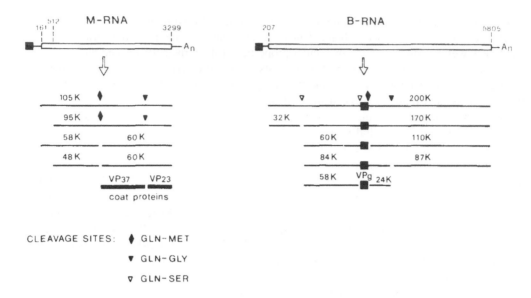

FIGURE 1.   Expression of M- and B-RNA of CPMV. M- and B-RNA each contain a single, open reading frame represented by the open bars. The position of the translational start and stop codons is indicated. Translation of M-RNA in vitro starts at the AUG codon at position 161, but more efficiently at the AUG codon at position 512. B-RNA is translated into a 200-K (Kilodalton) polyprotein and M-RNA into 105- and 95-K polyproteins which are subsequently processed by specific proteolytic cleavages at the indicated sites into smaller functional proteins. (K = kilodalton.)

60-kdalton protein into 58-kdalton with the release of VPg has never been observed in vitro. All five final cleavage products of the 200-kdalton polyprotein encoded by B-RNA (see Figure 1) are detectable in CPMV-infected protoplasts, if in varying amounts. Free VPg has not been detected in vivo, but it occurs either in precursor form or linked to the 5′ phosphate of the terminal uridyl-residue of B- and M-RNA. Beside the final cleavage products, the processing intermediates 170-, 110-, 84-, and 60-kdalton are also found in considerable amounts in infected cells which suggests that they may also represent functional molecules.[2,18]

The order of the cleavage products NH$_2$-32 kdaltons-58 kdaltons-4 kdaltons (= VPg)-24 kdaltons-87 kdaltons-COOH, in the 200-kdalton B-RNA-encoded protein, initially established by comparison of the different proteins by peptide mapping and by immunological techniques, has been confirmed by determining the amino terminal sequences of the various B-RNA-encoded proteins and locating the coding regions for these proteins on the B-RNA sequence.[21,25,27-29] Such sequence analysis, moreover, revealed the cleavage sites used in the proteolytic processing of the 200-kdalton polyprotein. It was found that three types of proteolytic cleavage sites were used: a glutamine-serine pair (2x), a glutamine-methionine pair (1x), and a glutamine-glycine pair (1x) (see Figure 2).[29] The proteolytic activity involved in cleaving at all three pairs is located in the 24-k/dalton protein encoded by B-RNA.[31,31a,31b] In contrast to what was thought previously, the 32-kdalton protein does not bear proteolytic activity.[30,32]

B. Expression of CPMV-M-RNA

The sequence of CPMV-M-RNA is 3481 nucleotides long, not including the poly(A) tail, and also contains a single, large, open reading frame running from the AUG codon at position 161 to the UAG stop codon at position 3299.[9] Nevertheless, M-RNA produces upon in vitro translation two polyproteins with molecular weights of 105 and 95-kdaltons. These two proteins have overlapping carboxy-terminal ends and arise because initiation of translation does not only start at the AUG codon at position 161, but also, and even to a greater

GENETIC MAP OF CPMV

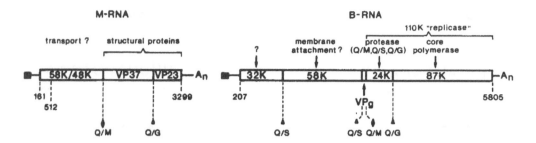

FIGURE 2.   Genetic map of CPMV. The single, open reading frames in M- and B-RNA are represented by the open bars. The indicated positions of the coding regions of the different functional protein domains in the reading frames are drawn to scale. (K = kilodalton.)

extent, at the AUG codon at position 512 in phase with the open reading frame.[9,20,21,30,33] Both M-RNA-encoded polyproteins are proteolytically cleaved and produce overlapping 58- and 48-kdalton proteins and a 60-kdalton protein which is the direct precursor of the two capsid proteins VP37 and VP23 (see Figure 1). A second cleavage generates the two coat proteins from the 60-kdalton protein.[20,30,32]

By locating the coding region of VP37 and VP23 on the nucleotide sequence of M-RNA, the proteolytic cleavage sites used to release the capsid proteins from the 105- and 95-kdalton primary translation products have been determined to be a glutamine-methionine pair and a glutamine-glycine pair (see Figure 2).[34] The cleavages at both sides are achieved by the B-RNA-encoded, 24-kdalton protease. Cleavage at the glutamine-methionine pair requires, moreover, the 32-kdalton encoded by B-RNA as a cofactor.[31a,31b]

The model for CPMV M-RNA expression depicted in Figure 1 has been derived from in vitro translation studies. In vivo, the capsid proteins are the only M-RNA-encoded products which are easily detectable. To verify whether this model also holds in vivo, a search in CPMV-infected cells was undertaken for other proteins occurring in the processing scheme. Using specific antibodies, it appeared possible to detect in CPMV-infected protoplasts small amounts of the 60-kdalton capsid precursor, and also of the 48-kdalton protein, but the 58-kdalton protein has not been found.[35] This demonstrates that in any case, the 95-kdalton polyprotein is synthesized in vivo, which is then rapidly cleaved into 48- and 60-kdalton products, followed by a second rapid cleavage of the 60-kdalton precursor to release the two capsid proteins. The 105-kdalton polyprotein is either not produced in vivo or only in amounts which are below the level of detection. It is therefore unclear if expression of M-RNA into a 105-kdalton protein has a role in vivo. In this connection it is striking that M-RNA of four other comoviruses, cowpea severe mosaic virus, bean pod mottle virus, red clover mottle virus, and squash mosaic virus, also direct in vitro translation of two poly-peptides of approximately the same size as CPMV M-RNA.[36-39] The occurrence of two AUG codons which give rise to translation in two large-sized proteins is apparently a common feature of comovirus M-RNA, which suggests that it may have biological significance.

IV. DISTINCTIVE FEATURES OF THE STRUCTURE OF CPMV RNA AND ITS REPLICATIVE FORM

The two genomic RNAs of CPMV are templates for both translation and replication. Once virus particles have invaded a host cell, the viral RNAs are released and first translated to produce proteins necessary for specific replication of virus RNA. From then onward the viral RNA also functions as template for the production of progeny RNA strands. Since in

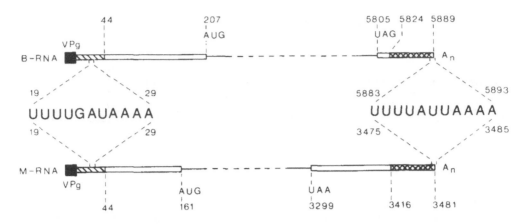

FIGURE 3.   Structural organization of the nontranslated regions of the CPMV genome. The 5' and 3' nontranslated regions in B- and M-RNA are represented as open bars. Within these bars two stretches with more than 80% nucleotide sequence homology between the RNAs are indicated (shaded areas); for further details see text.

CPMV-infection B- and M-RNA are multiplied by the same B-RNA-encoded replication machinery, it may be expected that both RNAs have features in common for their function in RNA replication. Indeed the 5'- and 3'-terminal noncoding regions of B- and M-RNA show conspicious sequence homology suggesting that these sequences contain recognition signals for various interactions with different viral and/or host protein involved in RNA replication (Figure 3).[9,10,40,41] The first 44 nucleotides in the two 5' leader sequences show 89% homology, and the last 65 nucleotides preceding the poly(A) tails show 82% homology. Particularly striking is a stretch of 11 nucleotides, UUUUGAUAAAA, in the homologous parts of the 5' leader sequences of both RNAs, which is complementary to the first four A's of the poly(A) tail and the last seven nucleotides before the beginning of the poly(A), allowing one G-U base pairing. Hence, it follows that the complementary negative strands of each genome segment have sequences at their 3' termini similar to those of the 3' ends of the positive viral RNAs. Such sequences may therefore constitute a recognition sequence for the viral RNA replicase.

The poly(A) tails of B- and M-RNA are heterogeneous in size and vary between 10 to 170 residues for B-RNA and between 20 to 400 residues for M-RNA.[42] It is not known whether the poly(A) tail is required for virus infectivity. The poly(A) tails are transcribed in the replication process, as is suggested by the absence of a polyadenylation recognition sequence AAUAAA in the region preceding the poly(A) tail, but has become apparent by the finding that poly(U) stretches are present at the 5' termini of negative strands of CPMV RNA replicative-form molecules.[9,10,40,43]

The functional significance of the protein VPg covalently linked to the 5' ends of B- and M-RNA is also not clear. Removal of VPg from the 5' termini by incubation with proteinase K does not lead to loss of infectivity of the viral RNA.[12] The protease treatment of isolated CPMV RNAs has equally been shown not to influence the messenger activity of the viral RNAs in cell-free systems.[12] Moreover, in rabbit reticulocyte lysates more than 90% of the VPg linked to CPMV RNA is removed and degraded within the first 10 min of incubation without any noticeable effect on the translational activity of the RNA.[44] A role of VPg in the translation of RNA is therefore not likely, and it appears that VPg has no function in establishing virus infection and replication.

Another possibility is that VPg, when linked to the virion RNAs, has no actual function, but represents a vestige of the process of RNA replication in which the RNA has been produced. VPg is encoded by B-RNA of CPMV as described in the previous section. It consists of a chain of 28 amino acid residues which is released from its precursors by cleavages at a glutamine-methionine pair and a glutamine-serine pair, respectively (Figure 2). VPg is linked to the 5'-terminal uridylic acid residue of the CPMV RNAs by a phosphodiester linkage with the −OH-side chain of its amino terminal serine.[45] Since B-RNA encodes proteins involved in viral RNA replication, it seems plausible that VPg is linked to the 5' ends during viral RNA synthesis. A role of VPg, or the generation of VPg, in viral RNA synthesis is substantiated by the finding that complementary negative strands in viral RNA replicative form are also provided with VPg at their 5' ends.[43] A possible role of VPg in viral RNA replication will be discussed in Section IX of this chapter.

V. REPLICATION OF CPMV RNA

Replication of CPMV RNA is associated with vesicular membranes of characteristic cytopathic structures in the cytoplasm of infected cells.[46] These cytopathic structures, consisting of arrays of vesicles surrounded by electron-dense material, the chemical nature of which is unknown, appear in cells early in infection with CPMV.[46-48] By fractionation of virus-infected cells it has been demonstrated that these structures contain CPMV-specific double-stranded RNAs and viral RNA replicase capable of synthesizing in vitro double-stranded RNA, and possibly some single-stranded viral RNA on endogeneous template RNA.[4,46,47] In cowpea protoplasts inoculated in vitro with B components of CPMV alone, B-RNA is expressed and replicated and the development of similar cytopathic membrane structures is observed.[49] This suggests that the induction of the membrane proliferation for the vesicular structures is achieved by a B-RNA-encoded function and may fulfill an essential role in viral RNA synthesis.

During the early stages of infection, CPMV replication is inhibited by actinomycin D, but no longer once virus replication is established at about 8 hr after infection.[8,50] Similar inhibition of virus replication by actinomycin D has been reported for bean pod mottle virus, another member of the comovirus group.[51,52] This inhibition of CPMV replication at an early stage of infection indicates that host DNA-dependent RNA synthesis is required to allow virus RNA replication and suggests that some host-specified component, the synthesis of which is induced by viral infection, is essential at that stage.

Involvement of a host factor in CPMV RNA replication is also demonstrated by a CPMV mutant that is no longer able to grow in cowpea, but still able to replicate in bean plants, *Phaseolus vulgaris* var. Pinto.[53] The mutation responsible for this behavior is located in B-RNA which codes for functions involved in viral RNA replication. The effect of the mutation may be interpreted as a defect in a specific interaction between a B-RNA encoded protein and a host factor necessary for the formation of a functional viral RNA replication complex. By the mutation, this interaction has become defective in cowpea, but is apparently still effective in beans.

Since addition of actinomycin D during active RNA replication does not affect virus synthesis and neither is there any effect on the activity of viral RNA replication complexes in vitro, actinomycin D does not seem to interfere with the process of viral RNA replication per se, but rather with the establishment of viral RNA synthesis in the infected cell. It is possible that host-dependent RNA synthesis is required for the development of the vesicular membrane structures as sites for virus RNA replication. Another possibility is that *de novo* synthesis of a host protein is necessary to supply an essential function for virus RNA replication.

In accordance with the location of viral RNA replication in the membranes of the cytopathic structures in CPMV-infected cells, the crude membrane fraction of CPMV infected leaves was found to contain RNA-dependent RNA polymerase activity capable in vitro of fully elongating nascent viral RNA chains initiated in vivo.[54] The completed chains are all positive strands and exclusively found in double stranded replicative forms. Beside the RNA replicase associated with negative strand viral RNA as template, the crude membrane fraction appeared to hold another RNA-dependent RNA polymerase. This second RNA-dependent RNA polymerase activity is also detectable in the membrane fraction of uninfected leaves in very small amounts, but is greatly increased in CPMV-infected leaves.[55] This host-encoded RNA polymerase transcribes endogenous plant RNA, and in infected plants also viral RNAs, into small (4 to 5 S) negative strand RNA molecules.[54] In CPMV-infected cowpea leaves the activity of the host-dependent RNA polymerase activity is enhanced at least 20-fold and almost overshadows the viral RNA replicase activity which represents less than 5% of the total RNA-dependent RNA polymerase activity in the crude membrane fraction. The viral replicase and the host RNA polymerase are, however, distinguished by virtue of the different products of their respective activities, which show that they are functionally different.[54]

The host RNA-dependent RNA polymerase activity can be separated from the CPMV RNA replication complex, as the binding of the host RNA polymerase to the membranes is much weaker than the binding of the replication complex. As a result, the host enzyme can be readily released from the membranes by washing with Mg^{++}-deficient buffer whereas the CPMV RNA replication complex remains firmly bound under these conditions. The host-encoded RNA polymerase has been purified by successive steps as outlined in Figure 4, and the purified enzyme proved to be a monomeric protein with a molecular weight of 130,000 daltons (130 kdaltons).[55] Using antibodies raised against purified enzyme preparations in an antibody-linked polymerase assay on nitrocellulose blots, the 130,000-dalton protein could definitely be identified as a RNA-dependent RNA polymerase.[56] It was also demonstrated, using the antibodies against the host RNA polymerase, that the increase of its activity in CPMV-infected leaves is indeed due to an increase in the amount of the 130-kdalton protein and not to activation of enzyme already present.[57] The increase has been further shown to be restricted to CPMV-infected cells within the leaf tissue and does not occur in leaf cells in which no virus multiplication takes place.[57]

Strikingly, no increase of 130-kdalton host-encoded, RNA-dependent RNA polymerase was found to accompany CPMV-RNA replication in cowpea mesophyll protoplasts upon inoculation with CPMV in vitro, indicating that the increased production of host RNA-dependent RNA polymerase is not a prerequisite for CPMV-RNA replication.[57]

VI. CHARACTERIZATION OF PURIFIED CPMV RNA REPLICATION COMPLEX

By definition, the native CPMV RNA replication complex consists of RNA replicase molecules bound to template viral RNA and is detectable by its capacity in vitro to elongate nascent in vivo-initiated viral RNA chains to full-length molecules. In CPMV-infected cowpea leaves this viral RNA replicase activity is first detectable in the crude membrane fraction one day after inoculation and then increases to reach a maximum 2 to 3 days later. At the time that the replicase activity has attained its maximum, the 130-kdalton host-encoded, RNA-dependent RNA polymerase has also strongly increased, and that makes the separation of the host-encoded RNA polymerase activity from the CPMV RNA replication complex the major problem to surmount in the purification of the viral RNA replicase.[54] The different steps in a procedure which has resulted in a highly purified CPMV RNA replication complex are summarized in Figure 4.[54,58] The main feature of this procedure is that in the purification the replication complex is maintained functionally intact, which allows distinguishing the viral RNA replicase activity from the host-encoded, RNA-dependent RNA polymerase. Washing of the crude membrane fraction of CPMV infected leaves with Mg^{++}-deficient buffer removes the vast majority of the 130-kdalton host RNA polymerase and leaves the viral replication complex intact in the membranes. The CPMV RNA replication complex can then be solubilized by treating the washed membranes with Triton®X-100,

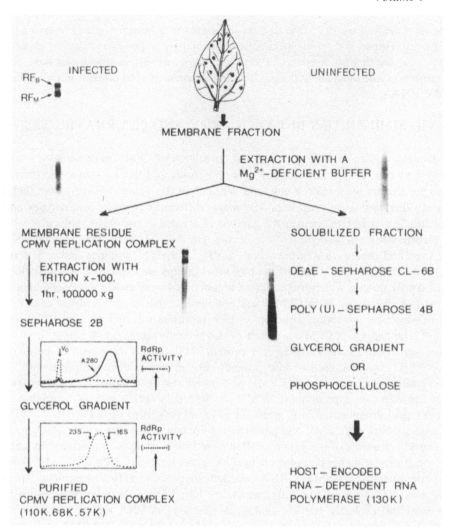

FIGURE 4.   Separation and purification of CPMV RNA replication complex and host-encoded, RNA-dependent RNA polymerase from CPMV infected cowpea leaves. The host-encoded activity produces short (4- to 55-base), negative strand RNAs from plant and viral RNA templates. The virus-encoded replicase elongates in vitro nascent chains initiated in vivo to full-length, plus-type viral RNA mainly found in double-strand form (RF$_B$ and RF$_M$). For details see text and references therein. (K = kilodalton.)

whereupon most of the contaminating proteins are removed by Sepharose®2B chromatography. After this step, the CPMV replication complex was free of the 130-kdalton host enzyme which was no longer detectable using Western blot analysis.[57] Further purification of the CPMV RNA replication complex was obtained by glycerol gradient centrifugation while the capacity of elongating nascent RNA chains to full-length viral RNAs was still preserved.[58] The active preparation after glycerol gradient centrifugation contained three major polypeptides of 110, 68, and 57 kdaltons. Using antisera against various viral proteins, the 110-kdalton, protein was proven to be a viral protein encoded by CPMV B-RNA. (Figure 2). The 68- and 57-kdalton polypeptides did not react with antibodies against viral-encoded proteins and presumably are host proteins which may either have a function in the viral RNA replication complex or be contaminating proteins. Since the amount of B-RNA encoded protein associated with the replication complex is correlated with the polymerase activity in the purified complex, the 110-kdalton protein has been assigned to represent the viral-encoded core polymerase in the CPMV RNA replication complex.

In the next section we shall first discuss the similarities between plant comoviruses and animal picornaviruses and, more specifically, the analogy in genome structure, expression, and replication mechanism between CPMV and poliovirus. These similarities have considerably influenced the ideas about the possible involvement of viral proteins in the replication of CPMV RNA.

VII. SIMILARITIES BETWEEN CPMV AND PICORNAVIRUSES

The detailed studies on the expression and replication of CPMV have revealed a striking similarity between the genome strategy of comoviruses and that of animal picornaviruses (see also the chapter by Semler, Kuhn, and Wimmer in this volume).[2] Apart from their host range and other biological properties, the major difference between comoviruses and picornaviruses is that in comoviruses the genome is divided in two RNA molecules whereas that in picornaviruses is a single RNA molecule. On the other hand, the genomic RNAs of comoviruses and picornaviruses have both a VPg at their 5' terminus and a 3' terminal poly(A) tail. The genomic RNAs of the two virus groups are expressed by translation into large-sized polyproteins which are processed to functional proteins by virus-encoded protease activities (this chapter).[59] Both CPMV and polioviruses induce in inoculated cells the formation of vesicular membrane structures which represent the sites of viral RNA replication.[46-48,60] The capsids of CPMV are built of two proteins present in 60 copies, each which are processed from a common precursor protein.[6-8] Similarly, the capsid of poliovirus is made up of 60 copies of each of four proteins derived from a common precursor.[60] More recent crystallographic studies on CPMV and poliovirus have shown that the polypeptide chain of the two capsid proteins of CPMV are folded in such a way as to produce three distinct β-barrel domains similarly arranged as in the poliovirus capsid.[61,62]

The analogy between CPMV and poliovirus is not confined to the structural organization of the capsids, features of the genomic RNAs, and their mode of expression. In addition, there is significant sequence homology between nonstructural (58-, VPg, 24-, and 87-kdalton) proteins contained in the CPMV B-RNA-encoded polyprotein and four nonstructural proteins of poliovirus, 2C, VPg, 3C, and 3D, which are found in similar relative positions in the genetic map and probably provide analogous functions in RNA replication (Figure 5).[63] Protein 3D has been identified as the core polymerase of the polioviral RNA replicase. The protein exhibits 20.9% amino acid sequence homology to the sequence of the 87-kdalton protein of CPMV. This homologous region of the 87-kdalton protein moreover contains a block of 14 amino acid residues consisting of a GDD (Gly–Asp–Asp) sequence flanked by hydrophobic residues at both sides, and a second block with the conserved sequence $\frac{S}{T}GxxxTxxxN\frac{S}{T}$ (in which x may be any amino acid residue) 37 residues upstream from the first conserved region, found in all viral RNA-dependent RNA polymerases characterized so far.[64]

This provides strong evidence that the 87-kdalton B-RNA encoded protein represents the core polymerase of CPMV replicase. This is in agreement with the occurrence of the B-RNA-encoded 110-kdalton protein in purified CPMV RNA replication complexes; for the 110-kdalton protein contains the sequences of the 87- and the 24-kdalton proteins.

The homology between the 24-kdalton CPMV protein, which has been shown to carry specific protease activity, and the poliovirus protease 3C adjoining the polymerase sequence is located in their C terminal sequences. In that part of the polypeptide chain both proteases have the features of the active site of a thiol protease, consisting of a Cys residue and a His residue approximately 14 to 18 residues apart.

The function of poliovirus protein 2C (formerly P2-x) has not yet been biochemically defined, but there is evidence that this protein, or derivatives thereof, is associated with the

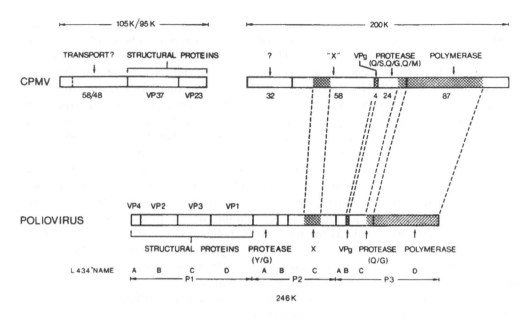

FIGURE 5.   Comparison of the functional organization in CPMV and polioviral polyproteins. Similar functions are found in similar relative positions in the genomes of CPMV and poliovirus. Moreover, in the B-RNA-encoded polyprotein, three regions show more than 20% amino acid sequence homology to regions in the polioviral polyprotein (hatched boxes). These homologous regions reside in proteins which probably have similar functions in viral RNA replication. (K = kilodalton). (From Franssen, H., Leunissen, J., Goldbach, R., Lomonossoff, G., and Zimmern, D., *EMBO J.*, 3, 855, 1984. With permission.)

vesicular membranes where polioviral RNA replication is located in infected cells.[65] Moreover, recent analyses of guanidine-resistant and guanidine-dependent isolates of poliovirus revealed that these contain a mutation in the central domain of about 140 amino acid residues of the protein 2C, which is strongly conserved among picornaviruses and shows 30% sequence homology to the central region of the 58-kdalton protein of CPMV.[63,66-68] Since the major effect of guanidine appears to be blocking of viral RNA synthesis, these observations provide additional evidence that the protein 2C of poliovirus plays a role in this process. By analogy, a similar role may be attributed to the 58-kdalton protein encoded by CPMV B-RNA.

The strong analogy between CPMV and poliovirus in structural organization and expression strategy, together with similar genetic organization and amino acid homology of viral proteins involved in RNA replication, prove that CPMV and poliovirus are somehow evolutionary related. Therefore, in view of the similarities between CPMV and poliovirus, it is natural to anticipate that the mechanism of CPMV RNA replication will be very similar to that of poliovirus RNA replication. In the following sections we shall take that line in further discussion of the replication of CPMV RNA.

VIII. CPMV RNA REPLICATION DIFFERS FROM THE RNA REPLICATION OF TMV, BMV, AlMV, AND CMV

If sequence conservation in nonstructural viral proteins which are involved in viral RNA replication defines a group of positive RNA viruses with a similar mechanism of viral RNA replication, then CPMV belongs to a different group of plant viruses than TMV, BMV, AlMV, and CMV. Although TMV is a monopartite genome virus and BMV, CMV, and AlMV are tripartite genome viruses, TMV is similar to BMV, CMV, and AlMV in containing cistrons for four viral proteins. Two of these proteins — the coat protein and a protein thought to be involved in transfer of virus from cell to cell in infected plants — are not required for RNA replication, whereas both other proteins appear to be involved in viral RNA synthesis. The latter two proteins of TMV, BMV, and AlMV contain distinct domains with sequence homology, but do not show homology with the virus-encoded proteins involved in replication of CPMV RNA beyond the GDD-associated sequences noted above.[63,69,70] Obviously then, the plant RNA viruses of which the genome structure and expression mechanism has been elucidated can so far be classified into two groups: one containing viruses like TMV, BMV, CMV, and AlMV, and another represented by CPMV and viruses such as the nepoviruses, which appear similar to CPMV in genome structure and translation strategy. Each group has a mechanism of viral RNA replication with characteristics based on similar functions of the proteins or the domains of proteins with conserved function. This would imply that functionally equivalent replication complexes are formed for the various viruses of each group. A major difference between the two groups could be that the viral RNA replicase of CPMV, just as the replicase of poliovirus, is a RNA-dependent RNA polymerase that requires for its activity not only a template, but also a primer.[60] In contrast, the replicase of the other group of viruses, as has been demonstrated for BMV replicase and AlMV replicase, do not require a primer, but can transcribe a template starting at a specific recognition sequence.[71-73] Such difference will imply a different mechanism of initiation of viral RNA replication for each group.

IX. INITIATION OF CPMV RNA REPLICATION

Although the CPMV RNA replication complex has been purified, and the virus-encoded RNA polymerase identified, virtually nothing is known about the mechanism of initiation of viral RNA synthesis. It is tempting to ascribe to the small protein, VPg, a role in the initiation of viral RNA replication, if only to explain its occurrence at the 5′ end of the viral RNA. There is, however, as yet no direct evidence for such a role of VPg. Neither for poliovirus is the mechanism of initiation of viral RNA replication and the possible role of VPg in this process completely understood. On the one hand it has been shown that a crude membrane fraction from poliovirus-infected HeLa cells is capable of synthesizing in vitro the uridylated proteins VPgpU and VPgpUpU, which under conditions of RNA synthesis are further elongated into much longer stretches of 5′ terminal poliovirus RNA.[74] Such data suggest that VPgpU might function as a primer for viral RNA synthesis. On the other hand, Flanegan and co-workers have reported that purified poliovirus replicase synthesizes in vitro dimer products with virus RNA as a template in the presence of a host factor.[75,76] According to these authors, the host factor represents a terminal uridylyl transferase that elongates the poly(A) tail at the 3′ end of the template RNA with a number of U residues which can form a hairpin with the poly(A) tail.[77,78] The virus-encoded replicase then starts elongation at the hairpin primer and further transcribes the viral RNA template producing RNA products up to twice the length of the genome. Addition of VPg subsequently resulted in cleavage of the dimer molecule at the hairpin and linkage of VPg to the 5′ end of the newly synthesized RNA strand. Such data conflict with a primer role for VPg, but, on the contrary, propose a role of VPg in the nucleolytic cleavage of the hairpin.

For CPMV there is no experimental evidence for either mechanism since no in vitro initiating system for CPMV RNA elongation has been developed. Free VPgpU or uridylation of VPg or uridylation of its direct 60-kdalton precursor has never been observed, nor has the occurrence of dimer length RNA molecules as possible intermediates in viral RNA synthesis so far been demonstrated. The only observation to suggest a role of VPg in an early stage of CPMV RNA replication is that VPg is found both at the 5′ end of negative and positive RNA strands in the replicative forms isolated from virus-infected leaves.[43]

X. ROLE OF PROTEIN PROCESSING IN STARTING VIRAL RNA REPLICATION

The occurrence of VPg at the 5' end of each progeny viral RNA strand raises more questions concerning the viral RNA replication mechanisms. VPg of CPMV is encoded by B-RNA which is translated into a 200-kdalton primary translation product. This polyprotein has to complete several successive processing steps (Figure 1) to make VPg available for RNA synthesis. Only a single VPg is released each time a 200-kdalton translation product is processed, and this must happen for each molecule of B and M-RNA to be produced. In this way protein processing appears to have a dominant role in the replication of CPMV RNA, and both processes seem to be closely connected. Indeed, the translation-expression mechanism used by CPMV implies the production of equimolar amounts of polymerase molecules and other B-RNA-encoded proteins involved in replication, i.e., since VPg, the core polymerases (110 or 87 kdaltons) and the 58-kdalton membrane protein, together with the 32-kdalton protein are produced from a common B-RNA-encoded polyprotein, the latter proteins are simultaneously produced with VPg. For that reason a viral polymerase molecule needs on the average to synthesize only a single RNA molecule to keep step with the production of VPg. On that line of reasoning it is conceivable that the viral RNA polymerase molecules lose their activity upon releasing their first template. For example, it may be imagined that the 110-kdalton protein is involved in binding to template RNA and initiation of transcription. If subsequently the 110-kdalton protein is processed into 24- and 87-kdalton proteins by an intramolecular cleavage carried out by the proteolytic activity in the 24-kdalton domain of the 110-kdalton protein, reinitiation will be prevented. Such restriction of the polymerase molecules would fit with the observation that isolated CPMV replication complexes contain polymerase molecules which are only able to complete nascent chains from their endogenous template, but are not capable of accepting added template molecules.

XI. A MODEL FOR CPMV RNA REPLICATION

In Figure 6 a model is depicted in which the possible linkage between processing and CPMV RNA replication is illustrated. In this model a complex is formed between the 60- and 110-kdalton B-RNA-encoded proteins, with the 60-kdalton protein tightly associated with the membranes of the cytopathic structure in infected cells. Template RNA and possibly a host factor are bound to the complex in such a way to allow initiation of transcription. The 60-kdalton protein has to supply VPg, while the 110-kdalton protein contains the 24-kdalton domain which is able to release VPg by proteolytic cleavage. VPg may become uridylated at its N-terminal serine prior to initiation. Since the 60- and 110-kdalton proteins are derived from the common 170-kdalton precursor, formation of the initiation complex may also start with the binding of 170-kdalton protein to the membrane, whereupon cleavage into 60- and 110-kdalton protein occurs. Following the release of VPg, the 87-kdalton polymerase domain in the 110-kdalton protein starts transcription of the template, possibly using uridylated VPg as a primer. It has been noticed that the 58-kdalton protein contains an amino acid sequence, $GxxxxGK\frac{T}{S}$, which is similar to the consensus sequence found in proteins with ATPase/GTPase activity and in virus-encoded proteins involved in viral RNA replication for several viruses.[79] This sequence may represent a binding site for nucleoside triphosphates. While remaining attached to the 58-kdalton protein, the 87-kdalton polymerase domain within the 110-kdalton proteins continues the elongation of the RNA strand which is transcribed from the template. When transcription is completed, template RNA is released from the protein complex. In the meantime, the 3' terminal end of the template RNA may be used in the formation of another initiation complex. This has to take place at another 60-

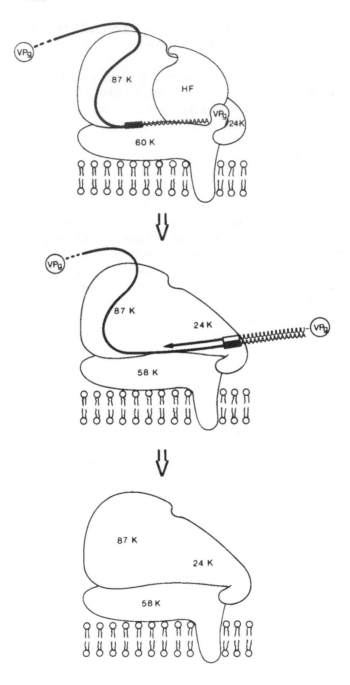

FIGURE 6. Model for membrane-bound CPMV RNA replication complex. Three stages in CPMV RNA replication are outlined: the replication complex just before RNA synthesis is initiated, an intermediate stage in which VPg has been linked to the 5' end of the nascent RNA chain, and the protein complex remaining when the newly synthesized viral RNA and the template have been released. For discussion of the model, see the text. (K = kilodalton.)

to 110-kdalton complex, as the site used (see Figure 6) is now devoid of VPg and, therefore, not capable of initiating the synthesis of a viral RNA strand.

The model may account both for the synthesis of negative and positive strands. It proposes an essential role for VPg in the initiation of viral RNA synthesis, which is speculative and needs to be tested. In the model are, however, incorporated the different observations which must be accounted for in CPMV RNA synthesis. The essential part of the model is the close linkage between protein processing and viral RNA replication. The host factor in the model may have a role in the uridylation of VPg for the priming of RNA synthesis. Alternatively, if the initiation of viral RNA synthesis proceeds by hairpin priming, as has been proposed for poliovirus, the host factor may represent the terminal uridylyl transferase activity for elongating the 3′ end of the template. The occurrence of terminal uridylyl transferase has been demonstrated in cowpea.[80]

The model does not explain the presence in infected cells of B-RNA-encoded 84- and 87-kdalton proteins, which are produced by alternative cleavage of the 170-kdalton polypeptides. Possibly it does not make much difference whether the 170-kdalton protein is first cleaved into 60- and 110-kdalton polypeptides or into 84- and 87-kdalton polypeptides if viral RNA replication as caricatured in Figure 6 can be established from either pair of polypeptides. This would imply that the 84-kdalton polypeptide also can supply VPg and the 87-kdalton polypeptide is active as polymerase. Alternatively, it must be seriously considered that a second pathway for cleaving the 170-kdalton polypeptide may be significant in regulating viral RNA replication in a way not yet understood.

XII. CONCLUSION

It will be clear from the discussion above that if we think we know the alphabet of CPMV RNA replication, we do not yet understand its grammar. Further biochemical identification of the activities of the various virus-encoded polypeptides and host factors involved is required to elucidate their role in CPMV RNA replication. It will be necessary to develop a system in which the initiation of CPMV RNA replication is reconstituted to resolve the speculations about the mechanism of viral RNA replication and to answer the questions raised by the model. For elucidating the functioning of the virus-encoded proteins involved in RNA replication, it may be equally helpful to have access to mutants of CPMV which each bear a single mutation at a well-defined site in the genomic RNA or in one of the virus-encoded proteins. Recently, full-size DNA copies of B- and M-RNA have been cloned in our laboratory and it was shown that such DNA copies can be transcribed in vitro into RNA molecules which were found to be infectious upon inoculation of cowpea mesophyll protoplasts (Vos et al., to be published).[31,33] With this, it seems that for CPMV a system has become available for producing site specific mutations in B-RNA, of which the effect on viral RNA replication can subsequently be tested. This genetic approach together with biochemical studies may result in further unravelling of the molecular mechanism of CPMV RNA replication.

ACKNOWLEDGMENT

We thank Gré Heitkönig, Jan Verver, and René Klein Lankhorst for their assistance in preparing this manuscript. This work was supported by the Netherlands Foundation for Chemical Research (S.O.N.), with financial aid from the Netherlands organization for the Advancement of Pure Research (Z.W.O.)

REFERENCES

1. **Dougherty, W. G. and Hiebert, E.,** Genome structure and gene expression of plant RNA viruses, in *Molecular Plant Virology,* Vol. 2, Davies, J. W., Ed., CRC Press, Boca Raton, Fl., 1985, chap. 2.
2. **Goldbach, R. and Van Kammen, A.,** Structure, Replication and Expression of the bipartite genome of cowpea mosaic virus, in *Molecular Plant Virology,* Vol. 2, Davies, J. W., Ed., CRC Press, Boca Raton, Fl., 1985, chap. 3.
3. **Goldbach, R.,** Comoviruses: molecular biology and replication, in *Plant Viruses: Viruses with Bipartite Genomes and Isometric Particles,* Harrison, B. D. and Murant, A. F., Eds., Plenum Press, New York, in press.
4. **Zabel, P., Weenen-Swaans, H., and Van Kammen, A.,** In vitro replication of cowpea mosaic virus RNA I. Isolation and properties of the membrane-bound replicase, *J. Virol.,* 14, 1049, 1974.
5. **Rottier, P. J. M., Rezelman, G., and Van Kammen, A.,** Protein synthesis in cowpea mosaic virus-infected cowpea protoplasts: detection of virus-related proteins, *J. Gen. Virol.,* 51, 359, 1980.
6. **Wu, G. J. and Bruening, G.,** Two proteins from cowpea mosaic virus, *Virology,* 46, 596, 1971.
7. **Geelen, J. L. M. C.,** Structuur en Eigenschappen van Cowpea-Mozaiekvirus, Ph.D. thesis, Agricultural University, Wageningen, 1974.
8. **Rottier, P. J. M., Rezelman, G., and Van Kammen, A.,** The inhibition of cowpea mosaic virus replication by actinomycin D, *Virology,* 92, 299, 1979.
9. **Van Weezenbeek, P., Verver, J., Harmsen, J., Vos, P., and Van Kammen, A.,** Primary structure and gene organization of the middle component RNA of cowpea mosaic virus, *EMBO J.,* 2, 941, 1983.
10. **Lomonossoff, G. P. and Shanks, M.,** The nucleotide sequence of cowpea mosaic virus B RNA, *EMBO J.,* 2, 2153, 1983.
11. **Daubert, S. D., Bruening, G., and Najarian, R. C.,** Protein bound to the genome RNAs of cowpea mosaic virus, *Eur. J. Biochem.,* 92, 45, 1978.
12. **Stanley, J., Rottier, P., Davies, J. W., Zabel, P., and Van Kammen, A.,** A protein linked to the 5' termini of both RNA components of the cowpea mosaic virus genome, *Nucleic Acids Res.,* 5, 4505, 1978.
13. **Daubert, S. D. and Bruening, G.,** Genome-associated protein of comoviruses, *Virology,* 98, 246, 1979.
14. **El Manna, M. M. and Bruening, G.,** Polyadenylate sequences in the ribonucleic acids of cowpea mosaic virus, *Virology,* 56, 198, 1973.
15. **Steele, K. P. and Frist, R. H.,** Characterization of the 3' termini of the RNAs of cowpea mosaic virus, *J. Virol.,* 26, 243, 1978.
16. **Van Kammen, A.** The relationship between the component of cowpea mosaic virus. I. Two ribonucleo-protein particles necessary for the infectivity of CPMV, *Virology,* 34, 312, 1968.
17. **De Jager, C. P.,** Genetic analysis of cowpea mosaic virus mutants by supplementation and reassortment tests, *Virology,* 70, 151, 1976.
18. **Goldbach, R., Rezelman, G., and Van Kammen, A.,** Independent replication and expression of B-component RNA of cowpea mosaic virus, *Nature (London),* 286, 297, 1980.
19. **De Varennes, A. and Maule, A. J.,** Independent replication of cowpea mosaic virus of bottom component RNA: in vivo instability of the viral RNAs, *Virology,* 144, 495, 1985.
20. **Pelham, H. R. B.,** Synthesis and proteolytic processing of cowpea mosaic virus proteins in reticulocyte lysates, *Virology,* 96, 463, 1979.
21. **Goldbach, R. W., Schilthuis, J. G., and Rezelman, G.,** Comparison of in vivo and in vitro translation of cowpea mosaic virus RNAs, *Biochem. Biophys. Res. Comm.,* 99, 89, 1981.
22. **Goldbach, R. and Rezelman, G.,** Orientation of the cleavage map of the 200-kilodalton polypeptide encoded by the bottom-component RNA of cowpea mosaic virus, *J. Virol.,* 46, 614, 1983.
23. **Franssen, H., Goldbach, R., and Van Kammen, A.,** Translation of bottom component RNA of cowpea mosaic virus in reticulocyte lysates: faithful proteolytic processing of the primary translation product, *Virus Res.,* 1, 39, 1984.
24. **Peng, X. X. and Shih, D. S.,** Proteolytic processing of the proteins translated from the bottom component RNA of cowpea mosaic virus. The primary and secondary cleavage reactions, *J. Biol. Chem.,* 259, 3197, 1984.
25. **Rezelman, G., Goldbach, R., and Van Kammen, A.,** Expression of bottom component RNA of cowpea mosaic virus in cowpea protoplasts., *J. Virol.,* 36, 366, 1980.
26. **Wellink, J., Jaegle, M., and Goldbach, R.,** Detection of a 24K protein encoded by B-RNA of cowpea mosaic virus using antibodies raised against a synthetic peptide, *J. Virol.,* submitted.
27. **Zabel, P., Moerman, M., Van Straaten, F., Goldbach, R., and Van Kammen, A.,** Antibodies against the genome linked protein VPg of cowpea mosaic virus recognize a 60,000-Dalton precursor polypeptide, *J. Virol.,* 41, 1083, 1982.
28. **Goldbach, R., Rezelman, G., Zabel, P., and Van Kammen, A.,** Expression of the bottom-component RNA of cowpea mosaic virus. Evidence that the 60-kilodalton VPg precursor is cleaved into single VPg and a 58-kilodalton polypeptide, *J. Virol.,* 42, 630, 1982.

29. **Wellink, J., Rezelman, G., Goldbach, R., and Beyreuther, K.,** Determination of the sites of proteolytic processing in the polyprotein encoded by the B-RNA of cowpea mosaic virus, *J. Virol.,* 59, 50, 1986.

30. **Franssen, H., Moerman, M., Rezelman, G., and Goldbach, R.,** Evidence that the 32,000-Dalton protein encoded by bottom-component RNA of cowpea mosaic virus is a proteolytic processing enzyme, *J. Virol.,* 50, 183, 1984.

31. **Verver, J., Goldbach, R., Garcia, J. A., and Vos, P.,** In vitro expression of a full-length DNA copy of cowpea mosaic virus B-RNA: identification of the B RNA-encoded 24-kilodalton protein as a viral protease, *EMBO J.,* 6, 549, 1987.

31a. **Vos, P.** Molecular Cloning and Expression of Full-Length DNA Copies of the Genomic RNAs of Cowpea Mosaic Virus, Ph. D. thesis, Agricultural University, Wageningen, 1987.

31b. **Vos, P., Verver, J., Jaegle, M., Wellink, J., van Kammen, A., and Goldbach, R.,** Role of the cowpea mosiac virus B RNA-encoded 24K and 32K polypeptides in the proteolytic processing of the viral proteins, submitted.

32. **Franssen, H., Goldbach, R., Broekhuijsen, M., Moerman, M., and Van Kammen, A.,** Expression of middle-component RNA of cowpea mosaic virus: in vitro generation of a precursor to both capsid proteins by a bottom-component RNA-encoded protease from infected cells, *J. Virol.,* 41, 8, 1982.

33. **Vos, P., Verver, J., Van Weezenbeek, P., Van Kammen, A., and Goldbach, R.,** Study of the genetic organisation of a plant viral RNA genome by in vitro expression of a full length DNA copy, *EMBO J.,* 3, 3049, 1984.

34. **Franssen, H., Roelofsen, W., and Goldbach, R.,** Mapping of coding regions for the capsid proteins of cowpea mosaic virus on the nucleotide sequence of middle component RNA, *Chin. J. Virol.,* 2, 46, 1984.

35. **Wellink, J., Jaegle, M., Prinz, H., Van Kammen, A., and Goldbach, R.,** Expression of the middle component RNA of cowpea mosaic in vivo, *J. Gen. Virol.,* 68, 2577, 1987.

36. **Hiebert, E. and Purcifull, D. E.,** Mapping of the two coat protein genes on the middle RNA component of squash mosaic virus (comovirus group), *Virology,* 113, 630, 1981.

37. **Beier, H., Issinger, D. G., Deuschle, M., and Mundry, K. W.,** Translation of the RNA of cowpea severe mosaic virus in vitro and in cowpea protoplasts, *J. Gen. Virol.,* 54, 379, 1981.

38. **Gabriel, C. J., Derrick, K. S., and Shih, D. S.,** The synthesis and processing of the proteins of bean pod mottle virus in rabbit reticulocyte lysates, *Virology,* 122, 476, 1982.

39. **Goldbach, R. and Krijt, J.,** Cowpea mosaic virus-encoded protease does not recognize primary translation products of M-RNA from other comoviruses, *J. Virol.,* 43, 1151, 1982.

40. **Davies, J. W., Stanley, J., and Van Kammen, A.,** Sequence homology adjacent to the 3'-terminal poly(A) of cowpea mosaic virus RNAs, *Nucleic Acids Res.,* 7, 493, 1979.

41. **Najarian, R. C. and Bruening, G.,** Similar sequences from the 5' end of cowpea mosaic virus RNAs, *Virology,* 106, 301, 1980.

42. **Ahlquist, P. and Kaesberg, P.,** Determination of the length distribution of poly(A) at the 3'-terminus of the virion RNAs of EMC virus, poliovirus, rhinovirus, RAV-61 and CPMV and of mouse globin mRNA, *Nucleic Acids Res.,* 7, 1195, 1979.

43. **Lomonossoff, G., Shanks, M., and Evans, D.,** The structure of cowpea mosaic virus replicative form RNA, *Virology,* 144, 351, 1985.

44. **Goldbach, R.,** Unpublished results.

45. **Jaegle, M., Wellink, J., and Goldbach, R.,** The genome-linked protein of cowpea mosaic virus is bound into the 5'-terminus of virus RNA by a phosphodiester linkage to serine, *J. Gen. Virol.,* 68, 627, 1987.

46. **De Zoeten, G. A., Assink, A. M., and Van Kammen, A.,** Association of cowpea mosaic virus-induced double stranded RNA with a cytopathological structure in infected cells, *Virology,* 59, 341, 1974.

47. **Assink, A. M., Swaans, H., and Van Kammen, A.,** The localisation of virus specific double-stranded RNA of cowpea mosaic virus in subcellular fractions of infected Vigna leaves, *Virology,* 53, 384, 1973.

48. **Hibi, T., Rezelman, G., and Van Kammen, A.,** Infection of cowpea mesophyll protoplasts with cowpea mosaic virus, *Virology,* 64, 308, 1975.

49. **Rezelman, G., Franssen, H. J., Goldbach, R. W., Ie, T-S., and Van Kammen, A.,** Limits to the independence of bottom component RNA of cowpea mosaic virus, *J. Gen. Virol.,* 60, 335, 1982.

50. **De Varennes, A., Davies, J. W., Shaw, J. G., and Maule, A. J.,** A reappraisal of the effect of actinomycin D and cordycepin on the multiplication of cowpea mosaic virus in cowpea protoplasts, *J. Gen. Virol.,* 66, 817, 1985.

51. **Lockhart, B. E. L. and Semancik, J. S.,** Inhibition of a plant virus by actinomycin D, *Virology,* 36, 504, 1968.

52. **Lockhart, B. E. L. and Semancik, J. S.,** Differential effect of actinomycin D on plant virus multiplication, *Virology,* 39, 62, 1969.

53. **Evans, D.,** Isolation of a mutant of cowpea mosaic virus which is unable to grow in cowpeas, *J. Gen. Virol.,* 66, 339, 1985.

54. **Dorssers, L., Van der Meer, J., Van Kammen, A., and Zabel, P.,** The cowpea mosaic virus RNA replicase complex and the host-encoded RNA-dependent RNA polymerase-template complex are functionally different, *Virology,* 125, 155, 1983.

55. **Dorssers, L., Zabel, P., Van der Meer, J., and Van Kammen, A.,** Purification of a host-encoded RNA-polymerase from cowpea mosaic virus-infected cowpea leaves, *Virology,* 116, 236, 1982.

56. **Van der Meer, J., Dorssers, L., and Zabel, P.,** Antibody-linked polymerase assay on protein blots: a novel method for identifying polymerase following SDS-polyacrylamide gel electrophoresis, *EMBO J.,* 2, 233, 1983.

57. **Van der Meer, J., Dorssers, L., Van Kammen, A., and Zabel, P.,** The RNA-dependent RNA polymerase of cowpea is not involved in cowpea mosaic virus RNA replication: immunologic evidence, *Virology,* 132, 413, 1984.

58. **Dorssers, L., Van der Krol, S., Van der Meer, J., Van Kammen, A., and Zabel, P.,** Purification of cowpea mosaic virus RNA replication complex: identification of a virus-encoded 110,000 dalton polypeptide responsible for RNA chain elongation, *Proc. Natl. Acad. Sci. U.S.A.,* 81, 1951, 1984.

59. **Toyada, H., Nicklin, M. J. H., Murray, M. G., Anderson, C. W., Dunn, J. J., Studier, F. W., and Wimmer, E.,** A second virus encoded proteinase involved in proteolytic processing of poliovirus polyprotein, *Cell,* 45, 761, 1986.

60. **Strauss, E. G. and Strauss, J. H.,** Replication strategies of the single-stranded RNA viruses of eukaryotes, in *Current Topics in Microbiology and Immunology 105,* Cooper, M., Hofschneider, M., Koprowski, H., Melchers, F., Rott, R., Schweiger, H. G., Vogt, P. K., and Zinkernagel, R., Eds., Springer Verlag, Berlin, 1983, 1.

61. **Hogle, J., Chow, M., and Filman, D. J.,** The high resolution structure of the Mahoney strain of type 1 poliovirus, *Science,* 229, 1359, 1985.

62. **Stauffacher, C. V., Usha, R., Harrington, M., Schmidt, T., Hosur, M. V., and Johnson, J. E.,** The structure of cowpea mosaic virus at 3.5 Å resolution, in *Crystallography in Molecular Biology,* Mora, D., Ed., Plenum Press, New York, in press.

63. **Franssen, H., Leunissen, J., Goldbach, R., Lomonossoff, G., and Zimmern, D.,** Homologous sequences in non-structural proteins from cowpea mosaic virus and picornaviruses, *EMBO J.,* 3, 855, 1984.

64. **Kamer, G. and Argos, P.,** Primary structural comparison of RNA-dependent polymerases from plant, animal and bacterial viruses, *Nucleic Acids Res.,* 12, 7269, 1984.

65. **Tershak, D. R.,** Association of poliovirus proteins with the endoplasmic reticulum, *J. Virol.,* 52, 777, 1984.

66. **Emini, E. A., Schleif, W. A., Colonno, R. J., and Wimmer, E.,** Antigenic conservation and divergence between the viral-specific proteins of poliovirus type 1 and various picornaviruses, *Virology,* 140, 13, 1985.

67. **Pincus, S. E., Diamond, D. C., Emini, E. A., and Wimmer, E.,** Guanidine-selected mutants of poliovirus: mapping of point mutations to polypeptide 2C, *J. Virol.,* 57, 638, 1986.

68. **Argos, P., Kamer, G., Nicklin, M. J. H., and Wimmer, E.,** Similarity in gene organization and homology between proteins of animal picornaviruses and a plant comovirus suggest common ancestry of these virus families, *Nucleic Acids Res.,* 12, 7251, 1984.

69. **Haseloff, J., Goelet, P., Zimmern, D., Ahlquist, P., Dasgupta, R., and Kaesberg, P.,** Striking similarities in amino acid sequence among nonstructural proteins encoded by RNA viruses that have dissimilar genomic organization, *Proc. Natl. Acad. Sci. U.S.A.,* 81, 4358, 1984.

70. **Goldbach, R. W.,** Molecular evolution of plant RNA viruses, *Annu. Rev. Phytopathol.,* 24, 289, 1986.

71. **Miller, W. A., Bujarski, J. J., Dreher, T. W., and Hall, T. C.,** Minus-strand initiation by Brome mosaic virus replicase within the 3' tRNA-like structure of native and modified RNA templates, *J. Mol. Biol.,* 187, 537, 1986.

72. **Houwing, C. J. and Jaspars, E. M. J.,** Coat protein blocks the in vitro transcription of the virion RNAs of alfalfa mosaic virus, *FEBS Lett.,* 209, 284, 1986.

73. **Berna, A., Doré, J.-M. Collendavelloo, J., Mugnier, M.-A., and Pinck, L.,** EMBO Workshop, Molecular Plant Virology, (Abstr.), 52, 1986.

74. **Takeda, N., Kuhn, R. J., Yang, C.-F., Takegami, T., and Wimmer, E.,** Initiation of poliovirus plus-strand RNA synthesis in a membrane complex of infected HeLa cells, *J. Virol.,* 60, 43, 1986.

75. **Young, D. C., Tuschall, D. M., and Flanegan, J. B.,** Poliovirus RNA-dependent RNA polymerase and host cell protein synthesize product RNA twice the size of poliovirion RNA in vitro, *J. Virol.,* 54, 256, 1985.

76. **Young, D. C., Dunn, B. M., Tobin, G. J., and Flanegan, J. B.,** Anti-VPg antibody precipitation of product RNA synthesized in vitro by the poliovirus polymerase and host factor is mediated by VPg on the poliovirion RNA template, *J. Virol.,* 58, 715, 1986.

77. **Andrews, N. C., Levin, D. H., and Baltimore, D.,** Poliovirus replicase stimulation by terminal uridylyl transferase, *J. Biol. Chem.,* 260, 7628, 1985.

78. **Andrews, N. C. and Baltimore, D.,** Purification of a terminal uridylyltransferase that acts as host factor in the in vitro poliovirus replicase reaction, *Proc. Natl. Acad. Sci. U.S.A.*, 83, 221, 1986.
79. **Gorbalenya, A. E., Blinov, V. M., and Koomin, E. V.,** Prediction of nucleotide-binding properties of virus-specific proteins from their primary structure, *Mol. Genetika*, N11, 30, 1985.
80. **Zabel, P., Dorssers, L., Wernars, K., and Van Kammen, A.,** Terminal uridylyl transferase of Vigna unguiculata: purification and characterization of an enzyme catalyzing the addition of a single UMP residue to the 3′-end of an RNA primer, *Nucleic Acids Res.*, 9, 2433, 1981.

Chapter 4

REPLICATION OF THE RNAs OF ALPHAVIRUSES AND FLAVIVIRUSES

James H. Strauss and Ellen G. Strauss

TABLE OF CONTENTS

I.   Introduction 72

II.   The Alphaviruses 72
       A.   The Genome Organization of Alphaviruses 72
       B.   Alphavirus RNA Replication 75
       C.   Evolution of Alphaviruses 78
       D.   Relationship of Alphaviruses to Certain Plant Viruses 81

III.   The Flaviviruses 82
       A.   The Genome Organization of Flaviviruses 82
       B.   Comparison of Flavivirus Sequences 83
       C.   Flavivirus RNA Replication 86
       D.   Comparison of Yellow Fever Strains 17D and Asibi 86

IV.   Concluding Remarks 87

Acknowledgments 88

References 88

I. INTRODUCTION

The alphaviruses and flaviviruses are two important groups of positive strand animal viruses which contain a number of human and veterinary pathogens. Members of each group (about 25 for alphaviruses and 60 for flaviviruses) have evolved from common ancestors in response to different environmental factors. Furthermore, each virus species consists of a number of distinguishable strains that exhibit different geographical distributions and host ranges, indicating that these viruses are continuing to evolve in response to external selection pressures.[1,2] Although alphaviruses and flaviviruses are similar in structure,[3] their replication strategies are distinct. Their genome organizations have been deduced from the complete nucleotide sequences of the RNAs of the type alphavirus, Sindbis virus,[4] and the type flavivirus, yellow fever virus.[5] In addition, there exists a considerable body of comparative RNA sequence data from different viruses[6,7] which has been useful for identifying conserved functional domains and in determining the evolutionary relationships among these viruses. Unfortunately biochemical studies of RNA replication, either in vitro or in vivo, have lagged behind the sequencing efforts. Therefore, this chapter will focus primarily on the genome organization and translation strategy of alpha and flaviviruses and their consequences for viral replication.

Until recently, both of these groups were considered to be genera within the family Togaviridae. Both groups contain plus-stranded RNA of about the same size (approximately 12 kilobases in the case of alphaviruses and 11 kilobases in the case of flaviviruses) that is infectious after extraction from the virion and deproteinization. The RNA is present in a nucleocapsid with apparent icosahedral symmetry, which is in turn enveloped by a lipoprotein envelope containing proteins encoded by the viral genome. Most members of both groups are transmitted by arthropod vectors; mosquitoes in the case of alphaviruses and either mosquitoes or ticks (depending upon the virus) for flaviviruses.[8] Once the genome organizations were determined, it became clear that the flaviviruses were quite distinct and they were reclassified as the monogeneric family Flaviviridae.[9] The family Togaviridae currently contains three genera, Alphavirus, Rubivirus, and Pestivirus, as well as a number of incompletely characterized viruses still tentatively classified as Togaviridae.[10]

II. THE ALPHAVIRUSES

A. The Genome Organization of Alphaviruses

The genome organization and translation strategy of the alphaviruses is illustrated in Figure 1 for the type alphavirus, Sindbis virus. Sindbis RNA is 11,703 nucleotides in length exclusive of a 60 ($\pm$)-nucleotide poly(A) tract at the 3' end and it has a 5' terminal cap. The genome is comprised of two regions, encoding the nonstructural and structural proteins, respectively. The nonstructural proteins are translated as polyprotein precursors from the virion RNA and processed to form four final products known as nsP1, nsP2, nsP3, and nsP4 in order of their location within the genome from 5' to 3'. The proteolytic activity responsible for processing the nonstructural precursor is believed to be virus encoded and probably autocatalytic. This protease cleaves (at the arrow) in the sequence of Gly–Ala ↓ Ala or Gly–Ala ↓ Gly (in five alphaviruses sequenced in this region) to separate nsP1 from nsP2, and in the sequence Gly–Ala ↓ Ala or Gly–Cys ↓ Ala to separate nsP2 and nsP3; the cleavage site for producing nsP4 is Gly–Gly ↓ Tyr or Gly–Ala ↓ Tyr.[11]

In Sindbis virus, Middelburg virus,[12] and Ross River virus[13] there is an opal termination codon (UGA) at the COOH terminus of nsP3. The predominant polyprotein made thus contains nsP1, nsP2, and nsP3 (Figure 1); readthrough of this termination codon is required to produce nsP4. The mechanism of readthrough is not understood, but nsP4 is only produced in small quantities relative to the other three products. On the other hand, in Semliki Forest

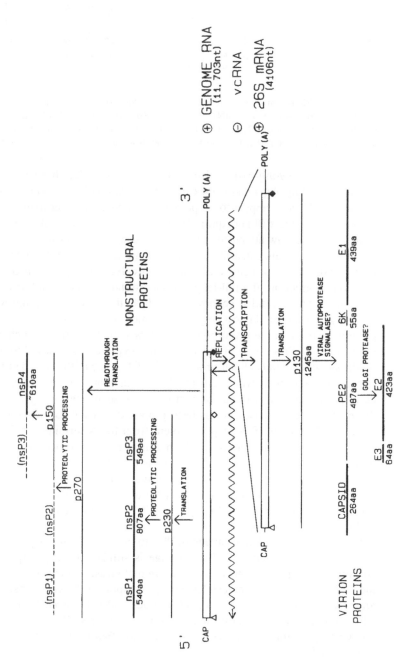

FIGURE 1. Replication strategy of Sindbis virus. Untranslated regions of the genomic RNA are shown as single lines, and the translated region as an open box. The subgenomic RNA region is expanded below using the same convention. Minus strand RNA (vc RNA) is shown as a wavy line. All translation products are indicated and the final protein products, both virion and nonstructural, are shown as bold lines. (△) indicates

```
                                          10                    →nsP4        20
                                          I                     ┌─            I
         R   R   S   R   R   T   E   Y  [X]  L   T   G   V   G   G │ Y   I   F   S   T   D
SIN     CGC AGG AGC AGG AGG ACU GAA UAC UGA CUA ACC GGG GUA GGU GGG│UAC AUA UUU UCG ACG GAC

         E   F   E   R   L   T   S   A  [X]  L   D   R   A   G   A │ Y   I   F   S   S   D
MID     GAG UUC GAA CGU UUA ACG UCA GCA UGA CUA GAC CGG GCG GGG GCC│UAC AUA UUC UCA UCG GAU

         G   D   I   D   F   D   Q   F  [X]  L   G   R   A   G   A │ Y   I   F   S   S   D
RR      GGC GAU AUU GAU UUU GAC CAA UUC UGA CUA GGC AGA GCG GGG GCG│UAC AUC UUC UCG UCU GAU

         F   G   D   F   D   D   V   L   R   L   G   R   A   G   A │ Y   I   F   S   S   D
SF      UUC GGA GAC UUC GAC GAC GUC CUG CGA CUA GGC CGC GCG GGU GCA│UAU AUU UUC UCC UCG GAC

         C   S   D   T   D   E   E   L   R   L   D   R   A   G   G │ Y   I   F   S   S   D
ONN     UGC UCG GAC ACA GAC GAA GAG UUA CGA CUA GAC AGA GCA GGG GGU│UAC AUA UUC UCC UCU GAC
```

FIGURE 2.   Alignment of sequences at the stop codon. Translated sequences of Semliki Forest virus (SF),[14] Middelburg (MID) and Sindbis (SIN) viruses,[12] and Ross River (RR) and O'Nyong-nyong (ONN) viruses[13] around the stop codon are shown. The stop codon (X) is boxed and the beginning of nsP4 (as determined from amino acid sequencing of the isolated protein for SFV and from protein homology in the other cases) is indicated.

virus[14] and O'Nyong-nyong virus,[13] the opal codon has been replaced by a sense codon (CGA, encoding Arg) and nsP4 can be produced without readthrough (Figure 2). For SFV at least, nsP4 does appear to be produced in amounts larger than that found during Sindbis infection, but it is still not equimolar with nsP1, nsP2, and nsP3, suggesting that production of nsP4 is somehow attenuated in this virus as well. It is unclear why nsP4 is underproduced, but it has been suggested that it serves some regulatory role during virus replication. All four nonstructural proteins are probably involved in replication of the viral RNA, which involves production of a full-length minus strand template and transcription of this template to produce plus-stranded genomes as well as the 26S subgenomic RNA. The subgenomic 26S RNA is translated into a polyprotein which is ultimately cleaved to form the nucleocapsid protein, two envelope glycoproteins, and two small peptides which are usually not present in virions.[15-17] Processing of the structural polyprotein is believed to occur by a combination of virus-encoded and cellular proteases.[11,17] Cleavage of the capsid protein from the precursor occurs by autoproteolytic cleavage.[18-20] From mapping of temperature-sensitive mutants defective in this activity and comparison of the sequences of alphavirus capsid proteins with those of serine proteases, it has been proposed that the capsid protein is a serine autoprotease.[21,22] Removal of the capsid protein from the N terminus of the polyprotein precursor reveals a new amino terminus which functions as a signal sequence to insert precursor glycoprotein PE2 into the rough endoplasmic reticulum. The 6K protein located between glycoproteins E2 and E1 functions as an internal signal sequence leading to the insertion of glycoprotein E1.[23] The cleavages after E2 and before E1 which liberate the 6K protein are postulated to be performed by signalase,[16,17] a cellular enzyme active in the lumen of the endoplasmic reticulum.

Finally, there is a late cleavage of PE2 to produce E2 and remove a small amino terminal domain of PE2 called E3. This cleavage is similar in many respects to cleavages of glycoprotein precursors of other enveloped animal viruses which occur during the transit through the Golgi apparatus or in vesicles during transport to the cell surface. This cleavage occurs after two or more basic amino acid residues and is believed to be due to a Golgi protease which processes a number of cellular components as well.[16,17] Alphavirus glycoproteins are known to pass through the Golgi and are modified by a number of other Golgi enzymatic functions.[24]

About 90% of the polysome-associated, virus-specific mRNA in infected cells is 26S, and structural proteins are produced in great molar excess over nonstructural proteins. Since more than 200 copies of the structural proteins are needed to package a single RNA molecule, such a mechanism to produce predominantly structural proteins is clearly efficient and advantageous. It is notable that utilization of the coding capacity of the alphavirus genome is quite efficient even though overlapping reading frames are not used, as is the case for many of the minus strand viruses. Thus, in a genome almost 12 kilobases in size, only about 80 nucleotides at the 5' end, about 50 nucleotides in the junction region, and 100 to 500 nucleotides at the 3' end are not translated into protein, and at least part of these nontranslated regions play a role in replication of the viral RNA.

B. Alphavirus RNA Replication

During the alphavirus replication cycle, three distinct RNA synthetic activities are needed: early in infection, a replicase is needed to synthesize a full-length minus strand template from the genomic RNA; in addition, starting early in infection, a replicase transcribes full-length genomic RNAs from this minus strand template, leading to amplification of the infecting RNA (later in infection, when large amounts of genomic RNA are produced for incorporation into progeny virions, this mode predominates); and an efficient transcriptase is produced that initiates internally on the full-length minus strand template and produces a large quantity of 26S messenger RNA. The various modes of virus-specific RNA synthesis are regulated both quantitatively and temporally (see the time course shown in Figure 3). By comparative sequence analysis of several alphaviruses, a number of conserved sequence elements have been identified which may serve as promoters for these various synthetic functions.[25-28] These conserved sequences are mapped on a schematic representation of the genome in Figure 4.

The conserved 19-nucleotide sequence element adjacent to the poly(A) tail at the 3' terminus of the RNA is postulated to be a promoter sequence for initiation of minus strand synthesis.[26] However, since 26S mRNA (which is 3' coterminal with the 49S genomic RNA) is not a template for minus strand synthesis, some additional non-26S sequence must also be required. There is a conserved double-loop sequence found near the 5' terminus of the genomic RNA, marked "51 nt" in Figure 4, that could conceivably play a role in discriminating between genomic RNA and 26S RNA.[27] The alphaviruses also contain a conserved 21 nucleotide domain just upstream of the beginning of 26S RNA which may form the recognition sequence for initiation of transcription of this subgenomic message.[28] Finally, there is a stem and loop structure, as well as a short nucleotide sequence at the 5' end which is conserved, which is postulated to be a promoter (in the minus strand) for initiation of plus-stranded genomic RNA synthesis.[27] Several lines of evidence indicate that it is the structure rather than the linear sequence that is important (also see below).[25] It is of interest that for alphaviruses, like most plus stranded RNA viruses, the 3' ends of the minus strand templates are quite different from the 3' ends of the plus strand templates, unlike the case for the minus strand viruses where both plus and minus strands contain the same short 3' terminal sequences.

The function of these conserved sequence elements in RNA replication has been tested in a defective interfering RNA system.[29,30] In these experiments DI RNA was cloned as cDNA and inserted into a bacterial plasmid. DI RNA was then transcribed from this plasmid with bacteriophage SP6 RNA polymerase, and the resulting RNA used to transfect cells that had been infected with wild-type Sindbis virus. If the in vitro transcribed RNA functioned as DI RNA it would replicate and be packaged, and within two or three passages become established as a dominant RNA which interfered with wild-type virus replication. The function of these promoter elements was tested by deletion analysis in which different domains were removed from the cloned DNA copy of the DI RNA. The most straightforward results tested the function of the 3' terminal 19-nucleotide sequence element. Deletions upstream of the sequence element, which left exactly 19 nucleotides of the RNA at the 3' terminus adjacent to the poly(A) tract intact, resulted in a functionally active DI RNA; but the deletion of even two nucleotides into this domain, leaving only the 17 3' terminal nucleotides, resulted in an inactive RNA. Thus, the original postulate that this 19-nucleotide sequence element served as a promoter required for viral RNA replication seems to be valid.

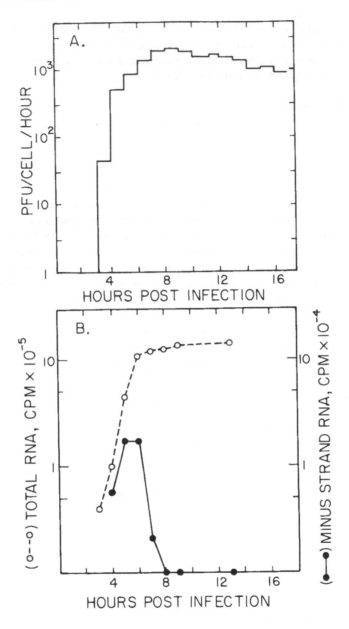

FIGURE 3.   Growth curve of Sindbis virus in chicken embryo fibroblasts at 30°C. (A) Release of progeny virus into the extracellular fluid. (B) Cells were infected with Sindbis virus and pulsed for 1 hr at the times specified with radioactive uridine. Open symbols (○) are total radioactivity incorporated into acid-insoluble form. Solid symbols (●) indicate incorporation into minus strand RNA (note that the scale for minus strand differs by tenfold from that for total RNA). (From Strauss, E. G. and Strauss, J. H., in *The Togaviridae and Flaviviridae*, Schlesinger, S. and Schlesinger, M., Eds., Plenum Press, New York, 1986, chap. 3. With permission.)

On the other hand, the function of the 51-nucleotide sequence element could not be determined from DI studies. Deletion of this sequence element resulted in a DI which was viable,[29] but less efficient, and whose efficiency of replication depended upon the 5′ terminal sequence element present in the DI RNA.[31] Thus, it appears that this 51-nucleotide sequence element is important for viral RNA replication, but the exact nature of its function is unclear.

GENOME ORGANIZATION OF ALPHAVIRUSES

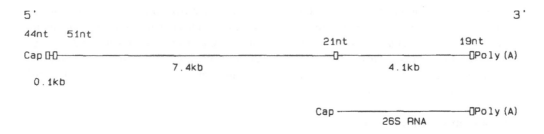

FIGURE 4.   Location of conserved sequences in alphavirus RNAs. The four boxes show the location of four regions of conserved sequence or structure. These are (right to left) the 19-nucleotide (19nt) sequence adjacent to the 3' poly(A) tract, the 21nt sequence at the junction region, the 51nt domain near the 5' end, and the 5' terminal 44 nucleotides. (From Strauss, E. G. and Strauss, J.H., in *The Togaviridae and Flaviviridae*, Schlesinger, S. and Schlesinger, M., Eds., Plenum Press, New York, 1986, chap. 3. With permission.)

This 51-nucleotide sequence is also not necessary for encapsidation, since DI RNAs lacking it can be packaged.

DI RNAs have been isolated that have a number of sequences as their 5' termini.[32,33] These include DI RNAs with the standard viral sequence at the 5' end, those with sequences from the 5' end of 26S RNA as their 5' end, and others which have as their 5' end a host cell transfer RNA. Thus, it is clear that a number of sequence elements at the 5' end suffice for replication and encapsidation of DI RNAs. The deletion experiments performed in the cDNA system showed that in fact, one or some combination of these sequences were required for DI replication because deletion of these 5' terminal sequences resulted in an inactive product.[29]

Finally, the function of the 21-nucleotide sequence element junction region has not been tested to date in such an in vitro system.

We know that since the naked RNA is infectious, the products of translation of the incoming 49S genomic RNA are sufficient to replicate the RNA and transcribe 26S RNA. However, at the current time, individual enzymatic activities cannot be unambiguously assigned to nsP1, nsP2, nsP3, or nsP4. Genetic analysis suggests that all four contain domains involved in RNA replication, since four complementation groups (groups A, B, F, and G) of temperature-sensitive mutants have been isolated which fail to produce RNA at the nonpermissive temperature.[34] Although these mutants have not been mapped, the phenotypes of a number of them have been examined by temperature shift experiments in which infection is initiated at the permissive temperature, followed by shift up to nonpermissive temperature and examination of the RNA phenotype. After shift, the group F mutants fail to make any form of RNA, either plus or minus, 49S or 26S, and are presumably defective in a major elongation function. After shift, the group B mutants stop making minus strands, possibly due to disfunction of a minus strand initiation factor. Groups A and G both make proportionally less 26S mRNA after shift, suggesting a defect in transcription rather than replication. A number of group A mutants have been examined and they appear to have diverse phenotypes,[35] thus implying that the group A polypeptide contains a number of different functional domains which can be independently modified. Indeed, part of the difficulties with the genetic analysis may be that all of the activities enumerated above are due to complexes involving more than one polypeptide. In addition, host cell factors have been implicated in alphavirus replication.[6,25]

As alluded to in the introduction, attempts to purify an active alphavirus RNA replicase have been only partially successful; no laboratory has succeeded in obtaining a preparation which is free of endogenous template RNA and capable of virus specific initiation of RNA synthesis. However, partially purified complexes have been characterized and it appears clear that nsP1 is a major component of them; minor amounts of both nsP2 and nsP4 are also seen in some preparations, but nsP3 has yet to be identified in a replicase complex.[6] Because the nonstructural proteins until recently have been characterized only by their apparent molecular weights on acrylamide gels, and these apparent molecular weights vary significantly from virus to virus and sometimes even between research groups, it is somewhat difficult to decipher the early literature in this field and determine which polypeptides are indeed present in the replication complexes isolated.

C. Evolution of Alphaviruses

Although a number of alphaviruses have been examined in the region encoding the structural proteins,[15-17,36,37] only two complete genomic sequences are known; that for Sindbis virus[4] and for Semliki Forest virus.[14] The deduced amino acid sequences of all the proteins encoded by these two viruses have been aligned and compared in Figure 5. Sindbis and Semliki Forest are two of the most distantly related alphaviruses, both on the basis of serological cross-reactions (which examine primarily external domains of the glycoproteins),[39] and on the evolutionary tree derived from the N terminal sequences of their E1 glycoproteins.[40] As seen in Figure 5 and Table 1, the proteins of the two viruses average overall ≈55% amino acid sequence homology. The homology is strikingly nonuniform, however; some proteins are more highly conserved than others, and within proteins there are conserved and nonconserved domains. The structural proteins share less amino acid sequence homology (46% on average) than the nonstructural proteins (59%). In the structural proteins, E1 is the most highly conserved virion glycoprotein (49%). The capsid protein exhibits two distinct domains, an N terminal domain exhibiting little conservation (31%) and a C terminal domain that is highly conserved (69%). Analogous considerations apply to the nonstructural proteins. nsP4 is the most highly conserved protein among alphaviruses, exhibiting 74% amino acid sequence homology between Sindbis and Semliki Forest viruses, and even here, the central core of the protein is more highly conserved than the ends. At the other extreme, the C terminus of nsP3 of Sindbis contains 80 amino acids which have no counterpart in Semliki Forest virus. Note that throughout the protein sequence, the hydrophobicity profile is conserved whether or not the amino aicd sequence is conserved (Figure 5).

Even though there is extensive amino acid sequence homology between different alphaviruses, the nucleotide sequence encoding any conserved amino acid domain has diverged markedly to the extent permitted by the degeneracy of the amino acid code. An example of this is shown in Figure 6 where a domain of nsP4 and the RNA encoding it is compared for four different alphaviruses. The nine amino acids in this domain are perfectly conserved among all four alphaviruses, but the nucleotide sequence has diverged, as indicated, to use many different codons for conserved amino acids. Quantitative comparisons have shown that codon usage for conserved amino acids is essentially completely randomized between any two alphaviruses.[6] Thus, the evolutionary separation of any two viruses has been extensive and selection for conservation of sequence has been at the level of amino acid sequence and protein function.

Such an analysis makes clear that when nucleotide sequences are found which are conserved among different alphaviruses, conservation is occurring at the level of the nucleotide sequence per se, suggesting that these conserved nucleotides play an essential role in viral RNA replication, transcription or encapsidation. The locations of several conserved RNA sequences in alphavirus genomes were shown in Figure 4, and their roles in RNA replication were discussed above.

The high degree of conservation of particular domains within the nonstructural proteins of alphaviruses may reflect differing evolutionary pressures upon different viral components.

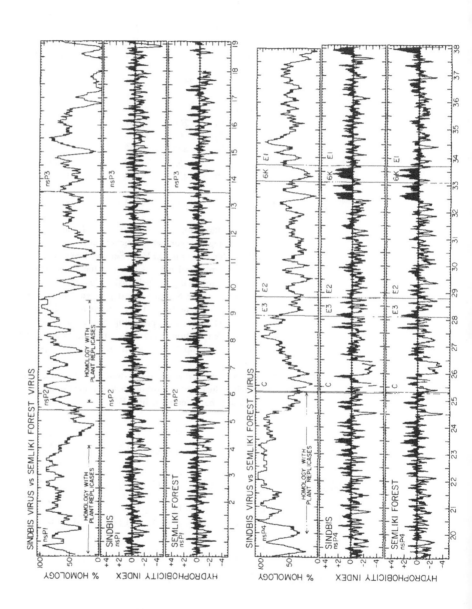

Table 1

Protein (total positions)[a]	Amino acids		Number identical (% homology)[b]
	SIN[c]	SFV[c]	
Nonstructural			
nsP1 (541)	540	537	339 (62.7)
nsP2 (810)	807	798	481 (59.4)
nsP3 (556)	549[d]	482[d]	208 (37.4)
nsP4 (614)	610	614	453 (73.8)
Totals 2521	2506[e]	2431	1481 (58.7)
Structural			
C (279)	264	267	149 (53.4)
E3 (67)	64	66	30 (44.8)
E2 (427)	423	422	170 (39.8)
6K (60)	55	60	17 (28.3)
E1 (439)	439	438	215 (49.0)
Totals 1272	1245	1253	581 (45.7)

[a]   Total positions include gaps introduced for alignment.

[b]   % Homology = number identical/total positions.

[c]   SIN = Sindbis virus; SFV = Semliki Forest virus.

[d]   The N terminus of nsP3 is taken as the amino acid immediately following the nsP2/nsP3 cleavage site (amino acid 1336 of the Semliki Forest polyprotein or 1348 of the Sindbis polyprotein). The C terminus is taken as the amino acid immediately preceding the nsP3/nsP4 cleavage site in SFV (position 1817) or the amino acid immediately preceding the opal termination codon in SIN (position 1896).

[e]   The SIN open reading frame totals 2513 amino acids, which include seven positions between the C terminus of nsP3 and the N terminus of nsP4.

```
            L       F       K       L       G       K       P       L
SIN        CUG     UUU     AAG     UUG     GGU     AAA     CCG     CUC

            L       F       K       L       G       K       P       L
MID        CUC     UUU     AAG     CUC     GGA     AAA     CCG     CUG

            L       F       K       L       G       K       P       L
RR         UUA     UUU     AAA     CUA     GGU     AAA     CCU     UUA

            L       F       K       L       G       K       P       L
SF         CUG     UUC     AAG     UUG     GGU     AAG     CCG     CUA
CODONS      3       2       2       3       2       2       2       4
```

FIGURE 6.   Randomization of codons used for conserved amino acids among alphaviruses. A short stretch of conserved protein sequence found in nsP4, about 80 amino acids from the C terminus, is shown for four alphaviruses along with the nucleotide sequence that encodes it. The first leucine is amino acid 2431 in the Sindbis nonstructural polyprotein. The bottom line gives the number of different codons used at each position. The single-letter amino acid code is used. Abbreviations for the virus names are the same as in Figure 2. The data are from Ou et al.[28] (From Strauss, E. G. and Strauss, J. H., in *The Togaviridae and Flaviviridae*, Schlesinger, S. and Schlesinger, M., Eds., Plenum Press, New York, 1986, chap. 3. With permission.)

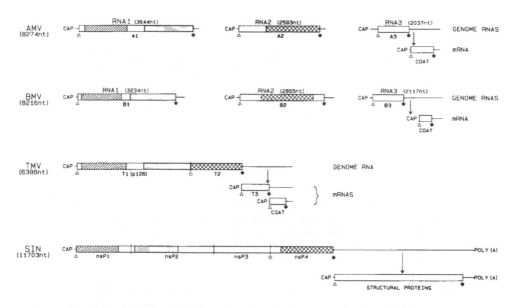

FIGURE 7. Genome organization and amino acid homologies of Sindbis virus and three plant viruses: alfalfa mosaic virus (AMV), bromegrass mosaic virus (BMV), and tobacco mosaic virus (TMV). The conventions are the same as those used in Figure 1. Within the translated regions there are three areas of homology indicated with different types of shading (hatching, stipling, and crosshatching); these regions were also indicated in Figure 5. All genomes are shown to scale. (From Ahlquist, P. E., Strauss, E. G., Rice, C. M., Strauss, J. H., Haseloff, J., and Zimmern, D., *J. Virol.*, 53, 536, 1985. With permission.)

Structural proteins, especially the glycoproteins, change relatively rapidly in response to the immunological defenses of the host. Changes in host range are probably also a source of alteration in the glycoproteins in that a change in affinity for host-cell receptors may permit infection of new cell types or new host species. Evolutionary pressures upon the nonstructural proteins, on the other hand, are of a different nature and appear to select against diversity. Replicase proteins have many interactions with one another to form different types of complexes, with host factors as yet unidentified, and with both genomic and virus-complementary RNA during RNA replication and transcription. Most alterations in the replicase are presumably deleterious and selected against in the absence of concommitant changes in the moieties with which they interact. Changes in proteins that interact with host cell factors or that are involved in RNA chain elongation would be expected to be particularly rare.

D. Relationship of Alphaviruses to Certain Plant Viruses

The replicase proteins of the alphaviruses have been found to share amino acid sequence homology with the corresponding proteins of three groups of plant viruses represented by tobacco mosaic virus (TMV), alfalfa mosaic virus (AMV), and brome mosaic virus (BMV).[41,42] These three groups of plant viruses have different morphologies: TMV is a helical rod; BMV is a simple icosahedron; and AMV is bacilliform. Three areas of sequence homology are illustrated in Figure 7 which also points out similarities in the replication and translation strategies used by these viruses. All of these viruses utilize a subgenomic messenger RNA to produce the structural protein(s) required for virion assembly. In addition, TMV has an amber termination codon in its replicase-encoding region which leads to synthesis of two proteins. The readthrough portion of this protein shows amino acid sequence homology with the readthrough portion of the Sindbis polyprotein and with the region encoded on a separate RNA segment (RNA2) in the case of BMV and AMV. Thus, in each case there is either demonstrated or potential regulation of the amounts produced of the nonstructural protein corresponding to nsP4 of Sindbis virus. The domains of homology to plant viruses were

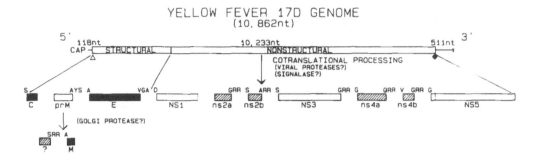

FIGURE 8.   Organization and processing of proteins encoded by the yellow fever virus genome. Conventions are the same as those in Figure 1. The single-letter amino acid code is used for sequences flanking assigned cleavage sites. Two other potential cleavage sites are shown as dotted vertical lines. Structural proteins, identified nonstructural proteins, and hypothesized nonstructural proteins are indicated by solid, open, and hatched boxes, respectively. (From Rice, C. M., Lenches, E. M., Eddy, S. R., Shin, S. J., Sheets, R. L., and Strauss, J. H., *Science*, 229, 726, 1985. With permission. Copyright 1985 AAAS.)

indicated also in Figure 5, in which Sindbis and Semliki Forest virus proteins were compared. It is of interest that the domains with homology to plant virus proteins also are the most conserved between Sindbis and Semliki Forest viruses (see Figure 5). A particularly striking example of this is the N terminal half of nsP2 where the Sindbis and Semliki Forest proteins demonstrate 74% amino acid sequence homology compared to an overall conservation of 59% for nsP2 as a whole.

The finding of three extensive regions of amino acid sequence homology among the replicase proteins of these four groups of viruses suggests that they have all descended from a common ancestor, perhaps an insect virus. If so, during evolution, as the viruses adpated to new hosts, the morphology of the virions and the sequence of the structural proteins have changed so dramatically that no detectable homology remains, although it is possible that detailed structural information from X-ray crystallography might reveal structural similarities in the proteins. For RNA viruses, replication strategy and genome organization may prove to be a more accurate predictor of evolutionary relationships than the morphological criteria currently used.

III. THE FLAVIVIRUSES

A. Genome Organization of Flaviviruses

The complete sequences of the type flavivirus, yellow fever virus,[5] and that of West Nile virus,[43-46] have been obtained as well as partial nucleotide sequences for Murray Valley encephalitis virus,[47] St. Louis encephalitis virus,[48] Japanese encephalitis virus,[49] and two serotypes of Dengue virus.[50-52] The genome organization and translation strategy utilized by the flaviviruses is illustrated in Figure 8 for yellow fever virus. The entire genome is 10,862 nucleotides in length and contains but a single, long, open reading frame of 10,233 nucleotides. It has been postulated that this open reading frame is translated into one large polyprotein which is proteolytically processed to produce both the structural and nonstructural flavivirus proteins. A combination of cellular proteases (to process the structural proteins in a manner analogous to alphaviruses) and virus-encoded proteases (to process the nonstructural proteins) have been postulated.[5,7,11] The structural proteins are found at the amino terminus of the precursor polyprotein, which allows overproduction of structural proteins relative to nonstructural proteins by attenuation and ribosome fall-off during translation. Nonetheless, in contrast to alphavirus infection, nonstructural proteins are made in significant quantities during flavivirus infection. Similar to the alphaviruses, there is an N terminal nucleocapsid protein, C, which is quite basic, and two envelope proteins. One envelope protein, M, which is not glycosylated, is derived from a precursor glycoprotein called prM. Cleavage of prM to produce M is an event very similar to the situation for alphavirus PE2; the same cleavage site is used and the same cellular protease is probably involved. The second envelope protein, E, is glycosylated in many, but not all, flaviviruses. This protein is the major envelope protein and carries the hemagglutinating and neutralization activities and the cellular attachment site. There is also a nonstructural glycoprotein, NS1, whose function during virus replication and assembly is unknown. These proteins, C, prM, E, and NS1, are postulated to be separated from one another by cellular signalase, similar to the case for alphaviruses. The remaining nonstructural proteins have been called ns2, NS3, ns4 and NS5, and are thought to be components of the viral RNA replicase. These proteins appear to be separated from one another by a virus-encoded protease that cleaves after two basic residues in succession. Once again this is postulated to be an autoproteolytic activity, at least for the initial cleavage events; an autoprotease, once it released itself from the precursor, could then act as a diffusible protease for other cleavage events, as occurs with the major poliovirus protease. As was the case for the alphaviruses and is true for RNA viruses in general, utilization of the coding capacity of the genome is very efficient, although, like other plus-stranded RNA viruses, flaviviruses do not use overlapping reading frames to increase coding capacity.

A number of features of the flavivirus genome suggest that flaviviruses are only distantly related to other groups of RNA animal viruses and that they deserve their recent reclassification as a separate family, Flaviviridae. The gene order and translation strategy, with structural proteins 5' terminal in the genome and translation of the entire genome as a single polyprotein, is a feature they share with Picornaviruses. The cleavage sites to separate the polyprotein precursor are very different from those utilized by Picornaviruses, however, as are details of RNA replication and virus maturation. Flaviviruses also lack a 3' poly(A) tract, a feature which distinguishes them from all other plus-stranded animal RNA viruses. They have instead a structure at the 3' end, shown in Figure 9, which is probably important in RNA replication and is reminiscent of those found at the 3' end of a number of plant virus RNAs. Finally, flaviviruses possess at the 3' end of the plus strand a limited number of nucleotides which are also found at the 3' end of the minus strand. This feature is unique among plus-stranded RNA viruses, but characteristic of minus strand RNA viruses. These common sequence elements are probably important in initiation of RNA synthesis, implying that the replicases for both plus and minus strand synthesis are very similar if not identical, and suggest that flaviviruses may be more closely related to the minus strand viruses than to the other plus-stranded RNA viruses. No amino acid sequence homologies have been found between flaviviruses and other groups of RNA viruses, except for very limited sequence homology in NS5,[7] of the type reported by Kamer and Argos,[54] which these authors use to suggest that all RNA viruses have evolved from a common ancestor.

B. Comparison of Flavivirus Sequences

There are about 60 members of the flavivirus family and the sequence relationships among the members of the family resemble in many ways the situation with the alphaviruses. Individual proteins demonstrate 25 to 90% sequence homology between two flaviviruses, depending upon the protein and the flaviviruses being compared. A comparison of the amino acid sequences of yellow fever and West Nile viruses is shown in Figure 10; these viruses represent members of two different serological subgroups among the mosquito-borne flaviviruses, and thus they are well separated, but not the most distantly related of flaviviruses. The structural proteins, especially the capsid protein, demonstrate much less sequence homology than do the nonstructural proteins (in particular, NS3 and NS5 which are probably the major components of the replicase). Within the structural protein E, there are domains specific for a given virus, domains shared among related viruses (for example, between

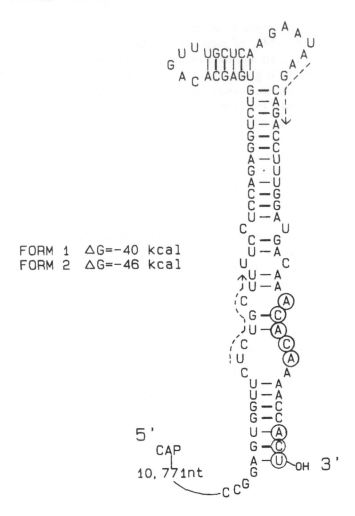

FORM 1   ΔG=-40 kcal
FORM 2   ΔG=-46 kcal

FIGURE 9.   Possible secondary structure at the 3' terminus of yellow fever virus 17D genomic RNA. Circled nucleotides are shared with the 3' terminus of the yellow fever minus strand. ΔG values were calculated according to Tinoco et al.[53] Form 2 assumes hydrogen bonding between the nucleotides indicated by the dashed arrows. (Adapted from Rice, C. M., Lenches, E. M., Eddy, S. R., Shin, S. J., Sheets, R. L., and Strauss, J. H., *Science*, 229, 726, 1985. With permission. Copyright 1985 AAAS.)

Murray Valley encephalitis virus and West Nile virus or between Dengue 2 virus and Dengue 4 virus) and domains conserved among all flaviviruses. These may correspond to the type, subcomplex, and group specific antigens which have been described for glycoprotein E.[55] Once again, the hydrophobicity profiles are remarkably well conserved, being virtually superimposable, even in domains sharing little amino acid sequence homology. Note in particular the ns2 region, where sequence homology is low. This suggests that the overall structures of the various proteins are the same for all flaviviruses, and that within this conserved structure some domains can accommodate little sequence variability while retaining function, whereas in other domains, the linear amino acid sequence is not crucial as long as the hydrophobicity relationships are maintained.

Like the alphaviruses, the flaviviruses have virtually randomized the codon usage even within domains of high amino acid conservation. In Figure 11 a conserved domain in the E protein is shown for four flaviviruses. Although the amino acid sequence is perfectly conserved in all four sequences, the codon usage varies considerably.

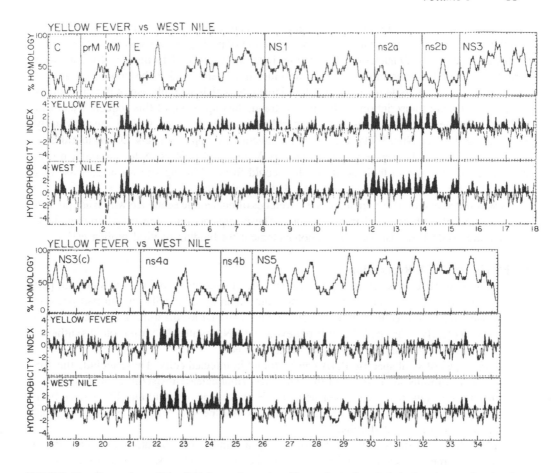

FIGURE 10.   Comparison of the flavivirus polyproteins. The method of analysis is the same as that for the alphaviruses shown in Figure 5. Polyproteins of yellow fever virus[5] and of West Nile virus[43-45] are compared. Boundaries of the individual polypeptides are indicated.

		P	P	F	G	D	S	Y	I
YF		CCA	CCU	UUU	GGA	GAC	AGC	UAC	AUU
		P	P	F	G	D	S	Y	I
WN		CCC	CCG	UUU	GGU	GAC	UCU	UAC	AUC
		P	P	F	G	D	S	Y	I
MVE		CCA	CCC	UUC	GGA	GAC	UCA	UAC	AUU
		P	P	F	G	D	S	Y	I
DEN2		CCU	CCA	UUC	GGA	GAC	AGC	UAC	AUC
CODONS		3	4	2	2	1	3	1	2

FIGURE 11.   Codon usage randomization in flavivirus proteins. A short stretch of nucleotide sequence encoding a portion of the E protein is shown for yellow fever virus (YF),[5] West Nile virus (WN),[43] Murray Valley encephalitis virus (MVE)[47] and Dengue 2 virus (Den 2)[52] along with the encoded amino acids. The number of codons used at any position is shown below. (Adapted from Rice, C. M., Strauss, E. G., and Strauss, J. H., in *The Togaviridae and Flaviviridae*, Schlesinger, S. and Schlesinger, M., Eds., Plenum Press, New York, 1986, chap. 10. With permission.)

A number of conserved sequence elements or structures have been found in flaviviruses, and these are probably important for RNA replication or packaging. As was shown in Figure 9, in the 3′ terminal 80 nucleotides of yellow fever are capable of forming a structure with a high calculated thermostability.[5] This structure has been found to be conserved among flaviviruses.[46,49,56] It could form a promoter element to which the replicase of the virus binds in order to initiate transcription of the minus strand. Just upstream of this sequence element are two stretches of nucleotides that are highly conserved between yellow fever, Murray Valley,[57] and West Nile.[46] These sequence elements are approximately 20 nucleotides in length, and one of them is repeated in Murray Valley and in West Nile, although not in yellow fever. Because of their position, they could form a binding site for a replicase that is bound to this sequence as well as to the 3′ terminal structure.

As was noted above, there are also short stretches of nucleotide sequence which are present at the 3′ end of both plus and minus strands. These sequences are conserved among flaviviruses and could form part of the promoter element for binding of viral replicases.[5]

C. Flavivirus RNA Replication

The intracellular events occurring during flavivirus infection are difficult to study, in part because host macromolecular synthesis is not shut off in infected cells, and in part because flaviviruses in general replicate to low titer. The intracellular localization of synthetic events has not been clearly delineated, but it is believed that RNA synthesis, viral protein synthesis, and virion morphogenesis are all membrane associated.[56]

Several types of viral-specific RNA can be found in infected cells. These include the 40S genomic RNA, minus strand RNA of the same size, ribonuclease-resistant replicative forms, and partially resistant replicative intermediates. However, no subgenomic RNAs have been described for flavivirus infection and it is believed that the 40S genomic RNA is the only mRNA in flavivirus-infected cells.

Crude cell-free extracts have been isolated by a number of laboratories which contain flavivirus polymerase activity, and in all cases, the polymerase appears to be a membrane-associated complex. Recent experiments have shown that for West Nile virus such complexes contain NS3 and NS5, and that antibodies to NS3 inhibit in vitro polymerase activity.[56] Although the 3′ termini of both plus and minus strands have short sequence elements in common, regulation of synthesis of the two strands is different since only approximately 10% of flavivirus-specific RNA found in cells is of minus polarity.[58] At this time it is not known what distinguishes the plus strand replicase complex from the minus strand replicase.

D. Comparison of Yellow Fever Strains 17D and Asibi

Two strains of yellow fever virus have now been sequenced in their entirety. One is the 17D vaccine strain[5] which has been widely used to inoculate people against yellow fever. The second is the parental virulent Asibi strain,[59] from which the avirulent 17D strain was derived. These two strains differ by some 240 passages in chicken cells in culture, and a comparison of the sequences is of interest both in order to quantitate the rate of change in RNA genomes with passage and also to understand the molecular basis of virulence. The two viruses differ in 68 nucleotides (0.63% difference) which lead to a change in 32 different amino acids (0.94% difference). Thus, approximately 0.25 nucleotide changes and 0.12 amino adic changes became fixed on average per passage in cell culture. These changes are not randomly distributed over the genome, but cluster in certain areas. Figure 12 illustrates the positions of both nucleotide and amino acid substitutions between the two strains. Proteins which are highly conserved among flaviviruses, such as NS3 and NS5 (compare Figure 10), show few changes between Asibi and 17D, and many nucleotide changes that do occur are silent or result in conservative amino acid substitutions. Presumably, changes in this region are detrimental to virus growth and are selected against. On the other hand, amino acid

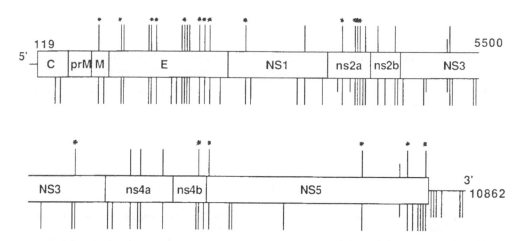

FIGURE 12.   Location of differences between the Asibi strain of yellow fever virus and the 17D vaccine strain. Lines above and below the schematic of the yellow fever genome indicate changes in amino acids (above) and nucleotides (below). Half-length lines show clonal differences in the Asibi strain. Asterisks indicate nonconservative changes. Data are from Hahn et al.[59]

changes in proteins that show great sequence divergence among flaviviruses, such as ns2 and ns4, are more frequent between the two strains. The envelope protein also shows a disproportionate number of changes and a large fraction (12 of 15) of the nucleotide changes in the E protein region lead to amino acid substitutions. Because the E protein carries the viral hemagglutinin and neutralization sites and binds to cellular receptors to initiate infection, it seems reasonable that changes occurring in E might lead to avirulence due to altered tissue tropisms. Growth of the virus in tissue culture would select for strains of virus that are able to bind rapidly and efficiently to chicken cells in culture and might lead to loss of or reduction in the ability to bind and replicate in cells from primates, the sole natural vertebrate hosts of yellow fever in nature. Further work will be required to determine if the changes in the E protein have in fact been responsible for the loss of virulence or whether changes in other regions of the genome are also involved.

IV. CONCLUDING REMARKS

In this brief review we have outlined the genome organization and translation strategies used by two important groups of RNA animal viruses, the alphaviruses and the flaviviruses, and have tried to describe some aspects of what is known about variation in RNA sequences among these viruses. It is clear that RNA genomes are quite mutable and thus able to diverge rapidly. In nature, alphaviruses and flaviviruses have both evolved into a fairly large number of family members which differ sufficiently to be considered as separate viruses and show limited serological cross reaction, but which all have essentially the same structure and share considerable amino acid sequence homology. These different viruses are simultaneously present in the field and can have either overlapping or distinct geographical distributions. This is similar to the situation for a number of other groups of RNA animal viruses, such as the picornaviruses, and in marked contrast to the case of influenza viruses, where despite their ability to undergo extensive genetic alteration, only one or a few strains of the virus predominate in the human population at any given time (although other strains are present in other hosts such as birds). The necessity for alpha- and flaviviruses to undergo an obligate alternation between the arthropod vectors (very often distinct for different members of the group) and a wide range of mammalian or avian hosts may affect their evolution in ways which are unclear at present.

ACKNOWLEDGMENTS

Work of the authors is supported by grants AI 10793 and AI 20612 from the National Institutes of Health and grand DMB 83 16856 from the National Science Foundation. The assistance of Tim Hunkapiller is gratefully acknowledged in the preparation of the figures, using the computer facility of Dr. Leroy Hood.

REFERENCES

1. **Griffin, D. E.,** Alphavirus pathogenesis and immunity, in *The Togaviridae and Flaviviridae,* Schlesinger, S. and Schlesinger, M., Eds., Plenum Press, New York, 1986, chap. 8.
2. **Monath, T. P.,** Pathology of flaviviruses, in *The Togaviridae and Flaviviridae,* Schlesinger, S. and Schlesinger, M., Eds., Plenum Press, New York, 1986, chap. 12.
3. **Murphy, F. A.,** Togavirus morphology and morphogenesis, in *The Togaviruses,* Schlesinger, R. W., Ed., Academic Press, New York, 1980, chap. 8.
4. **Strauss, E. G., Rice, C. M., and Strauss, J. H.,** Complete nucleotide sequence of the genomic RNA of Sindbis virus, *Virology,* 133, 92, 1984.
5. **Rice, C. M., Lenches, E. M., Eddy, S. R., Shin, S. J., Sheets, R. L., and Strauss, J. H.,** Nucleotide sequence of yellow fever virus: implications for flavivirus gene expression and evolution, *Science,* 229, 726, 1985.
6. **Strauss, E. G. and Strauss, J. H.,** Structure and replication of the alphavirus genome, in *The Togaviridae and Flaviviridae,* Schlesinger, S. and Schlesinger, M., Eds., Plenum Press, New York, 1986, chap. 3.
7. **Rice, C. M., Strauss, E. G., and Strauss, J. H.,** Structure of the flavivirus genome, in *The Togaviridae and Flaviviridae,* Schlesinger, S. and Schlesinger, M., Eds., Plenum Press, New York, 1986, chap. 10.
8. **Chamberlain, R. W.,** Epidemiology of arthropod-borne Togaviruses. The role of arthropods as hosts and vectors and of vertebrate hosts in natural transmission cycles, in *The Togaviruses,* Schlesinger, R. W., Ed., Academic Press, 1980, chap. 6.
9. **Westaway, E. G., Brinton, M. A., Gaidamovich, S. Ya., Horzinek, M. C., Igarashi, A., Kääriäinen, L., Lvov, D. K., Porterfield, J. S., Russell, P. K., and Trent, D. W.,** Flaviviridae, *Intervirology,* 24, 183, 1985.
10. **Matthews, R. E. F.,** Classification and nomenclature of viruses. Fourth Report of the International Committee on Taxonomy of Viruses, *Intervirology,* 17, 1, 1982.
11. **Strauss, J. H., Strauss, E. G., Hahn, C. S., Hahn, Y. S., Galler, R., Hardy, W. R., and Rice, C. M.,** Replication of alphaviruses and flaviviruses: proteolytic processing of polyproteins, in *Positive Strand RNA Viruses,* Vol. 54, U.C.L.A. Symp. on Molecular and Cellular Biology, Brinton, M. A. and Rueckert, R., Eds., Alan R. Liss, New York, 1986.
12. **Strauss, E. G., Rice, C. M., and Strauss, J. H.,** Sequence coding for the alphavirus nonstructural proteins is interrupted by an opal termination codon, *Proc. Natl. Acad. Sci. U.S.A.,* 80, 5271, 1983.
13. **Strauss, E. G. and Strauss, J. H.,** Unpublished data, 1986.
14. **Takkinèn, K.,** Complete nucleotide sequence of the nonstructural protein genes of Semliki Forest virus, *Nucleic Acids Res.,* 14, 5667, 1986.
15. **Garoff, H., Frischauf, A. M., Simons, K., Lehrach, H., and Delius, H.,** The capsid protein of Semliki Forest virus has clusters of basic amino acids and prolines in its amino terminal region, *Proc. Natl. Acad. Sci. U.S.A.,* 77, 6376, 1980.
16. **Garoff, H., Frischauf, A. M., Simons, K., Lehrach, H., and Delius, H.,** Nucleotide sequence of cDNA coding for Semliki Forest membrane glycoproteins, *Nature (London),* 288, 236, 1980.
17. **Rice, C. M. and Strauss, J. H.,** Nucleotide sequence of the 26S mRNA of Sindbis virus and deduced sequence of the encoded virus structural proteins, *Proc. Natl. Acad. Sci. U.S.A.,* 78, 2062, 1981.
18. **Simmons, D. T. and Strauss, J. H.,** Translation of Sindbis virus 26S RNA and 49S RNA in lysates of rabbit reticulocytes, *J. Mol. Biol.,* 86, 397, 1974.
19. **Aliperti, G. and Schlesinger, M.,** Evidence for an autoprotease activity of Sindbis virus capsid protein, *Virology,* 90, 366, 1978.
20. **Scupham, R. K., Jones, K. J., Sagik, B. P., and Bose, H. R.,** Virus directed posttranslational cleavage in Sindbis virus infected cells, *J. Virol.,* 22, 568, 1977.
21. **Hahn, C. S., Strauss, E. G., and Strauss, J. H.,** Sequence analysis of three Sindbis virus mutants temperature-sensitive in the capsid protein autoprotease, *Proc. Natl. Acad. Sci. U.S.A.,* 82, 4648, 1985.

22. **Boege, U., Wengler, G., Wengler, G., and Wittmann-Liebold, B.,** Primary structures of the core proteins of the alphaviruses, Semliki Forest virus and Sindbis virus, *Virology,* 113, 293, 1981.

23. **Hashimoto, K., Erdel, S., Keranen, S., Saraste, J., and Kaariainen, L.,** Evidence for a separate signal sequence for the carboxy-terminal envelope glycoprotein El of Semliki Forest virus, *J. Virol.,* 38, 34, 1981.

24. **Schlesinger, M. J. and Schlesinger, S.,** Formation and assembly of alphavirus glycoproteins, in *The Togaviridae and Flaviviridae,* Schlesinger, S. and Schlesinger, M., Eds., Plenum Press, New York, 1986, chap. 5.

25. **Strauss, E. G., and Strauss, J. H.,** Replication strategies of the single stranded RNA viruses of eukaryotes, *Curr. Top. Microbiol. Immunol.,* 105, 1, 1983.

26. **Ou, J.-H., Trent, D. W., and Strauss, J. H.,** The 3'-noncoding regions of alphavirus RNAs contain repeating sequences, *J. Mol. Biol.,* 156, 719, 1982.

27. **Ou, J.-H., Strauss, E. G., and Strauss, J. H.,** The 5'-terminal sequences of the genomic RNAs of several alphaviruses, *J. Mol. Biol.,* 168, 1, 1983.

28. **Ou, J.-H., Rice, C. M., Dalgarno, L., Strauss, E. G., and Strauss, J. H.,** Sequence studies of several alphavirus genomic RNAs in the region containing the start of the subgenomic RNA, *Proc. Natl. Acad. Sci. U.S.A.,* 79, 5235, 1982.

29. **Levis, R., Weiss, B. G., Tsiang, M., Huang, H., and Schlesinger, S.,** Deletion mapping of Sindbis virus DI RNAs derived from cDNAs defines the sequences essential for replication and packaging, *Cell,* 44, 137, 1986.

30. **Schlesinger, S. and Weiss, B. G.,** Defective RNAs of alphaviruses, in *The Togaviridae and Flaviviridae,* Schlesinger, S. and Schlesinger, M., Eds., Plenum Press, New York, 1986, chap. 6.

31. **Schlesinger, S.,** Personal communication, 1986.

32. **Monroe, S. and Schlesinger, S.,** RNAs from two independently isolated defective-interfering particles of Sindbis virus contain a cellular tRNA sequence at their 3' ends, *Proc. Natl. Acad. Sci. U.S.A.,* 80, 3279, 1983.

33. **Tsiang, M., Monroe, S., and Schlesinger, S.,** Studies of defective-interfering RNAs of Sindbis virus with and without tRNA[ASP] sequences at their 5' termini, *J. Virol.,* 54, 38, 1985.

34. **Strauss, E. G., and Strauss, J. H.,** Mutants of alphaviruses: genetics and physiology, in *The Togaviruses,* Schlesinger, R. W., Ed., Academic Press, New York, 1980, chap. 14.

35. **Sawicki, D. L. and Sawicki, S. G.,** Functional analysis of the A complementation group mutants of Sindbis HR virus, *Virology,* 144, 20, 1985.

36. **Dalgarno, L., Rice, C. M., and Strauss, J. H.,** Ross River virus 26S RNA: complete nucleotide sequence and deduced sequence of the encoded structural proteins, *Virology,* 129, 170, 1983.

37. **Kinney, R. M., Johnson, B. J. B., Brown, V. L., and Trent, D. W.,** Nucleotide sequence of the 26S mRNA of the virulent Trinidad donkey strain of Venezuelan equine encephalitis virus and deduced sequence of the encoded structural proteins, *Virology,* 152, 400, 1986.

38. **Kyte, J. and Doolittle, R. F.,** A simple method for displaying the hydropathic character of a protein, *J. Mol. Biol.,* 157, 105, 1982.

39. **Calisher, C. H., Shope, R. E., Brandt, W., Casals, J., Karabatsos, N., Murphy, F. A., Tesh, R. B., and Wiebe, M. E.,** Proposed antigenic classification of registered arboviruses. I. Togaviridae, alphavirus, *Intervirology,* 14, 229, 1980.

40. **Bell, J. R., Kinney, R. M., Trent, D. W., Strauss, E. G., and Strauss, J. H.,** An evolutionary tree relating eight alphaviruses based on amino terminal sequences of their glycoproteins, *Proc. Natl. Acad. Sci. U.S.A.,* 81, 4702, 1984.

41. **Ahlquist, P. E., Strauss, E. G., Rice, C. M., Strauss, J. H., Haseloff, J., and Zimmern, D.,** Sindbis virus proteins nsP1 and nsP2 contain homology to nonstructural proteins from several RNA plant viruses, *J. Virol.,* 53, 536, 1985.

42. **Haseloff, J., Goelet, P., Zimmern, D., Ahlquist, P., Dasgupta, R., and Kaesberg, P.,** Striking similarities in amino acid sequence among nonstructural proteins encoded by RNA viruses that have dissimilar genomic organization, *Proc. Natl. Acad. Sci. U.S.A.,* 81, 4358, 1984.

43. **Wengler, G., Castle, E., Leidner, U., Nowak, T., and Wengler, G.,** Sequence analysis of the membrane protein V3 of the flavivirus West Nile virus and of its gene, *Virology,* 147, 264, 1985.

44. **Castle, E., Nowak, T., Leidner, U., Wengler, G., and Wengler, G.,** Sequence analysis of the viral core protein and the membrane associated proteins VI and NV2 of the flavivirus West Nile virus and of the genome sequence for these proteins, *Virology,* 145, 227, 1985.

45. **Castle, E., Leidner, U., Nowak, T., Wengler, G., and Wengler, G.,** Primary structure of the West Nile flavivirus genome region coding for all nonstructural proteins, *Virology,* 149, 10, 1986.

46. **Wengler, G. and Castle, E.,** Analysis of structural properties which probably are characteristic for the 3'-terminal sequence of the genome RNA of flaviviruses, *J. Gen. Virol.,* 67, 1183, 1986.

47. **Dalgarno, L., Trent, D. W., Strauss, J. H., and Rice, C. M.,** Partial nucleotide sequence of Murray Valley encephalitis virus: comparison of the encoded polypeptides with yellow fever virus structural and nonstructural proteins, *J. Mol. Biol.,* 187, 309, 1986.

48. **Trent, D. W., Kinney, R. M., Johnson, B. J. B., Vorndam, A. V., Grant, J. A., Deubel, V., Rice, C. M., and Hahn, C.,** Partial nucleotide sequence of St. Louis encephalitis virus RNA: structural proteins, NS1, ns2a, and ns2b, *Virology,* 156, 293, 1986.

49. **Takegami, T., Washizu, M., and Yasui, K.,** Nucleotide sequence at the 3' end of Japanese encephalitis virus genomic RNA, *Virology,* 152, 483, 1986.

50. **Deubel, V., Kinney, R. M., and Trent, D. W.,** Nucleotide sequence and deduced amino acid sequence of the structural proteins of dengue type 2 virus, Jamaica genotype, *Virology,* 155, 365, 1986.

51. **Zhao, B., Mackow, E., Buckler-White, A., Markoff, L., Chanock, R. M., Lai, C.-J., and Makino, Y.,** Cloning full length dengue type 4 viral DNA sequence: analysis of genes coding for structural proteins, *Virology,* 155, 77, 1986.

52. **Hahn, Y. S., Galler, R., Hunkapiller, T., Dalrymple, J. M., Strauss, J. H., and Strauss, E. G.,** Nucleotide sequence of Dengue 2 RNA and comparison of the encoded proteins with those of other flaviviruses, *Virology,* 1987, in press.

53. **Tinoco, I., Borer, P. N., Dengler, B., Levine, M. D., Uhlenbeck, O. C., Crothers, D. M., and Gralla, J.,** Improved estimation of secondary structure in ribonucleic acids, *Nature (London) New Biol.,* 246, 40, 1973.

54. **Kamer, G. and Argos, P.,** Primary structural comparison of RNA-dependent polymerases from plant, animal, and bacterial viruses, *Nucleic Acids Res.,* 12, 7269, 1984.

55. **Roehrig, J. T.,** The use of monoclonal antibodies in studies of the structural proteins of togaviruses and flaviviruses, in *The Togaviridae and Flaviviridae,* Schlesinger, S. and Schlesinger, M. J., Eds., Plenum Press, New York, 1986, chap. 9.

56. **Brinton, M. A.,** Replication of flaviviruses, in *The Togaviridae and Flaviviridae,* Schlesinger, S. and Schlesinger, M., Eds., Plenum Press, New York, 1986, chap. 11.

57. **Rice, C. M., Dalgarno, L., Galler, R., Hahn, Y. S., Strauss, E. G., and Strauss, J. H.,** Molecular cloning of flavivirus genomes for comparative analysis and expression, in *Modern Trends in Virology,* Proc. Int. Symp., Bauer, H., Klenk, H.-D., and Scholtissek, C., Eds., Springer-Verlag, Berlin, 1986, in press.

58. **Cleaves, G. R., Ryan, T. E., and Schlesinger, R. W.,** Identification and characterization of type 2 dengue virus replicative intermediate and replicative form RNAs, *Virology,* 111, 73, 1981.

59. **Hahn, C. S., Dalrymple, J. M., Strauss, J. H., and Rice, C. M.,** Comparison of the virulent Asibi strain of yellow fever virus with the 17D vaccine strain derived from it, *Proc. Natl. Acad. Sci. U. S. A.,* 84, 2019, 1987.

Chapter 5

RNA REPLICATION OF BROME MOSAIC VIRUS AND RELATED VIRUSES

T. W. Dreher and T. C. Hall

TABLE OF CONTENTS

I.   Introduction..92

II.   A Perspective on Research on Plant Viral Replication.........................92

III.   RNA-Dependent RNA Synthesizing Activities in Plants........................93
        A.   Viral Replicase...93
        B.   Host RNA-Dependent RNA Polymerase Activity..........................94
        C.   Host Terminal Transferases 95

IV.   Genomic Organization of BMV and Related Viruses 95

V.   Function of Viral Genes in Replication 98
        A.   Role of Gene 2..98
        B.   Role of Gene 1..98
        C.   Role of Gene 3..99

VI.   Studies on the Replication of BMV RNA......................................99
        A.   The Replicative Cycle of BMV 99
        B.   Preparation of BMV Replicase from BMV-Infected
                Barley Leaves 99
        C.   Template Dependency and Specificity of BMV Replicase..............100
        D.   Synthesis of ($-$) Strands by BMV Replicase 101
        E.   Relationship Between Replicase and tRNA-Like Activities...........103
        F.   Use of ($-$) Sense BMV RNA Templates by BMV Replicase.............104
        G.   Synthesis of Subgenomic RNA4 by BMV Replicase 105

VII.   Studies on the Replication of TYMV RNA 106
        A.   Preparation of TYMV Replicase from TYMV-Infected Plants...........106
        B.   Polypeptide Components of TYMV Replicase..........................107

VIII.   Future Directions 108
        A.   The Form of Actively Replicating Intermediates 108
        B.   The Composition of the Active Replication Complex 108
        C.   The Role of the tRNA-Like Structures 108

IX.   Conclusions...109

References..109

I. INTRODUCTION

Despite a widespread awareness over the last 15 years or so of the importance of in vitro studies on the replication of plant RNA viruses, our understanding of the subject remains remarkably incomplete.[1-4] The ability to amplify viral RNA in vitro using an enzyme purified from infected plant tissue is still a long way off. In this article, we discuss the state of our knowledge of the replication of those (+)-stranded RNA viruses of plants, the RNAs of which have capped 5' termini and can be aminoacylated at the 3' end.[1,5] Emphasis is placed on brome mosaic virus (BMV) and, to a lesser extent, turnip yellow mosaic virus (TYMV); the status for related viruses has been reviewed recently.[3] Similar replication strategies are probably used by all of the viruses with aminoacylatable 3' ends, and also by alfalfa mosaic virus (AlMV) and the ilarviruses, which do not bind amino acids, but do have very similar genomic organization.[6] Table 1 lists these groups of viruses and their type members. Much of the discussion in this article can be cautiously applied to the entire group, although idiosynchratic differences between viruses in certain instances are to be expected. Abbreviations used throughout the chapter are as follows:

- BMV   brome mosaic virus
- TYMV   turnip yellow mosaic virus
- AlMV   alfalfa mosaic virus
- TMV   tobacco mosaic virus
- CMV   cucumber mosaic virus
- CCMV   cowpea chlorotic mottle virus
- 12-M   dodecyl-β-D-maltoside

II. A PERSPECTIVE ON RESEARCH ON PLANT VIRAL REPLICATION

An understanding of the reasons for our poor knowledge of the biochemistry of the replication of the (+)-stranded RNA plant viruses is important in order to delineate the most productive approaches for future research. Although the first attempts were made in the early 1970s to study replication in vitro, consistent efforts have been made by only a few research groups; several investigators have been drawn away from this important field of study to work on other aspects of viral function. Research on replication has been frustrated by the fact that extracts from many healthy plants are able to incorporate ribonucleotides into acid-precipitable form when viral (or other) RNAs are offered as templates. These activities are often enhanced by viral infection, sometimes to the point where they catalyze the incorporation of similar or even greater amounts of [^{32}P]UTP than do viral-specific activities. Unless products are carefully scrutinized, at least two activities (terminal transferase and RNA-dependent RNA polymerase) that are distinct from the activity responsible for the specific replication of viral RNA can confuse studies of viral replication.

Efforts to ascribe viral replicase function to host activities[7] have certainly obscured the study of true replicase, an enzyme characterized by its specificity for copying cognate viral RNA. Only recently has it become generally accepted that viral replication is indeed dependent on a unique and specific RNA-dependent RNA polymerase (replicase) present only in infected tissues. There is some evidence, discussed below, that the core polymerase of this replicase is of viral origin; other host- and virus-encoded proteins may well participate in the function of the enzyme. One of the intentions of this article is to demonstrate that a careful analysis of the products of in vitro polymerase reactions makes the separation of true replicase activities from the various host activities straightforward. With the application of powerful new techniques, such as RNA engineering[8-10] and the use of antibodies to help

Table 1
GROUPS OF (+) RNA PLANT VIRUSES RELATED TO BMV[a]

Virus group	Type member	Characteristics
Tripartite viruses		
Bromoviruses	Brome mosaic virus	RNAs bind tyrosine
Cucumoviruses	Cucumber mosaic virus	RNAs bind tyrosine
Hordeiviruses	Barley stripe mosaic virus	RNAs bind tyrosine
Alfalfa mosaic virus	Alfalfa mosaic virus ⎱	Coat protein needed for infectivity
Ilarviruses	Tobacco streak virus ⎰	
Monopartite viruses		
Tymoviruses	Turnip yellow mosaic virus	RNA binds valine
Tobamoviruses	Tobacco mosaic virus	RNA binds histidine

[a]   Classification after Matthews.[99]

define the function of the nonstructural, virally encoded proteins,[11] we see a bright future for research on plant virus replication. It is worth noting that progress has not only been slow in plant virology, but also with the (+)-stranded viruses of animal systems,[12] indicating that this field of research is inherently difficult. The replicase from the bacteriophage Qβ is the only system capable of catalyzing in vitro the entire cycle of viral replication,[13] that is, the amplification of genomic RNA. The Qβ replicase system remains a useful model for the plant viral replicases.[1]

III. RNA-DEPENDENT RNA SYNTHESIZING ACTIVITIES IN PLANTS

When cytoplasmic extracts of plants infected with virus are incubated with RNA templates, incorporation of nucleoside monophosphates (usually studied with radiolabeled UTP) into an acid-precipitable form can readily be demonstrated. Such incubations are usually done in the presence of actinomycin D to prevent DNA-dependent transcription. The activities obtained, typical of RNA polymerases, are insensitive to inorganic phosphate, but strongly inhibited by pyrophosphate. Studies from a number of virus-plant combinations show basic similarities in the types of activities found, although absolute levels and some properties of these activities vary, depending on the plant species. The extracts studied are the supernatants of low-speed centrifugation, which is designed to remove the nuclei and, hence, the bulk of the DNA-dependent transcriptional machinery.

Throughout this article we refer to the polymerase(s) responsible for specific viral RNA synthesis as the viral replicase(s). Synthesis of full-length (+) and (−) strands from the complementary viral RNA templates will be referred to as RNA replication, and synthesis of subgenomic RNA as viral transcription. The nature of the plant viral replicases is unknown, but a reasonable working hypothesis is that they exist as a complex consisting of a virally encoded core polymerase and one or more associated factors that may be important in template selection. The removal, addition, or substitution of such factors would enable the core polymerase to synthesize (+), (−), and subgenomic viral RNAs. The nonviral (host) RNA-dependent RNA polymerase activities produce their complementary products by a process referred to as transcription.

A. Viral Replicase
Activity responsible for the synthesis of full-length, double-stranded viral RNAs is in-

Table 2
METHODS USED TO SEPARATE VIRAL REPLICASES FROM OTHER MEMBRANE-ASSOCIATED, RNA-DEPENDENT RNA SYNTHESIZING ACTIVITIES

Example	Treatment	Comments
BMV-infected barley[15]	Nonionic detergent; 1% dodecyl-β-D-maltoside	Replicase remains in particulate fraction
TYMV-infected chinese cabbage[14]	Nonionic detergent; 5% Lubrol® WX	Replicase solubilized
CPMV-infected cowpeas[16]	Mg^{2+}-deficient buffer wash; 1 mM EDTA	Selective removal of host RNA-dependent RNA polymerase and terminal transferases from membranes
CCMV-infected cowpeas[22]	High salt wash; 0.5 M KCl	Selective, but incomplete removal of host activity from membranes

variably membrane associated, and pelletable by 10,000 g[14] to 100,000 g[15] centrifugation. Many plants have other RNA-dependent RNA-synthesizing activities of unknown function associated with cytoplasmic or organellar membranes, but these can be separated from the viral RNA-specific replicase. This has been accomplished by treatment of membranes with nonionic detergents, by washing membranes with salt, or with Mg^{2+}-deficient buffer (see Table 2); critical evaluation of the size, origin, and polarity of products is necessary in developing an appropriate protocol. The viral RNA-specific activity found in pellets represents preinitiated replication complexes, bearing (−) strand templates and initiated but incomplete nascent (+) strands, which can be elongated in vitro to produce complete or nearly complete complementary strands.[16-18] These extracts are therefore unable to accept exogenous templates without further treatment. The synthesis of all genomic viral RNAs is observed, and in some cases subgenomic RNA elongation also occurs.[18] The fact that most preinitiated replication complexes are engaged in (+) strand synthesis is consistent with their role in RNA amplification, and presumably reflects a preferential initiation on (−) strand templates. Some methods that have been used to further study these replication complexes, and to make them dependent on exogenous template, will be discussed below.

The association of viral replication with membranes has also been concluded from electron microscopic studies of infected plant tissues. The appearance of membrane vesicles containing fibrillar material, considered to be viral RNA, is characteristic of infections of these (+)-stranded RNA viruses.[19] The association of the vesicles with particular membranes or organelles occurs in a pattern characteristic of the particular virus. The best studied case is the association of TYMV replication with the outer chloroplast envelope.[20,21]

B. Host RNA-Dependent RNA Polymerase Activity

An activity exists in many plants that is capable of synthesizing double-stranded RNA products from a wide range of RNA templates.[7] This activity typically increases slightly after mock inoculation, but much more so after viral infection; total levels in extracts of infected leaves may be considerably higher than those of true replicase.[16,22] The literature contains many reports of such activities, but the relationship between those studied in different plants is not clear.

The best studied activity is that present in cowpea mosaic virus-infected cowpea leaves.[16,23,24] On extraction, this activity was membrane associated and existed in a complex with bound RNA template. The products after in vitro incubation represented both CPMV-specific and unidentified host RNA sequences of negative polarity and consisted mostly of single-stranded

RNAs with short, double-stranded segments.[16] The transcribed (−) strands were hetero-disperse, 4 to 5 S, and appeared to be initiated at multiple sites. Specific solubilization of this activity was achieved by washing the membranes with 1 mM EDTA in the absence of Mg^{2+}, a treatment leaving the viral RNA-specific replicase intact and still membrane associated.[16] The removal of endogenous template from the solubilized host polymerase by DEAE-Sepharose® chromatography yielded a template-dependent activity that was able to transcribe short sections of cowpea mosaic virus RNA. However, there was no specificity for viral RNA. On further purification, the enzyme was identified as a 130-kdalton monomeric protein. This activity is clearly functionally different from the viral replicase[16] and is definitively absent from the CPMV RNA replication complex.[24] The normal function of this enzyme in cowpea plants is not understood.

Host RNA-dependent RNA polymerase activities detected in other virus-infected plants make products similar to those from the cowpea enzyme, namely, polydisperse 4- to 7-S-sized RNAs transcribed from longer single-stranded RNAs.[22,25-28] In many cases, the activity is distributed between the soluble and particulate compartments,[7,22,27,29] but may also be predominantly soluble.[26,30]

All reports mentioned above cite increases between 2- and 20-fold in the host RNA-dependent RNA polymerase activities following viral infection. However, even the strongest former proponents of the notion that these enzymes are responsible for viral replication no longer believe this to be the case.[31] None of the earlier reports describing replicase-like activity dependent on a host enzyme have been followed up, and the biological significance of these observations remains uncertain. Indeed, there are no obvious roles for such an activity in healthy plants. We believe that insufficient efforts have been made to exclude the possibility that the observed RNA-dependent RNA polymerase activities may be due to an artifactual activity of enzyme(s) normally engaged in other functions in vivo. DNA-dependent RNA polymerase from *Escherichia coli* is able to utilize single-stranded RNA as a template in vitro, but there is no evidence for such an activity in vivo.[32] Further, purified RNA polymerases II from tomato and wheat germ are able to synthesize full-length copies of viroid RNA templates in vitro.[33] The fact that the host RNA-dependent RNA polymerase activity is stimulated on viral infection suggests that it may be due to one of a number of proteins, such as the so-called pathogenesis-related proteins,[34] involved in defense responses of the plant. It is interesting that RNA transcriptional factors have been identified among heat shock proteins in both prokaryotes and eukaryotes,[35] and their induction may in general be associated with stress responses; in many systems, viral infection induces some of the heat-shock proteins.[35] A critical reexamination to determine whether the host activities can be ascribed to DNA-dependent polymerases seems worthwhile.

C. Host Terminal Transferases

A number of different terminal transferase activities have been reported from various cytoplasmic plant extracts. A nonprocessive uridylyl transferase has been studied from cowpeas,[36] and processive transferases capable of adding long tracts of nucleoside mono-phosphate residues, not necessarily restricted to UMP, are apparently present in various plants. These activities have been reported from both soluble and particulate fractions in different species.[27,37,38] The terminal transferase activities can also be enhanced by viral infection,[27,36] but they are easily distinguished from true RNA-dependent RNA polymerases. Even though products appear full length on denaturing gels, all are single stranded, and radioactivity is only incorporated terminally.

IV. GENOMIC ORGANIZATION OF BMV AND RELATED VIRUSES

The viruses of Table 1 have similar replication and gene expression strategies, and similar

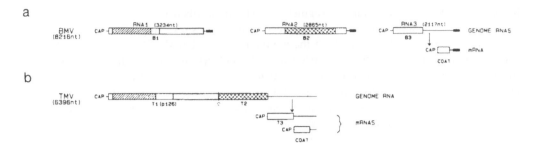

FIGURE 1.   Genomic organization of (a) brome mosaic virus and (b) tobacco mosaic virus, drawn to scale. Boxed regions are the protein-coding sequences, with homologous domains represented by similar shading. The open diamond represents a stop codon that is partially readthrough in TMV. The thickened line segments at the 3' ends of the BMV RNAs represent the circa 200-nucleotide-long conserved regions that include the tRNA-like structure. TMV RNA also possesses a tRNA-like structure, not indicated. (Adapted from Ahlquist, P. E., Strauss, E. G., Rice, C. M., Strauss, J. H., Haseloff, J., and Zimmern, D., *J. Virol.*, 53, 536, 1985.)

coding capacity. However, there are two different types of genomic organization: that of the tripartite viruses, including BMV,[39] and that of the monopartite viruses, including TMV and TYMV.[40,41] The genomic makeup of BMV is depicted in Figure 1a. The three genomic RNAs encode four proteins, with RNA3 being dicistronic. Three of the viral genes encode nonstructural proteins: protein B1 (109 kdaltons), protein B2 (94 kdaltons),[42] and protein B3 (32.5 kdaltons).[43] The single structural gene is a silent cistron in the 3' half of RNA3 and encodes the coat protein which assembles into capsids composed of 180 identical subunits; the coat protein is translated only from the subgenomic RNA4, whose entire sequence is present within RNA3.[43] RNA4 is thus not required for infection.[44] The expression of coat protein via a subgenomic RNA is a strategy used by both the tripartite and monopartite viruses, and is presumably a mechanism by which structural gene expression is delayed until after replication of the genomic RNAs has been established.

All four BMV RNAs are messenger RNAs with 7-methyl guanosine 5'-ppp-5' guanosine caps, but lacking methylation of penultimate nucleotides.[45] A distinctive feature at the 3' end of each of the four RNAs is the presence of a highly conserved sequence of about 200 nucleotides.[46] This region has long been recognized as being responsible for the specific esterification of tyrosine at the −CCA terminus by host aminoacyl tRNA synthetase.[47] Viral RNAs are tyrosylated during an infection in barley,[48] and virion RNAs can be stoichiometrically tyrosylated in vitro by wheat germ[49] or yeast (Dreher, unpublished results) synthetases. Aminoacylatability is a common feature of this group of viruses (except for AlMV and the ilarviruses; Table 1), but its role has not yet been elucidated. The tRNA mimicry of plant viral RNAs is indeed a remarkable phenomenon, extending to a number of other tRNA-associated activities,[5] the most important of which are the ability to form a ternary complex with GTP and elongation factor, and the ability of CTP,ATP:tRNA nucleotidyl transferase to repair truncated or incomplete 3' −CCA termini.

Recent advances in our understanding of the structure of BMV and other plant viral RNAs have explained the structural basis for the tRNA-like properties by describing the formation of a plausible aminoacyl acceptor stem. In tRNAs, this stem is formed by the base-pairing of both 5' and 3' termini, with the 3' −CCA terminus free and available for aminoacylation. For the plant viral RNAs, recent structural models have suggested a novel solution to the problem of aminoacyl acceptor stem formation in the presence of long tracts of genomic RNA to the 5' side of the tRNA-like regions. In these elegant models, construction of the acceptor stem is achieved by coaxial stacking of disconnected helical segments[50] (Figure 2); the continuation in the 5' direction of the antiparallel strand that is base-paired to nucleotides near the 3' terminus is accomplished by the sharp turn of a "pseudoknot".[51,52] Thus, the

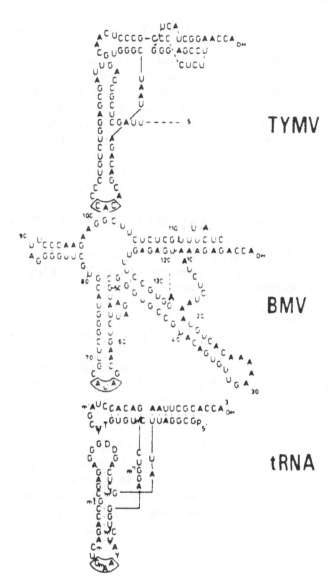

FIGURE 2.   Representations of the tertiary conformations of the tRNA-like structures of BMV and TYMV RNAs and of tRNA^{Phe}. The viral structures are based on solution structural studies; pseudoknots are essential features in the construction of an aminoacyl acceptor stem in these RNAs. Anticodon triplets are boxed. (Adapted from Rietveld, K., Linschooten, K., Pleij, C. W. A., and Bosch, L., *EMBO J.*, 3, 2613, 1984. With permission.)

overall structures of the plant viral RNAs have analogies to that of tRNA (Figure 2), although there are only very restricted sequence homologies to the isoaccepting tRNA. The tRNA-like structures shown in Figure 2 are the basic sequences required for the tRNA-associated functions.[53,54] It is known for BMV that the tRNA-like structure also functions as the specific promoter used by BMV replicase to initiate (−) strand synthesis.[55] The relationship between the tRNA-like activities and replication is discussed below.

In TMV, the only monopartite virus of Table 1 that has been completely sequenced, there are also four encoded proteins[40,41,56] (Figure 1b). Both coat protein and the 30-kdalton protein

(T3) that is possibly analogous to protein B3 of BMV are translated from subgenomic RNAs. Two large proteins (T1, 126 kdaltons and T2, 183 kdaltons)[56] are translated from the genomic RNA, the larger as a readthrough product beyond a leaky termination codon.[57] The genomic organization of TYMV RNA is not yet as well characterized as that of TMV and BMV, and only a part of the sequence has been reported. The primary translation product of 150 kdaltons and a readthrough protein of 195 kdaltons may correspond to the 126- and 183-kdalton proteins of TMV, but there is some evidence for proteolytic processing of the readthrough protein.[58] No gene product analogous to protein 3 of the other viruses has been reported, but the coat protein gene is situated at the 3' end of the genome and translated from a subgenomic RNA as in the other viruses.[59] The RNAs of the monopartite viruses have 5' caps and tRNA-like structures analogous to that of BMV.[40]

V. FUNCTION OF VIRAL GENES IN REPLICATION

Very little is known about the function of the virally encoded nonstructural proteins. The completion of the sequences of the entire genomes of a number of RNA plant viruses over the last few years has permitted some refinement of our knowledge about the proteins they encode. Amino acid homologies have been observed in proteins 2 and in two domains of proteins 1 between the tripartite viruses BMV, CMV, and AlMV; these homologies extend to the large proteins of TMV[60-62] and indeed also to nonstructural proteins of the alphaviruses of animals[63] (see Chapter 4, Section II.D. of this Volume). The functions of these proteins are hence almost certainly related in the different viruses. Although there is no clear biochemical or genetic data ascribing detailed functions to any of the three nonstructural proteins of the viruses of Table 1, some observations permit conclusions suggesting likely functions.

A. Role of Gene 2

It is now established for BMV that RNA1 and RNA2 together, but not alone, are able to replicate in barley protoplasts in the absence of RNA3.[65,87] The same is true for AlMV,[66] except for a requirement for coat protein bound to inoculum RNA. The two large viral proteins are thus the only ones directly involved in RNA replication, presumably in concert with host proteins. Computer-assisted comparisons of amino acid sequences have shown strongly conserved regions among the RNA-dependent polymerases from two main groups of animal viruses, the reverse transcriptases of retroviruses, and the RNA-dependent RNA polymerases of picornaviruses.[67] Conserved sequences related to known RNA-dependent polymerases were detected in the sequence of protein B2 of BMV, and in the analogous proteins of other viruses of Table 1. Proteins 2 of BMV, CMV, and AlMV, and the 183-kdalton readthrough protein of TMV may thus be RNA-dependent RNA polymerases. It is not known whether these putative enzymes are able to replicate RNA without other viral or host factors, or whether they are in fact the core polymerases of replication complexes.

B. Role of Gene 1

There are no concrete observations relating a detailed function to protein 1. One possibility is that this protein has a role in the capping of viral RNA. Many viruses that replicate in the cytoplasm and that rely on the translation of capped viral mRNAs supply their own enzymes to accomplish RNA capping. This is necessary because the host capping system is localized within the nucleus.[68,69] Cytoplasmically replicating viruses with (−)-stranded RNA, double stranded RNA, or DNA (especially Vaccinia) genomes have been shown to encode capping activities comprising both guanylyl transferase and guaninine-7-methyl transferase.[68,69] A unique methyl transferase has recently been detected in cells infected with the alphavirus Semliki Forest virus.[70] Similar activities should be encoded by the (+)-stranded RNA plant viruses. Similar arguments have been made by van Kammen.[2] The capping enzymes of Vaccinia exist as a 120- to 130-kdalton complex of two or three polypeptides, with the guanylyl and methyl transferase activities residing in separate proteins.[69] However, the separation of the various capping functions, and indeed of the polymerase activity, to individual proteins is not obligatory. Vesicular stomatitis virus, for example, encodes a 241-kdalton RNA-dependent RNA polymerase which also functions in capping, methylation, and polyadenylation.[71] We suggest that gene 1 of the (+)-stranded RNA plant viruses of Table 1 may encode a capping enzyme. The two conserved domains[60] within this protein may represent domains responsible for guanylyl and methyl transfer that are present on the same protein. It follows that the 183-kdalton protein of TMV could thus carry capping and polymerase activities on the one polypeptide chain. This suggested role for protein 1 is consistent with the requirement for RNA1 for replication in BMV and AlMV; uncapped mRNAs are known to be very unstable in both plant and animal cytoplasms.[10,72,73]

C. Role of Gene 3

Since RNA3 is not required for the replication of BMV RNA (see above), protein B3 of BMV apparently does not have a direct role in replication, although an involvement in the regulation of replication or of subgenomic RNA synthesis cannot be excluded. The analogous proteins in the other viruses are probably also not obligatory for replication, and do not show sequence homologies.[60,62] With AlMV,[66] the absence of RNA3 from an inoculum causes an alteration in (−) and (+) strand ratios, suggesting a regulatory effect on replication by either protein A3 or the coat protein. The observations may well reflect the requirement for the binding of coat protein to the viral (+) RNA. The 30-kdalton protein of TMV and the proteins 3 of the tripartite viruses are implicated in the cell-to-cell spread of virus, but the mechanism of this effect is not understood.[4]

VI. STUDIES ON THE REPLICATION OF BMV RNA

A. The Replicative Cycle of BMV

In BMV, three distinct phases of RNA synthesis are necessary in completing a replicative cycle: (−) strand synthesis using the genomic RNAs as template; full length (+) strand synthesis on (−) strand templates to generate daughter genomic RNAs; and subgenomic RNA synthesis. Each of these replicative modes have been studied in vitro with the objective of understanding the characteristics of the replicase responsible for viral replication and subgenomic transcription. Ultimately, one would hope to achieve an amplification of virion RNAs in vitro using purified components of the replicase complex. Using the BMV replicase preparation described below, we have observed in vitro the initiation and synthesis of (−) strands and of the subgenomic RNA, but have to date not achieved initiation and transcription of full-length genomic (+) strands using (−) strand templates.

B. Preparation of BMV Replicase from BMV-Infected Barley Leaves

An activity capable of specifically transcribing full-length (−) strand complementary RNAs on BMV (+) strand RNA templates exists in the particulate, presumably membranous, fraction of cytoplasmic extracts prepared from leaves of BMV-infected barley plants.[74] Symptomless second leaves of barley (*Hordeum vulgare* cv. Dickson) are harvested 4 to 6 days after inoculation of the primary leaves with BMV. The variety of barley is important because many crop cultivars are selected for resistance to BMV. Leaves are ground on ice in buffer A (50 mM Tris-HCl, pH 7.4, containing 10 mM KCl, 1 mM EDTA, 10 mM Mg(OAc)$_2$, 10 mM DTT, and 15% [v/v] glycerol). The extract is centrifuged at 1100 g to remove debris and nuclei, and the supernatant is then recentrifuged at high speed (50,000 to 250,000 g) to obtain a pellet containing the viral-specific RNA-dependent RNA polymerase activity. The high-speed pellet is then suspended in buffer containing detergent in order to solubilize many of the membrane-associated proteins present in the replicase extract. Do-decyl-β-D-maltoside (12-M), used at 1% (w/v) in buffer A,[15] has given the most satisfactory results of a number of detergents tested. We have not been able to use detergent to solubilize the replicase in a template-dependent and template-specific form, and have thus preferred to study the properties of the particulate template-specific activity obtained after 12-M extraction of membranes. Following extraction of the membranes with 12-M by stirring at 4°C for 90 min, the extract is recentrifuged at high speed and the pellet is resuspended in buffer A containing 0.1% 12-M. The resuspended activity is finally centrifuged through a 40% sucrose (in buffer A containing 0.1% 12-M) pad at 45,000 g in order to completely remove solubilized proteins,[15] including nucleases. All of the polymerase activity is found in the almost colorless pellet which is resuspended in buffer B (50 mM Tris-HCl, pH 8.0, containing 0.5 mM MgCl$_2$, 10 mM DTT, and 0.1% 12-M).

When stored in aliquots at −70°C, the activity is stable for 6 months or longer, but the template specificity is gradually lost during storage. There is a small amount of endogenous BMV RNA present in these extracts, which can serve as template during incubations. This is routinely removed by micrococcal nuclease treatment of the replicase preparation prior to assay.[75] Low levels of Mg^{2+} in buffer B are important for efficient nuclease digestion of the endogenous template, but the total absence of Mg^{2+} at this stage results in the loss of replicase activity. Equivalent preparations from mock-inoculated barley plants show no activity, and the levels of activity in the first high-speed pellet from such plants are negligible compared with the levels in pellets from BMV-infected plants. Because it is unique to BMV-infected tissue and is able to specifically synthesize full length complementary strands on BMV (+) strand RNA templates, we refer to the activity present in the 12-M extracted pellets as BMV replicase. These preparations still contain many proteins; we have thus far concentrated on a description of the catalytic activities of the replicase, rather than to attempt purification of an activity lacking specificity.

A typical assay for in vitro replicase activity uses 20 μℓ of micrococcal nuclease-treated replicase in a 25-μℓ reaction containing 1 to 5 μg of BMV RNA (1.5 to 7.5 pmol). The final reaction conditions are: 40 mM Tris-HCl, pH 8.0, containing 10 mM Mg(OAc)$_2$, 8 mM DTT, 0.08% 12-M, 0.5 mM ATP, GTP, and CTP, 80 μg/mℓ actinomycin D, and 2 to 5 μM radiolabeled UTP. Reactions are incubated at 30°C, and complete elongation of the complementary strands of genomic RNAs takes less than 25 min.[76] The products are full-length complementary strands which are double stranded even before the phenol ex-traction, used to terminate reactions, as evidenced by their resistance to ribonuclease digestion in 0.3 M salt. There is no significant reinitiation on (−) strands produced in vitro.

C. Template Dependency and Specificity of BMV Replicase

After 12-M treatment, BMV replicase has rather low levels of endogenous activity, that is, elongation of preinitiated nascent strands to make full-length viral RNA products; no discrete products other than that of viral RNA have been observed. Treatment of the replicase with micrococcal nuclease renders the preparation essentially totally dependent on exoge-nously supplied template. Figure 3 illustrates the above points. The addition of individual BMV RNA components to an incubation with replicase results in synthesis only of its complement, and not the complements of the other RNAs; endogenous activity is absent. This high template dependence has made possible many types of experiments aimed at studying the template specificity of the replicase, the characterization of the promoters recognized within BMV RNA, and the initiation of complementary strand synthesis.

BMV replicase shows a strong preference for BMV RNAs as templates, over both viral and nonviral RNAs. Neither nonviral RNAs nor RNAs of unrelated viruses such as Qβ and pea enation mosaic virus[74,76] are copied. RNAs of viruses closely related to BMV are copied, but with efficiencies well below that for BMV RNA. CCMV RNA is about one third as

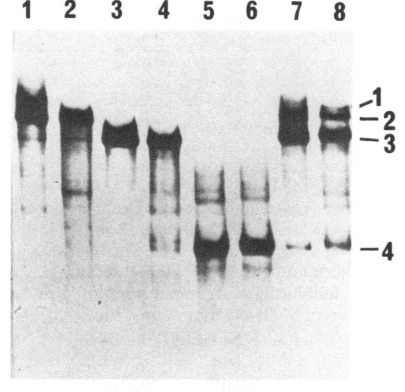

FIGURE 3. Products (³H-labeled) synthesized by BMV replicase, analyzed by electrophoresis and fluorography. The replicase was micrococcal nuclease-treated to make it template dependent. The products are full-length, ribonuclease-resistant duplex RNAs. Products were synthesized using sucrose density gradient-purified BMV RNA components as templates: RNAs 1 and 2 (lanes 1 and 2), RNA3 (lanes 3 and 4), RNA4 (lanes 5 and 6), and an equimolar mixture of all four RNAs (lanes 7 and 8). The products in even-numbered lanes were ribonuclease-treated prior to analysis. (From Miller, W. A. and Hall, T. C., *Virology*, 125, 236, 1983. With permission.)

good a template as BMV RNA.[75] This is in accord with results obtained from experiments in which the replication of pseudorecombinant mixtures between BMV and CCMV RNAs was studied. An infection established by inoculation with BMV RNAs 1 and 2 was able to support the replication of CCMV RNA3,[77] but its accumulation was quite low.

The protocol we have used in preparing BMV replicase may have applicability to related viruses in different host backgrounds, although variations from the method may be necessary in certain instances. From CCMV-infected cowpea leaves, we have prepared CCMV replicase that has very analogous properties to BMV replicase.[37] As others have reported, cowpeas have fairly high activities of other RNA-dependent, RNA-synthesizing activities, especially terminal transferase activities,[22,36] but these did not interfere with the replicase preparation. The BMV replicase protocol has recently been applied to AlMV-infected bean tissue, and a replicase activity of similar properties, also remaining particulate after 12-M treatment, was prepared.[64]

D. Synthesis of (−) Strands by BMV Replicase

The template dependence of micrococcal nuclease-treated BMV replicase has permitted many studies on the characteristics of the synthesis of (−) strand when (+) sense BMV RNAs are offered as templates The complementary strands are synthesized following true *de novo* initiation, as shown by the incorporation of γ-³²P-labeled nucleoside triphosphate

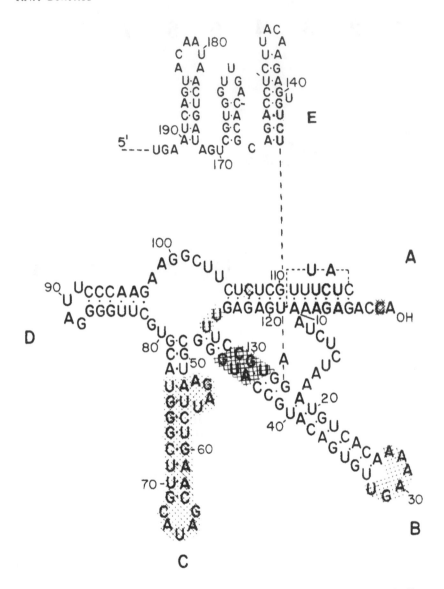

FIGURE 4. Representation of the structure of the 3′ region of (+) BMV RNA3. The tRNA-like structure, 134 nucleotides in length, functions in vitro as the (−) strand promoter for BMV replicase and as the core structure recognized by the tRNA-associated activities. The regions in which sequence alterations have resulted in large decreases in promoter activity are crosshatched; those in which sequence alterations have resulted in large decreases in tyrosylation are stippled. Nucleotides are numbered from the 3′ terminus.

at their 5′ termini. Initiation occurs with a GTP which is incorporated opposite C2 of the (+) strand template[55] (Figure 4), as judged by sizing fragments synthesized in the presence of the chain terminator cordycepin triphosphate. Two other observations are consistent with initiation at this site: firstly, nucleotide substitutions at C2 as well as A1 were far more detrimental to in vitro (−) strand synthesis than were substitutions at A1 alone;[8] secondly, removal of A1 did not affect (−) strand synthesis, but removal of both C2 and A1 decreased the template activity to a level commensurate with the proportion of residual C2 present.[55]

The identity of the (−) strand product has been verified by hybridization studies using single-stranded M13 probes, and by the use of cordycepin triphosphate as a chain terminator during strand synthesis by the replicase activity. In this way, a ladder analogous to that obtained in normal dideoxy sequencing and corresponding to the sequence of the (−) strand was obtained.[55]

The template characteristics of BMV (+) sense RNA and the nature of the signals required for template recognition by the replicase have been extensively investigated. All of the signals necessary for template recognition, for initiation, and for the elongation of complementary (−) strand are present in the 3′ terminal 134 nucleotides of each BMV RNA that comprise the tRNA-like structure (Figure 4). A number of experimental approaches have been used to define this promoter. 3′ terminal fragments of BMV RNA can be prepared either by partial ribonuclease T1 cleavage to produce a 161-nucleotide-long fragment,[78] or by ribonuclease H cleavage directed to specific sites by short complementary deoxyoligonucleotides.[55] A 3′ fragment 134 nucleotides in length was an active template for the replicase, while shorter ones were not. The function of the (−) strand promoter has also been studied in hybrid-arrest experiments, where complementary single-stranded DNAs cloned in bacteriophage M13 were annealed to BMV RNAs offered as templates for replicase.[79] The inactivity of double-stranded hybrids as templates suggests that conformational features of the folded RNA structure are used by the replicase in template selection.

By far the most versatile and powerful way to study promoter function has been to produce native (wild type) and specifically altered templates by in vitro transcription from cDNA clones of BMV RNA.[8] Sufficient quantities of RNA for most biochemical applications (1 to 20 μg) can easily be synthesized using the RNA polymerases from the bacteriophages SP6, T7, or T3 after the cDNA has been cloned under the control of the appropriate phage promoter.[80] Transcripts with correct 3′ termini and representing the 200-nucleotide-long homologous region at the 3′ end of the BMV RNAs (Figure 1a) can be made in vitro with SP6 RNA polymerase by transcribing DNA templates linearized at a point corresponding exactly to the 3′ end of the RNA sequence.[8] Such transcripts are active as templates for replicase and can be tyrosylated by tyrosyl tRNA synthetase. Numerous alterations have been made in the promoter region, including point substitutions and deletions, to further study the properties of the promoter used by BMV replicase in (−) strand synthesis. These results are discussed below in relation to the other activities of the mutant RNAs. All results have been consistent with the proposition that the 134-nucleotide-long tRNA-like structure includes all the information required for promoter function, and that the folded conformation of the RNA in this region is important in promoter function.

E. Relationship Between Replicase and tRNA-Like Activities

The properties of the tRNA-like structure of BMV RNA in (−) strand promotion and in aminoacylation have been studied in parallel with the use of specific mutant RNAs generated in vitro from altered cDNA-clones as described above. A comprehensive set of mutations within the 3′ region of BMV RNA has been assembled[8,81] (Dreher and Hall, in preparation) with two aims in mind. First, we wanted to map the regions of the RNA sequence and structure required for the replicase template and tRNA-associated activities, and thereby study the interrelationship of these activities. Second, we aimed to construct one or more mutants which had specifically lost the capacity to be tyrosylated, but which retained high in vitro replicase template activity. Such mutants could then be used in infection experiments in barley protoplasts[82] to learn whether aminoacylatability is an absolute necessity for efficient replication. Although our understanding of the tRNA-like structure has increased greatly, we still do not have an answer to that question.

It has not been difficult to obtain mutant RNAs where one of the above in vitro activities has selectively been decreased, indicating that the replicase and tyrosyl tRNA synthetase are unlinked enzymes and that the synthetase is unlikely to be a subunit of the replicase. It is not known whether the presence or absence of tyrosine alters template activity in vitro, but it is highly unlikely that the RNAs serving as templates in a normal assay are tyrosylated since no tyrosine is supplied in the reaction. The correct initiation of (−) strand synthesis by BMV replicase is apparently not dependent on aminoacylation, but at present there is insufficient evidence to conclude that tyrosylation is not required in vivo to promote high rates of replication. This question will need to be answered by studying appropriate mutants in vivo and by identifying the subunits of BMV replicase and their function. A limited number of mutants with rather large deletions in the tRNA-like region have been studied in vivo.[83] All mutants that were inactive in either or both of the above in vitro assays failed to replicate in vivo, but the large size of the deletions makes these results difficult to interpret. However, one mutant which lacked arm D (Figure 4) and was active in the in vitro assays was able to replicate in barley plants, demonstrating that viable mutations are indeed possible in the 3′ region of the BMV RNAs.

The study of a number of mutations in the tRNA-like region has enabled the identification of locations necessary for (−) strand promoter and aminoacylation activities. Regions in which small sequence alterations result in the loss of activity may represent regions of interaction between the RNA and the enzyme. Neither activity was inhibited by the removal of arm D; indeed, replicase template activity was increased.[81] Mutations in all other regions of the tRNA-like structure caused loss of replicase template activity (Figure 4), reflecting an interaction of the replicase complex with most parts of the structure. Tyrosine tRNA synthetase, on the other hand, appears to interact with more limited parts of the structure (Figure 4). Mutations in two entire stems are clearly selective. Arm E, to the 5′ side of the tRNA-like structure, is important for efficient tyrosylation, but lies outside the (−) strand promoter. Arm C, Commonly but erroneously called the anticodon stem because of the presence in the loop of a tyrosine (AUA) anticodon, is not necessary for aminoacylation, but is critical for promoter function (Figure 4).[8] Indeed, the −AUA− triplet appears to be a very important sequence-specific determinant, which may be used by the replicase in specific template recognition. Point mutations in this sequence drastically decreased template activity, while leaving aminoacylatibility unaffected.[8] The overall tRNA-like conformation of the RNA is clearly vital in the recognition by both enzymes, as evidenced by the loss of both activities when substitutions were introduced into the pseudoknot to prevent formation of a complete aminoacyl acceptor stem (Dreher and Hall, in preparation).

A few mutants in which small sequence alterations have resulted in the selective loss of aminoacylation have been obtained. The effect that these mutations have on replication in vivo is being studied, as is their effect on elongation factor binding and on 3′ adenylation in vitro. These studies should allow an assessment of the role of aminoacylation in the replication of the virus.

F. Use of (−) Sense BMV RNA Templates by BMV Replicase

The BMV replicase preparation is able to use (−) strand templates in the synthesis of viral (+) sense RNA. As is the case for many aspects of current RNA virus research, the demonstration of the template activity of (−) BMV RNA3 has been dependent on the preparative transcription of RNAs in vitro since it has not been feasible to isolate (−) sense free of (+) sense RNA from infected tissue. BMV replicase preparations are able to synthesize the subgenomic RNA4 from (−) sense RNA3,[9] but we have not yet detected full-length (+) strand synthesis from (−) strand BMV RNA templates. We do not currently know whether this deficiency lies in the activities present in the replicase preparation, or in the form of the (−) strand template used. There is some evidence that (−) strands of BMV and related viruses may have an extra G residue at their 3′ termini,[55,84] which may be necessary for (+) strand initiation.

There have been no direct studies permitting the definition of regions necessary for promoter function in (+) strand genomic RNA synthesis from (−) strand templates. Studies of the replicatability in barley protoplasts of RNA3 variants carrying large deletions indicate that portions of all three noncoding regions of the RNA are required for replication.[85] Because of the autonomy of the tRNA-like structure as promoter for (−) strand synthesis in vitro, the 5′ and central untranslated regions may be needed for (+) strand synthesis. It is of interest that a conserved sequence, GUUCPuAPyPyCC, is present in both of these regions of RNA3, at the 5′ ends of RNAs 1 and 2, as well as at the 5′ ends of CMV RNAs.[61] These sequences may have a role in the synthesis of (+) strand by replicase.

G. Synthesis of Subgenomic RNA4 by BMV Replicase

Incubation of (−) sense BMV RNA3 templates in vitro with BMV replicase results in the synthesis of subgenomic RNA4.[9] The activity level for subgenomic synthesis was similar to that for (−) strand synthesis. As for (−) strand synthesis, *de novo* initiation occurs with GTP at a specific site, and the product strand remains annealed to the template. The initiation event occurs internally on the (−) strand template at the point corresponding to the 5′ end of virion RNA4. This mode of subgenomic RNA formation is probably used by all the viruses related to BMV, as well as by the alphaviruses of animals.[12] Although the same BMV replicase preparation is able to catalyze (−) and subgenomic (+) RNA synthesis, we do not know whether the same related or different polymerase complexes are involved in these two reactions. We have also not addressed the question of capping of (+)-stranded viral RNAs. Since incorporation of γ-labeled GTP into subgenomic products occurs,[9] at least some of the molecules are uncapped. This suggests that transcription and capping may not be obligately linked (as in cytoplasmic polyhedrosis virus)[69] and may be dependent on separate enzymes.

The promoter used by BMV replicase in subgenomic RNA synthesis has been studied using altered (−) RNA3 templates produced by in vitro transcription.[86] Core sequences necessary for promoter function in vitro are located between the end of the poly(A) stretch, 20 nucleotides upstream of the subgenomic start site, and the start of translation of the coat protein gene, 9 to 11 nucleotides within the subgenomic RNA sequence (Figure 5). Very conveniently, these points correspond to Bgl II and Sal I restriction sites, respectively, in the cDNA of RNA3. Studies with (−) RNA3 templates truncated at points close to the subgenomic initiation site (especially at the position of the Bgl II site) have demonstrated that the regions upstream of and including the poly(A) stretch (referring to the (+) strand sequence) are not necessary for normal subgenomic RNA synthesis in vitro.[9] Removal of the poly(U) from a full-length (−) RNA3 template, however, resulted in a substantial loss of activity in subgenomic RNA synthesis (Marsh, Dreher, and Hall, in preparation). These apparently contradictory results suggest that the poly(A) region may be important structurally in providing a spacer and ensuring that the subgenomic promoter remains accessible. According to computer-generated predictions of the RNA folding pattern of (−) RNA3, the poly(U) region is not involved in secondary structure and probably facilitates an open RNA structure in the region of the subgenomic promoter. It is interesting that homologies exist in the regions of subgenomic RNA initiation points from the plant viruses of Table 1 and from the alphaviruses.[86] The similarities between putative subgenomic promoters among the plant viruses include the presence of surrounding A-U-rich tracts, in place of the poly(A) present in BMV RNA3.

Studies using barley protoplasts have supported the localization of the subgenomic promoter sequence to the region upstream of the Sal I site in the BMV cDNA3.[87] Substitution of the coat protein gene with a gene from another source did not prevent subgenomic RNA production, with initiation apparently occurring at the normal site. In a number of constructions studied either in vitro[86] or in vivo,[87] sequence substitutions abolished the potential to form the normal stem and loop predicted by folding programs in the wild-type subgenomic promoter. This structure positions the subgenomic initiation site at an accessible point in

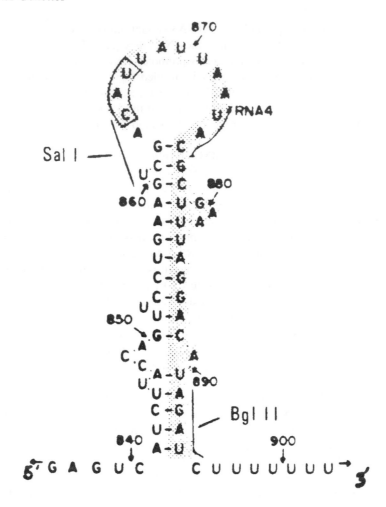

FIGURE 5. Part of the intergenic region of (−) BMV RNA3. Sequences comprising the core of the promoter that is used in vitro for the generation of subgenomic RNA4 are indicated (stippled). The site of initiation of RNA4 is marked, and the positions of Bgl II and Sal I restriction sites present in the cDNA are indicated. The position corresponding to the AUG codon that initiates the synthesis of the coat protein is boxed. The postulated secondary structure shown is based on computer folding predictions,[101] but analysis of the sequences necessary for subgenomic RNA synthesis suggest that the stem and loop are not involved in promoter function.

the loop (Figure 5), but is apparently not involved in the regulation of subgenomic RNA synthesis. Recognition of the subgenomic promoter by the polymerase may thus be sequence specific rather than secondary structure dependent.

VII. STUDIES ON THE REPLICATION OF TYMV RNA

A. Preparation of TYMV Replicase from TYMV-Infected Plants

Extracts of TYMV-infected chinese cabbage leaves carry a membrane-bound, RNA-dependent RNA polymerase activity capable of synthesizing double-stranded, ribonuclease-resistant products. The products appear by sucrose gradient ultracentrifugation and gel electrophoresis to be in the size range of TYMV genomic RNA.[17,88] Preinitiated nascent strands are apparently elongated on bound template during incubation in vitro since the reaction is not template dependent.[14] Hybridization studies performed with excess virion or double-stranded TYMV RNAs suggested that the product is (+) sense TYMV RNA, and that the activity is typical of template-replicase complexes engaged in virion RNA synthesis. The replicase activity is associated with chloroplast membranes,[14] which correlates well with cytological studies indicating the association of TYMV replication with chloroplast outer membranes.[20,21] The activity was found only in extracts from plants infected with TYMV.[14]

Using a procedure partially adapted from the purification method of Qβ replicase, it was possible both to dissociate the replicase activity from the membranes and to remove the bound template in order to make the enzyme template dependent.[14] Removal of the intact replicase-template complex from the membranes was achieved by treatment with 5% Lubrol® WX, a nonionic detergent. Bound RNA was then removed from the solubilized complex by treatment with a two-phase system consisting of dextran T500 and polyethylene glycol 6000 in the presence of 4 M NaCl, yielding a template-dependent activity in the polyethylene glycol phase. The soluble, template-dependent activity was then purified by a series of chromatographic separations, using DEAE-cellulose, phosphocellulose, TYMV RNA affinity chromatography, and sucrose density gradient centrifugation.[89] The resultant enzyme was *template dependent, and showed specificity for TYMV RNA; relative to the activity with* TYMV RNA, an activity of 71% was observed with RNA from the related eggplant mosaic virus, but only 14 and 18% for TMV and BMV RNAs, respectively. We are unaware of any reviewed publication giving a detailed characterization of the activities and properties of this preparation. However, the unpurified enzyme from the polyethylene glycol phase was shown to synthesize (−) sense TYMV RNA on (+) strand TYMV RNA templates.[17] This enzyme preparation apparently was also able to use (−) strand TYMV RNAs as templates for (+) strand synthesis,[88] but this activity has not been characterized. It is unfortunate that the available reports do not show high-resolution autoradiographs of products of the TYMV replicase analyzed on gels, and that the earlier work investigating the polarity of products has not been repeated with strand-specific cDNA probes. It is clear that there is a great need for biochemical characterization of the TYMV replicase in order to prove convincingly that it does indeed represent at least part of the enzyme complex involved in TYMV RNA replication in vivo.

B. Polypeptide Components of TYMV Replicase

Study of the subunits present in the purified TYMV replicase has provided evidence that the replicase complex may contain a virally encoded protein. Two polypeptide subunits, of molecular weights about 115 and 45 kdaltons purified with the TYMV RNA-specific polymerase activity.[89] The 115-kdalton protein cross-reacted with unidentified ^{35}S-methionine-labeled products of an in vitro translation of TYMV RNA in a reticulocyte lysate. Surprisingly, results from direct immunoprecipitation of TYMV RNA translation products with the antireplicase serum were not reported. The 115-kdalton protein of the replicase could correspond to a 120-kdalton fragment thought to be derived by proteolytic cleavage from the 195-kdalton readthrough protein.[58] In contrast, the 45-kdalton protein of the replicase appears to be of host origin.[89] Evidently, TYMV encodes at least one subunit of its replicase, presumably the core polymerase, as suggested by its similar size to other plant viral genes provisionally identified as polymerases by amino acid homologies.[67] The TYMV replicase was shown to be distinct from any host-encoded, RNA-dependent RNA polymerase by virtue of different template specificities, sedimentation coefficients, and chromatographic properties.[89]

The identity of host subunits in plant viral replicases is clearly a very important question. The 45-kdalton protein component of the TYMV replicase has not been identified, but recent results have shown immunologically that the elongation factor (EF-1α), which is able to bind valylated TYMV RNA, is not present in the template-specific replicase preparation.[90] Addition of excess elongation factor did not stimulate replicase activity.

The nature of the promoter signals used in the replication of TYMV virion RNA is not known. Some preliminary efforts in this direction have been reported,[17] but the study was apparently discontinued.

VIII. FUTURE DIRECTIONS

The above discussion has focused on the two best-studied examples of replicases from this group of plant viruses. Both systems have lacked attention in major areas and are poised for further detailed investigations. Some of the important questions remaining are discussed below.

A. The Form of Actively Replicating Intermediates

All plant viral replicase activities studied to date complete a single passage on the RNA template, resulting in a double-stranded RNA product. There has been speculation about the structure of replicative intermediates in the plant viruses and the number of replicase molecules transcribing a given template at the same time, but little definitive data exists.[4] Viral double-stranded RNAs can be extracted from probably all virus-infected plant tissue,[91] including plants infected with BMV[55,82] and TYMV,[20] but this may be an artifact of phenol extraction. The tremendous thermodynamic stability of long helices of double-stranded RNA makes such molecules inconceivable as replicative intermediates[92] unless replication involves the transcription of RNA duplexes as in the reoviruses.[93] It is probable that full-length, double-stranded RNAs are either an artifact of extraction or are present late in infection, where they presumably reflect the decreased metabolic capacity of sick and exhausted host cells.[94] By rather indirect cytological means, it has been concluded that replicating TYMV RNA is indeed predominantly single stranded.[20] The ability of BMV replicase to use free (−) strands to synthesize subgenomic RNA suggests that the replicative intermediates are largely single stranded. This problem clearly requires more direct attention to verify that replicative intermediates are not duplexes and to understand the way in which product and template strands are kept apart. This may permit improvements in the preparation of isolated replicases to permit the release of products synthesized in vitro and the completion of the replicative cycle.

B. The Composition of the Active Replication Complex

As discussed above, a variety of evidence supports the conclusion that the replicase responsible for viral replication includes one or more viral products, despite the absence of direct biochemical verification. This concept is in accord with the situation in all characterized RNA viruses which invariably are replicated by polymerase of viral origin. In the (+)-stranded RNA viruses, host factor(s) is frequently part of the replicase.[12] The best-characterized example is the replicase of bacteriophage Qβ,[13] which provides a useful model for the replicases of the plant viruses with aminoacylatable RNAs (see also Chapter 1 of this volume).[1] The host factors present in the Qβ replicase are proteins normally interacting with tRNA (elongation factors) or the ribosome (S1 protein).[13] An analogous protein of host cells (the elongation factor EF-1α) interacts with the known (−) strand promoter of BMV RNA[5] (the tRNA-like structure), and may be a component of the replicase, although this remains to be demonstrated. EF-1α has been immunologically detected in BMV replicase preparations (Ravel, J. M. and Dreher, T. W., unpublished observations), but may simply be a contaminant. EF-1α was not detected in the TYMV replicase,[90] and the identity of the 45-kdalton subunit is unknown. Clearly, much research is needed to identify the origin and function of polypeptides present in the plant viral replicases.

C. The Role of the tRNA-Like Structures

The tRNA-like structure of BMV RNA, being the (−) strand promoter, clearly has a vital role in replication. Whether the tRNA-associated activities also resident in the tRNA-like structure are necessary for efficient promoter function in vivo has not been established. As suggested above, the tRNA-like structures may be a device by which the tRNA mimicry permits the binding of host protein(s) and their use as replicase subunit(s). These may act analogously to the transcriptional factors of eukaryotic nuclear RNA polymerase II and polymerase III systems[95,96] in directing the viral polymerase (protein 2) specifically to viral RNA templates, and guiding its positioning on the template to permit correct initiation of complementary strand. Alternatively, the viral polymerase may itself provide the specificity for viral templates, but the host factors could function in stabilizing the complex, or facilitating the initiation of elongation.

Other roles for the tRNA-like structure are very likely. The tightly folded structure of the RNA is itself probably an advantage in the battle against nuclease degradation, and this stability could be enhanced by aminoacylation. Host CTP,ATP:tRNA nucleotidyl transferase probably also plays an important role in maintaining intact 3' termini, and indeed is likely to be an activity required for the completion of viral ($+$) strands.[55] In TYMV, most virion RNAs lack the 3' terminal A which must be added in vivo to permit their valylation.[5] In BMV, viral double-stranded RNAs isolated from infected barley tissue lacked a 3' terminal A, implying that this nucleotide is not normally added by the replicase.[55] It will be recalled that the complementary U is absent from ($-$) strand templates, since their initiation occurs at C2 of the ($+$) strands. Nucleotidyl transferase is thus probably a necessary activity in completing the replication cycle.

Additional roles for the tRNA-like structures have been proposed which do not necessarily exclude those discussed above. The binding of tRNA-associated proteins at the 3' end may accomplish a control over the relative rates of translation and replication.[42,97] Further analysis of this multifunctional region of the viral genome will undoubtedly uncover many surprises.

IX. CONCLUSIONS

The BMV and TYMV replicase systems described above provide good examples of techniques and approaches that can be applied to studies on the replication of related plant viruses. Frustration resulting from the confusing presence of various host DNA-independent, RNA-synthesizing activities, a great problem in the past, can be avoided by the use of rigorous analysis of products to focus on viral RNA-specific polymerases. One can expect significant and rapid advances in the development of replicase systems from other plant viruses, and there have recently been very encouraging results from research on CCMV[37] and AlMV[64,98] replicases. Studies on the replication of a number of different viruses will certainly stimulate the entire field.

REFERENCES

1. **Hall, T. C.,** Transfer RNA-like structures in viral genomes, *Int. Rev. Cytol.*, 60, 1, 1979.
2. **van Kammen, A.,** The replication of plant virus RNA, *Microbiol. Sci.*, 2, 170, 1985.
3. **Hall, T. C., Miller, W. A., and Bujarski, J. J.,** Enzymes involved in the replication of plant viral RNAs, *Adv. Plant Pathol.*, 1, 179, 1982.
4. **Hull, R. and Maule, A. J.,** Virus multiplication, in *The Plant Viruses: Polyhedral Virions with Tripartite Genomes*, Vol. 1, Francki, R. I. B., Ed., Plenum Press, New York, 1985, 83.
5. **Haenni, A.-L., Joshi, S., and Chapeville, F.,** tRNA-like structures in the genomes of RNA viruses, *Prog. Nucleic Acid Res. Mol. Biol.*, 27, 85, 1982.
6. **Francki, R. I. B., Ed.,** *The Plant Viruses: Polyhedral Virions with Tripartite Genomes*, Vol. 1, Plenum Press, New York, 1985, 1.

7. **Fraenkel-Conrat, H.,** RNA-dependent RNA polymerases of plants, *Trends Biochem. Sci.,* 4, 184, 1979.

8. **Dreher, T. W., Bujarski, J. J., and Hall, T. C.,** Mutant viral RNAs synthesized *in vitro* show altered aminoacylation and replicase template activities, *Nature (London),* 311, 171, 1984.

9. **Miller, W. A., Dreher, T. W., and Hall, T. C.,** Synthesis of brome mosaic virus subgenomic RNA *in vitro* by internal initiation on (−) sense genomic RNA, *Nature (London),* 313, 68, 1985.

10. **Ahlquist, P., French, R., Janda, M., and Loesch-Fries, L. S.,** Multicomponent RNA plant virus infection derived from cloned viral cDNA, *Proc. Natl. Acad. Sci. U.S.A.,* 81, 7066, 1984.

11. **Berna, A., Briand, J.-P., Strussi-Garaud, C., and Godefroy-Colburn, T.,** Kinetics of accumulation of the three non-structural proteins of alfalfa mosaic virus in tobacco plants, *J. Gen. Virol.,* 67, 1135, 1986.

12. **Strauss, E. G. and Strauss, J. H.,** Replication strategies of the single stranded RNA viruses of eukaryotes, *Curr. Top. Microbiol. Immunol.,* 105, 1, 1983.

13. **Blumenthal, T. and Carmichael, G. G.,** RNA replication: function and structure of Qβ-replicase, *Annu. Rev. Biochem.,* 48, 525, 1979.

14. **Mouchès, C., Bové, C., and Bové, J. M.,** Turnip Yellow Mosaic Virus RNA replicase: Partial purification of the enzyme from the solubilized enzyme-template complex, *Virology,* 58, 409, 1974.

15. **Bujarski, J. J., Hardy, S. F., Miller, W. A., and Hall, T. C.,** Use of dodecyl-β-D-maltoside in the purification and stabilization of RNA polymerase from brome mosaic virus-infected barley, *Virology,* 119, 465, 1982.

16. **Dorssers, L., van der Meer, J., van Kammen, A., and Zabel, P.,** The cowpea mosaic virus RNA replication complex and the host-encoded RNA-dependent RNA polymerase-template complex are functionally different, *Virology,* 125, 155, 1983.

17. **Mouchès, C., Bové, C., Barreau, C., and Bové, J. M.,** TYMV replicase: formation of a complex between the purified enzyme and TYMV RNA, *Ann. Microbiol. (Paris),* 127A, 75, 1976.

18. **Watanabe, Y. and Okada, Y.,** *In vitro* viral RNA synthesis by a subcellular fraction of TMV-inoculated protoplasts, *Virology,* 149, 64, 1986.

19. **Martelli, G. P. and Russo, M.,** in *The Plant Viruses: Polyhedral Virions with Tripartite Genomes,* Vol. 1, Francki, R. I. B., Ed., Plenum Press, New York, 1985, 163.

20. **Garnier, M., Mamoun, R., and Bové, J. M.,** TYMV RNA replication *in vivo:* replicative intermediate is mainly single-stranded, *Virology,* 104, 357, 1980,

21. **Garnier, M., Candresse, T., and Bové, J. M.,** Immunocytochemical localization of TYMV-coded structural and nonstructural proteins by the protein A-gold technique, *Virology,* 151, 100, 1986.

22. **White, J. L. and Dawson, W. O.,** Characterization of RNA-dependent RNA polymerases in uninfected and cowpea chlorotic mottle virus-infected cowpea leaves: selective removal of host RNA polymerase from membranes containing CCMV RNA replicase, *Virology,* 88, 33, 1978.

23. **Dorssers, L., Zabel, P., van der Meer, J., and van Kammen, A.,** Purification of a host-encoded RNA-dependent RNA polymerase from cowpea mosaic virus-infected cowpea leaves, *Virology,* 116, 236, 1982.

24. **van der Meer, J., Dorssers, L., van Kammen, A., and Zabel, P.,** The RNA-dependent RNA polymerase of cowpea is not involved in cowpea mosaic virus RNA replication: immunological evidence, *Virology,* 132, 413, 1984.

25. **Chifflot, S., Sommer, P., Hartmann, D., Stussi-Garaud, C., and Hirth, L.,** Replication of alfalfa mosaic virus RNA: evidence for a soluble replicase in healthy and infected tobacco leaves, *Virology,* 100, 91, 1980.

26. **Duda, C. T.,** Synthesis of double-stranded RNA. II. Partial purification and characterization of an RNA-dependent RNA polymerase in healthy tobacco leaves, *Virology,* 92, 180, 1979.

27. **Evans, D. M. A., Bryant, J. A., and Fraser, R. S. S.,** Characterization of RNA-dependent RNA polymerases in healthy and tobacco mosaic virus-infected tomato plants, *Ann. Bot.,* 54, 271, 1984.

28. **Kumarasamy, R. and Symons, R. H.,** Extensive purification of the cucumber mosaic virus-induced RNA replicase, *Virology,* 96, 622, 1979.

29. **Gill, D. S., Kumarasamy, R., and Symons, R. H.,** Cucumber mosaic virus-induced RNA replicase: solubilization and partial purification of the particulate enzyme, *Virology,* 113, 1, 1981.

30. **Romaine, C. P. and Zaitlin, M.,** RNA-dependent RNA polymerases in uninfected and tobacco mosaic virus-infected tobacco leaves: viral-induced stimulation of a host polymerase activity, *Virology,* 86, 253, 1978.

31. **Khan, Z. A., Hiriyanna, K. T., Chavez, F., and Fraenkel-Conrat, H.,** RNA-directed RNA polymerases from healthy and from virus-infected cucumber, *Proc. Natl. Acad. Sci. U.S.A.,* 83, 2383, 1986.

32. **Krakow, J. S., Rhodes, G., and Jovin, T. M.,** RNA Polymerase: catalytic mechanisms and inhibitors, in *RNA Polymerase,* Losick, R. and Chamberlin, M., Eds., Cold Spring Harbor Laboratory, Cold Spring Harbor, New York, 1976, 127.

33. **Rackwitz, H. R., Rohde, W., and Sänger, H. L.,** DNA-dependent RNA polymerase II of plant origin transcribes viroid RNA into full-length copies, *Nature (London),* 291, 297, 1981.

34. **van Loon, L. C.,** Pathogenesis-related proteins, *Plant Mol. Biol.*, 4, 111, 1985.
35. **Neidhardt, F. C., VanBogelen, R. A., and Vaughn, V.,** The genetics and regulation of heat-shock proteins, *Annu. Rev. Genet.*, 18, 295, 1984.
36. **Zabel, P., Dorssers, L., Wernars, K., and van Kammen, A.,** Terminal uridylyl transferase of *Vigna unguiculata:* purification and characterization of an enzyme catalyzing the addition of a single UMP residue to the 3'-end of an RNA primer, *Nucleic Acids Res.*, 9, 2433, 1981.
37. **Miller, W. A. and Hall, T. C.,** RNA-dependent RNA polymerase isolated from cowpea chlorotic mottle virus-infected cowpeas is specific for bromoviral RNA, *Virology*, 132, 53, 1984.
38. **Boege, F.,** Simultaneous presence of terminal adenylyl, cytidylyl, guanylyl and uridylyl transferase in healthy tomato leaf tissue: separation from RNA-dependent RNA polymerase and characterization of the terminal transferases, *Biosci. Rep.*, 2, 379, 1982.
39. **Symons, R. H.,** Viral genome structure, in *The Plant Viruses: Polyhedral Virions with Tripartite Genomes*, Vol. 1, Francki, R. I. B., Ed., Plenum Press, New York, 1985, 57.
40. **Joshi, S. and Haenni, A.-L.,** Plant RNA viruses: strategies of expression and regulation of viral genes, *FEBS Lett.*, 177, 163, 1984.
41. **Hirth, L. and Richards, K. E.,** Tobacco mosaic virus: model for structure and function of a simple virus, *Adv. Virus Res.*, 26, 145, 1981.
42. **Ahlquist, P., Dasgupta, R., and Kaesberg, P.,** Nucleotide sequence of the brome mosaic virus genome and its implications for viral replication, *J. Mol. Biol.*, 172, 369, 1984.
43. **Ahlquist, P., Luckow, V., and Kaesberg, P.,** Complete nucleotide sequence of brome mosaic virus RNA3, *J. Mol. Biol.*, 153, 23, 1981.
44. **Lane, L. C.,** The bromoviruses, *Adv. Virus Res.*, 19, 151, 1974.
45. **Dasgupta, R., Harada, F., and Kaesberg, P.,** Blocked 5' termini in brome mosaic virus RNA, *J. Virol.*, 18, 260, 1976.
46. **Ahlquist, P., Dasgupta, R., and Kaesberg, P.,** Near identity of 3' RNA secondary structure in bromoviruses and cucumber mosaic virus, *Cell*, 23, 183, 1981.
47. **Hall, T. C., Shih, D. S., and Kaesberg, P.,** Enzyme-mediated binding of tyrosine to brome mosaic virus ribonucleic acid, *Biochem. J.*, 129, 969, 1972.
48. **Loesch-Fries, L. S. and Hall, T. C.,** *In vivo* aminoacylation of brome mosaic and barley stripe mosaic virus RNAs, *Nature (London)*, 298, 771, 1982.
49. **Kiberstis, P. A. and Hall, T. C.,** *N*-acetyl-tyrosine at the 3' end of brome mosaic virus RNA has little effect on infectivity, *J. Gen. Virol.*, 64, 2073, 1983.
50. **Rietveld, K., van Poelgeest, R., Pleij, C. W. A., van Boom, J. H., and Bosch, L.,** The tRNA-like structure at the 3' terminus of turnip yellow mosaic virus RNA. Differences and similarities with canonical tRNA, *Nucleic Acids Res.*, 10, 1929, 1982.
51. **Rietveld, K., Pleij, C. W. A., and Bosch, L.,** Three-dimensional models of the tRNA-like 3' termini of some plant viral RNAs, *EMBO J.*, 2, 1079, 1983.
52. **Pleij, C. W. A., Rietveld, K., and Bosch, L.,** A new principle of RNA folding based on pseudoknotting, *Nucleic Acids Res.*, 13, 1717, 1985.
53. **Joshi, R. L., Joshi, S., Chapeville, F., and Haenni, A.-L.,** tRNA-like structures of plant viral RNAs: conformational requirements for adenylation and aminoacylation, *EMBO J.*, 2, 1123, 1983.
54. **Joshi, S., Chapeville, F., and Haenni, A.-L.,** Length requirements for tRNA-specific enzymes and cleavage specificity at the 3' end of turnip yellow mosaic virus RNA, *Nucleic Acids Res.*, 10, 1947, 1982.
55. **Miller, W. A., Bujarski, J.J., Dreher, T. W., and Hall, T. C.,** Minus-strand initiation by brome mosaic virus replicase within the 3' tRNA-like structure of native and modified RNA templates, *J. Mol. Biol.*, 187, 537, 1986.
56. **Goelet, P., Lomonossoff, G. P., Butler, P. J. G., Akam, M. E., Gait, M. J., and Karn, J.,** Nucleotide sequence of tobacco mosaic virus RNA, *Proc. Natl. Acad. Sci. U.S.A.*, 79, 5818, 1982.
57. **Pelham, H. R. B.,** Leaky UAG termination codon in tobacco mosaic virus RNA, *Nature (London)*, 272, 469, 1978.
58. **Morch, M.-D. and Benicourt, C.,** Post-translational proteolytic cleavage of *in vitro*-synthesized turnip yellow mosaic virus RNA-coded high-molecular-weight proteins, *J. Virol.*, 34, 85, 1980.
59. **Guilley, H. and Briand, J.-P.,** Nucleotide sequence of turnip yellow mosaic virus coat protein mRNA, *Cell*, 15, 113, 1978.
60. **Haseloff, J., Goelet, P., Zimmern, D., Ahlquist, P., Dasgupta, R., and Kaesberg, P.,** Striking similarities in amino acid sequence among nonstructural proteins encoded by RNA viruses that have dissimilar genomic organization, *Proc. Natl. Acad. Sci. U.S.A.*, 81, 4358, 1984.
61. **Rezaian, M. A., Williams, R. H. V., and Symons, R. H.,** Nucleotide sequence of cucumber mosaic virus RNA1: Presence of a sequence complementary to part of the viral satellite RNA and homologies with other viral RNAs, *Eur. J. Biochem.*, 150, 331, 1985.
62. **Cornelissen, B. J. C. and Bol, J. F.,** Homology between the proteins encoded by tobacco mosaic virus and two tricornaviruses, *Plant Mol. Biol.*, 3, 379, 1984.

63. **Ahlquist, P., Strauss, E. G., Rice, C. M., Strauss, J. H., Haseloff, J., and Zimmern, D.,** Sindbis virus proteins nsP1 and nsP2 contain homology to nonstructural proteins from several RNA plant viruses, *J. Virol.,* 53, 536, 1985.

64. **Huisman, C. J. and Jaspars, E. M. J.,** Coat protein blocks the in vitro transcription of the virion RNAs of alfalfa mosaic virus, *FEBS Lett.,* 209, 285, 1986.

65. **Kiberstis, P. A., Loesch-Fries, L. S., and Hall, T. C.,** Viral protein synthesis in barley protoplasts inoculated with native and fractionated brome mosaic virus RNA, *Virology,* 112, 804, 1981.

66. **Nassuth, A. and Bol, J. F.,** Altered balance of the synthesis of plus- and minus-strand RNAs induced by RNAs 1 and 2 of alfalfa mosaic virus in the absence of RNA3, *Virology,* 124, 75, 1983.

67. **Kamer, G. and Argos, P.,** Primary structural comparison of RNA-dependent polymerases from plant, animal and bacterial viruses, *Nucleic Acids Res.,* 12, 7269, 1984.

68. **Banerjee, A. K.,** 5′-Terminal cap structure in eucaryotic messenger ribonucleic acids, *Microbiol. Rev.,* 44, 175, 1980.

69. **Keith, J. M.,** 5′-Terminal modification of mRNAs by viral and cellular enzymes, in *Enzymes of Nucleic Acid Synthesis and Modification,* Vol. 2, Jacob, S. T., Ed., CRC Press, Boca Raton, Florida, 1983, 111.

70. **Cross, R. K.,** Identification of a unique guanine-7-methyl transferase in Semliki Forest Virus infected cell extracts, *Virology,* 130, 452, 1983.

71. **Schubert, M., Harmison, G. G., Richardson, C. D., and Meier, E.,** Expression of a cDNA encoding a functional 241-kilodalton vesicular stomatitis virus RNA polymerase, *Proc. Natl. Acad. Sci. U.S.A.,* 82, 7984, 1985.

72. **Green, M. R., Maniatis, T., and Melton, D. A.,** Human β-globin pre-mRNA synthesized *in vitro* is accurately spliced in *Xenopus* oocyte nuclei, *Cell,* 32, 681, 1983.

73. **Loesch-Fries, L. S., Jarvis, N. P., Krahn, K. J., Nelson, S. E., and Hall, T. C.,** Expression of alfalfa mosaic virus RNA4 cDNA transcripts *in vitro* and *in vivo, Virology,* 146, 177, 1985.

74. **Hardy, S. F., German, T. L., Loesch-Fries, L. S., and Hall, T. C.,** Highly active template-specific RNA-dependent RNA polymerase from barley leaves infected with brome mosaic virus, *Proc. Natl. Acad. Sci. U.S.A.,* 76, 4956, 1979.

75. **Miller, W. A. and Hall, T. C.,** Use of micrococcal nuclease in the purification of highly template-dependent RNA-dependent RNA polymerase from brome mosaic virus-infected barley, *Virology,* 125, 236, 1983.

76. **Miller, W. A.,** Replication of Brome Mosaic Virus RNA *In Vitro,* Ph.D. thesis, University of Wisconsin, Madison, 1984.

77. **Bancroft, J. B.,** A virus made from parts of the genome of brome mosaic and cowpea chlorotic mottle viruses, *J. Gen. Virol.,* 14, 223, 1972.

78. **Dasgupta, R. and Kaesberg, P.,** Sequence of an oligonucleotide derived from the 3′ end of each of the four brome mosaic viral RNAs, *Proc. Natl. Acad. Sci. U.S.A.,* 74, 4900, 1977.

79. **Ahlquist, P., Bujarski, J. J., Kaesberg, P., and Hall, T. C.,** Localization of the replicase recognition site within brome mosaic virus RNA by hybrid-arrested RNA synthesis, *Plant Mol. Biol.,* 3, 37, 1984.

80. **Melton, D. A., Krieg, P. A., Rebagliati, M. R., Maniatis, T., Zinn, K., and Green, M. R.,** Efficient *in vitro* synthesis of biologically active RNA and RNA hybridization probes from plasmids containing a bacteriophage SP6 promoter, *Nucleic Acids Res.,* 12, 7035, 1984.

81. **Bujarski, J. J., Dreher, T. W., and Hall, T. C.,** Deletions in the 3′-terminal tRNA-like structure of brome mosaic virus RNA differentially affect aminoacylation and replication *in vitro, Proc. Natl. Acad. Sci. U.S.A.,* 82, 5636, 1985.

82. **Loesch-Fries, L. S. and Hall, T. C.,** Synthesis, accumulation and encapsidation of individual brome mosaic virus RNA components in barley protoplasts, *J. Gen. Virol.,* 47, 323, 1980.

83. **Bujarski, J. J., Ahlquist, P., Hall, T. C., Dreher, T. W., and Kaesberg, P.,** Modulation of replication, aminoacylation and adenylation *in vitro* and infectivity *in vivo* of BMV RNAs containing deletions within the multifunctional 3′ end, *EMBO J.,* 5, 1769, 1986.

84. **Collmer, C. W. and Kaper, J. M.,** Terminal sequences of the double-stranded RNAs of cucumber mosaic virus and its satellite: implications for replication, *Virology,* 145, 249, 1985.

85. **Ahlquist, P., French, R., and Bujarski, J. J.,** Molecular studies of brome mosaic virus using infectious transcripts from cloned cDNA, *Adv. Virus. Res.,* 32, 215, 1987.

86. **Marsh, L. E., Dreher, T. W., and Hall, T. C.,** Mutational analysis of the internal promoter for transcription of the subgenomic RNA4 of BMV, in *Proc. UCLA Symp. Positive Strand RNA Viruses,* Brinton, M. A. and Rueckert, R., Eds., Alan R. Liss, New York, 1987, 327.

87. **French, R., Janda, M., and Ahlquist, P.,** Bacterial gene inserted in an engineered RNA virus: efficient expression in monocotyledonous plant cells, *Science,* 231, 1294, 1986.

88. **Ricard, B., Barreau, C., Bové, C., Charron, A., Mouchès, C., Latrille, J., and Bové, J. M.,** Plant virus replication: *in vitro* synthesis of turnip yellow mosaic virus coat protein and RNA, in *Acides Nucléiques et Synthèse des Protéines chez les Végétaux,* No. 261, Bogorad, L. and Weil, J. H., Eds., Colloques Internationaux du C.N.R.S., Paris, 1976, 43.

89. **Mouchès, C., Candresse, T., and Bové, J. M.,** Turnip yellow mosaic virus RNA-replicase contains host and virus-encoded subunits, *Virology*, 134, 78, 2984.
90. **Joshi, R. L., Ravel, J. M., and Haenni, A.-L.,** Interaction of turnip yellow mosaic virus Val-RNA with eukaryotic elongation factor EF-1α. Search for a function, *EMBO J.*, 5, 1143, 1986.
91. **Dodds, J. A., Morris, T. J., and Jordan, R. L.,** Plant viral double-stranded RNA, *Annu. Rev. Phytopathol.*, 22, 151, 1984.
92. **Reanney, D. C.,** The evolution of RNA viruses, *Annu. Rev. Microbiol.*, 36, 47, 1982.
93. **Joklik, W. K., Ed.,** *The Reoviridae*, Plenum Press, New York, 1983.
94. **Ralph, R. K.,** Double-stranded viral RNA, *Adv. Virus Res.*, 15, 61, 1969.
95. **Dynan, W. S. and Tjian, R.,** Isolation of transcription factors that discriminate different promoters recognized by RNA polymerase II, *Cell*, 32, 669, 1983.
96. **Lassar, A. B., Martin, P. L., and Roeder, R. G.,** Transcription of class III genes: formation of preinitiation complexes, *Science*, 222, 740, 1983.
97. **Florentz, C., Briand, J. P., and Giegé, R.,** Possible functional role of viral tRNA-like structures, *FEBS Lett.*, 176, 295, 1984.
98. **Berna, A.,** The Non-Structural Proteins of Alfalfa Mosaic Virus: Immunodetection and Possible Role in the Multiplication of the Virus, Ph.D. thesis, Université Louis Pasteur, Strasbourg, France, 1986.
99. **Matthews, R. E. F.,** Classification and nomenclature of viruses, *Intervirology*, 17, 1, 1982.
100. **Rietveld, K., Linschooten, K., Pleij, C. W. A., and Bosch, L.,** The three-dimensional folding of the tRNA-like structure of tobacco mosaic virus RNA. A new building principle applied twice, *EMBO J.*, 3, 2613, 1984.
101. **Zuker, M. and Stiegler, P.,** Optimal computer folding of large RNA sequences using thermodynamics and auxiliary information, *Nucleic Acids Res.*, 9, 133, 1981.

Chapter 6

REPLICATION OF CORONAVIRUS RNA

Michael M. C. Lai

TABLE OF CONTENTS

I.   Introduction 116

II.   Structure and Organization of Coronavirus 116
     A.   Structural Proteins 116
     B.   RNA Genome 117
     C.   Lipid 117

III.   Replication of Coronavirus RNA 117
     A.   General Pathway of Coronavirus Replication 117
     B.   Negative Strand RNA 119
     C.   mRNA Structure and Coding Functions 120
     D.   Mechanism of Subgenomic mRNA Transcription 122
     E.   Replicative Intermediate (RI) and Replicative Form (RF) RNA 125
     F.   RNA Replication 126

IV.   RNA-Dependent RNA Polymerases and Regulatory Proteins 128
     A.   RNA-Dependent RNA Polymerases 128
     B.   RNA Regulatory Proteins 130

V.   Temperature-Sensitive Mutants 130

VI.   RNA Recombination 131

VII. Defective-Interfering (DI) RNA 131

VIII. Perspectives 131

Acknowledgments 132

References 132

I. INTRODUCTION

Coronaviruses are a group of enveloped viruses with a nonsegmented, positive-sensed RNA genome. These viruses have common morphology with characteristic petal-shaped spikes, and were grouped together as Coronaviridae in 1974.[1] Coronaviruses have been isolated from many animal species. Among these are several economically and clinically important viruses which infect various species, causing a variety of respiratory and gastrointestinal illnesses. For instances, porcine transmissible gastroenteritis virus (TGEV), bovine coronavirus (BCV), avian infectious bronchitis virus (IBV), and feline infectious peritonitis virus (FIPV) all cause severe infections in livestocks and other domestic animals. Human coronaviruses are responsible for approximately 20% of common colds in winter,[2] and may play an increasingly important role in acute or chronic diarrhea.[3] Furthermore, human coronaviruses have been linked to multiple sclerosis.[4]

In addition to morphological similarity, coronaviruses share many biochemical features which distinguish these viruses as a unique taxonomic group. Many of these biochemical characteristics have only recently begun to be understood. Studies have revealed that coronaviruses have unique pathways of virus replication, particularly with regard to the mechanisms of RNA replication and transcription. The scope of this review is to discuss the mechanisms of RNA synthesis of coronaviruses and the implication of these mechanisms to RNA genetics. For other aspects of coronaviruses, the readers should refer to recent reviews by Sturman and Holmes,[5] Wege et al.,[6] and Siddell et al.[7]

II. STRUCTURE AND ORGANIZATION OF CORONAVIRUSES

A. Structural Proteins

Coronavirus virions are spherical, enveloped particles ranging from 80 to 160 nm in diameter. On the surface of virus particles are spikes, or peplomers, of roughly 20 nm in length.[8] The spikes have an appearance of a crown, or corona, thus giving the name for the virus. Inside the virus particle is a helical nucleocapsid of 6 to 8 nm in diameter, which could be released from the virion by mild detergents such as NP®-40 or Triton® X-100.[9]

All of the coronaviruses contain three or four structural proteins in the purified virion. Most of the viruses have two classes of glycoproteins, one with a molecular weight of approximately 180,000 daltons, and the other in the vicinity of 23,000 daltons. The 180-kdalton protein constitutes the peplomers on the viral envelope and is designated E2.[9] The peplomers are usually, but not always, cleaved into two different 90-kdalton subunits in virions,[10,11] one of which contains covalently linked palmitic acid.[10] This cleavage appears to be carried out by a cellular protease and is required for at least some of the biological activities of the peplomers.[10] The peplomers are required for the virus binding to the cellular receptors, induction of cell fusion, induction of neutralizing antibody and cell-mediated cytotoxicity, and may also contribute to the target cell specificity (for review, see Reference 5).

The second glycoprotein of 23 kdaltons, designated as E1, constitutes the matrix protein of the envelope. It usually consists of several molecules with different degrees of glycosylation,[12,13] ranging in molecular weight from 21 to 36 kdaltons. The gene sequence of E1 has so far been determined for mouse hepatitis virus (MHV)[14] and IBV,[15] and suggests that E1 is a transmembrane molecule which spans the membrane several times.[16a] The extracellular domain is glycosylated. In MHV, the carbohydrate side chains are linked through an O-glycosidic bond; thus, the synthesis of MHV E1 protein is not inhibited by tunicamycin.[17,18] However, in IBV, only N-glycosidic bond is present.[13] The intracellular domain of E1 seems to interact with the virion nucleocapsid directly.[8] This interaction might provide a focal point for assembly of virion particles.

The third structural protein is the nucleocapsid protein, N, which is a phosphorylated protein of 50 to 60 kdaltons.[19] The phosphorylation occurs on serine residues. The N protein appears to interact directly with the RNA of the virus, forming a helical nucleocapsid.[19a] It is the most abundant protein of the virion and in the infected cells. In an in vitro assay, the N protein has been shown to bind to various RNAs without apparent specificity,[20] probably as a result of the basic nature of the protein. A cAMP-independent protein kinase activity has been detected in virion, which can phosphorylate the N protein in vitro.[21] However, it is not known whether this activity is responsible for the phosphorylation of the N protein in the infected cells. Several N-related proteins of slightly different molecular weights have also been observed in the infected cells,[20] which are apparently the cleavage products of the nucleocapsid protein.

A fourth structural protein, gp65, has been detected in some coronaviruses, such as BCV[22,23] and human coronavirus OC43.[24] This glycoprotein is an envelope protein and appears to be responsible for the hemagglutinating activity of the virus.[23] It is significant that only the coronaviruses with hemagglutinating activity possess this protein.

B. RNA Genome

Coronaviruses contain a single piece of positive-stranded, nonsegmented RNA genome with a molecular weight of 5 to 8×10^6 daltons.[25-29] The RNA genome is infectious.[28,30,31] It contains polyadenylated sequences of roughly 100 nucleotides at the 3' end[27,32] and a cap structure at the 5' end,[33] and can serve as template for in vitro translation.[34,35] Furthermore, oligonucleotide fingerprinting studies[36,37] indicate that the genomic RNA and mRNAs in the infected cells are of the same sense. Thus, it is a positive-sensed RNA by definition. The RNA genome is contained within a helical nucleocapsid,[38,39] which is unusual for a positive-stranded RNA virus. Attempts have been made to search for negative-stranded RNA in the virion; no such RNA was detected.[33] Analysis of the genetic complexity and oligonucleotide fingerprinting studies[26,37] suggest that the coronavirus RNA genome does not contain significant redundancy of sequences. Complete sequences have been obtained for the IBV genomic RNA, which is 27.6 kilobases long,[39a] considerably larger than the previous estimates. Other coronaviruses are likely to have an RNA genome of comparable size. Thus, coronaviruses contain the largest RNA genome among all of the RNA viruses.

C. Lipid

The viral envelope consists of a lipid bilayer which is probably derived from the host cell membrane. Pike and Gawes[40] have found that the lipid composition of the virus particles reflects that of host cells in which the virus was grown. However, cholesterol and fatty acid esters are selectively deficient in virions. No information is yet available concerning the specific membrane origin of the lipids. This information would be particularly interesting since the virus matures by budding into endoplasmic reticulum, instead of plasma membrane (see Section III.).

III. REPLICATION OF CORONAVIRUS RNA

A. General Pathway of Coronavirus Replication

Coronaviruses generally infect only animal species of their origin. However, some coronaviruses could cross species barriers and result in epizootic infections. This possibility is suggested by the remarkable sequence homology between BCV and human coronavirus OC43.[41] This specificity is also reflected in vitro in that the viruses generally grow only in cells of the same species. Again, exceptions exist for some coronaviruses, such as BCV, which grows in bovine, human, and mouse cell lines. The species specificity is probably controlled at the level of cellular receptors for the viruses. In addition, coronaviruses also

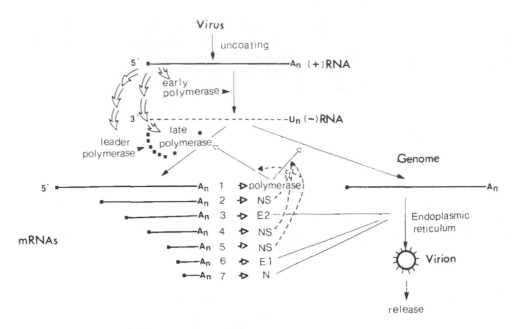

FIGURE 1.   General pathway of coronavirus replication. In this diagram, the three polymerases are identified as separate and distinct. They could be the modified forms of the same polymerase. The polymerase made from the newly synthesized mRNA1 contributes to the pool of the leader and late polymerases. The NS (nonstructural) proteins are depicted as regulating either transcription of mRNAs or replication of genomic RNA. The solid squares represent the leader RNAs. The nomenclature of MHV is used.

show marked tissue tropism which may also be determined by the distribution of cellular receptors.

The viruses initiate infection by binding to receptors on the surface of susceptible cells, apparently mediated by the E2 peplomers of the viruses. Recently, receptors for MHV have been identified in the liver and gastrointestinal tract of susceptible mouse strains.[41a] After virus adsorption, the viruses penetrate either by viropexis[42,43] or cell fusion.[44,45] The process of penetration and subsequent uncoating is poorly understood at this time.

Throughout the replication cycle, the virus is restricted to the cytoplasm of the infected cells. This fact is supported by ultrastructural studies[43] and by the observations that MHVs replicate in enucleated cells[46,47] and that actinomycin D and α-amanitin do not inhibit viral replication.[46,47] However, it has been suggested that some coronaviruses, such as IBV, may require some nuclear functions.[48]

The major biochemical events in the replication cycle of coronaviruses are summarized in Figure 1. The first biosynthetic event after virus penetration and uncoating is probably the synthesis of an RNA-dependent RNA polymerase, since this enzyme is not carried in the virion[49,50] and is required for the replication of viral RNA. Indeed, the requirement for continued protein synthesis has been shown to begin immediately following virus adsorption.[49,51,52] The incoming RNA is presumably transcribed by this enzyme into a full-length, negative-stranded RNA species[53] which then serves as the template for the synthesis of subgenomic and genomic mRNA species by a new or modified RNA polymerase.[49] These mRNAs are used to code for the structural and nonstructural viral proteins. The structural proteins are utilized for the assembly of virus particles, while the nonstructural proteins may participate in regulation of viral RNA transcription and replication and possibly, inhibition of host macromolecular synthesis.

After sufficient virus-specific mRNAs and proteins are synthesized, viral transcriptional machinery is probably shifted to the replication of viral genomic RNA. The switch from the transcription of mRNAs to replication of viral genome is not very clear in MHV[54] or IBV,[26] but is quite evident in BCV-infected cells (Keck et al., unpublished observations). The viral genome interacts with the viral structural proteins to form virus particles which bud into endoplasmic reticulum in the perinuclear region.[38,55,56] The mature virus particles travel through smooth endoplasmic reticulum and are released into the extracellular space.[5] The E2 envelope protein is cleaved probably by cellular proteases into two subunits during the maturation process.[10] The infected cells develop cytopathic effects (CPE), which may include cell fusion, and are eventually killed.

The kinetics of virus growth vary considerably among different coronaviruses and cell lines used. The MHVs in mouse cell lines generally start detectable RNA synthesis 3 to 4 hr after infection and reach the peak level of RNA synthesis 7 to 8 hr postinfection.[37,54,57] The CPE, such as cell fusion, becomes apparent 5 to 6 hr after infection. The production of virus particles coincides with the appearance of CPE and does not reach the peak until 10 to 12 hr postinfection. In contrast, other mammalian coronaviruses have a much slower growth rate. For instance, BCV or HCV takes 3 to 4 days to complete its replication cycle in HRT cell lines. In this review, the growth curve of MHV will be used throughout, unless otherwise indicated.

The following sections will examine in detail the steps involved in RNA transcription and replication.

B. Negative Strand RNA

Like most of positive strand RNA viruses, the replication of coronavirus RNA goes through a negative strand RNA. The negative strand RNA in MHV-infected cells has been examined by conventional "Northern" blot analysis using a positive-sensed virion genomic RNA as the probe.[53] This study showed that only a single species of negative strand RNA of genomic size could be detected late in the infection, and that no subgenomic-sized negative strand RNA were detected. The negative strand RNA was diffuse in agarose gel electrophoresis, suggesting possible heterogeneity or extensive secondary structure.[53] The negative strand RNA binds to poly(A) Sepharose® column, suggesting that it contains poly(U) sequences (Leibowitz and Weiss, personal communication); thus, it is very likely that the negative strand RNA synthesis initiates by copying the poly(A) tails at the 3' end of the virion genomic RNA. Recently the 3' ends of several coronavirus RNAs have been sequenced. A stretch of conserved sequences (GGGAAGAGCT) located within the 3'-noncoding region of genome (75 to 90 nucleotides from the 3' end) has been found in MHV, IBV, and TGEV.[61-64] This sequence corresponds to the 5' end of the negative strand RNA, and thus may provide the common recognition signal for regulation of negative strand RNA synthesis.

The kinetics of synthesis of negative strand RNA has been studied for MHV.[52] The negative-strand RNA was identified as the RNase-resistant fractions of the intracellular RNA after hybridization with an excess of unlabeled virion RNA. This study showed that negative strand RNA synthesis begins at about the same time (4 hr postinfection) as the total RNA synthesis and peaks at 5 to 6 hr postinfection, which is earlier than the peak (7 to 8 hr) of the total RNA synthesis.[52] However, there is still a considerable amount (about 20% of the peak level) of negative strand RNA synthesis even late in the infection. This result is in contrast to the indirect data of Brayton et al.[58] who, by using an in vitro RNA polymerase assay, suggest that negative strand RNA synthesis begins earlier than the positive strand RNA synthesis and stops before any significant amount of mRNAs are synthesized. These differences could be due to the different experimental conditions used. However, one piece of experimental data suggests that, if the negative strand RNA is synthesized late in the infection, it would not be likely to play a significant role in mRNA transcription: the UV transcriptional mapping studies performed by irradiating the infected cells at 6 hr postinfection indicated that the target sizes of subgenomic mRNAs are smaller than the genomic RNA.[59,60]

If the negative strand RNA synthesized late in the infection were utilized as the template for mRNA transcription, then the UV target sizes of MHV mRNAs (see Section III.D.) would be expected to be equal to that of the genome-length RNA. Thus, it is not clear whether the negative strand RNA synthesized late in the infection has any biological function.

The negative strand RNA template appears to be very stable, so that the plus strand RNA continues to be synthesized even though the synthesis of negative strand RNA is decreased or stopped. This may be due to the fact that all of the negative strand RNA exists in replicative intermediate (RI) form and no free negative strand RNA was detected in the infected cells.[52] The negative strand RNA does not constitute more than 1 to 2% of the total virus-specific RNA;[52,60a] thus, each negative strand RNA molecule is utilized for multiple rounds of mRNA transcription. A recent study using direct measurement of negative-strand RNA indeed shows that the MHV negative-strand RNA accumulates early in the infection, and then remains stable, but does not increase during the rest of the replication cycle.[60a]

C. mRNA Structure and Coding Functions

During infection, positive-sensed RNAs comprise the majority of the virus-specific RNAs. Thus, at any given time point after infection, the in vivo labeled RNA from the coronavirus-infected cells will represent the poly(A)-containing genomic RNA and mRNAs. The poly(A)-containing RNAs from the coronavirus-infected cells labeled in the presence of actinomycin D have consistently been resolved into six to seven species with molecular weights ranging from 0.6×10^6 to 6 to 7×10^6.[26,37,50,54,57,65-67] They are termed mRNAs 1 to 7 for MHVs and other mammalian coronaviruses, and F to A for IBV (Figure 2). The largest poly(A)-containing RNA (mRNA 1 for MHV and mRNA F for IBV) is equivalent to the genomic RNA, and may include both the genome-sized RNAs used as mRNAs and those used for packaging into the virion. The latter has been detected as an EDTA-resistant nucleocapsid structure,[57,68,60a] while the former is associated with polysomes, which can be dissociated with EDTA.[57,69] It is not clear whether there is any difference between these two classes of genomic RNA. The rest of the mRNAs are subgenomic in size. The genetic structure of these mRNAs has been determined by oligonucleotide fingerprinting studies,[36,37,54,70] which showed that these mRNAs have a 3′ coterminal, nested-set structure, i.e., all of the mRNAs represent sequences starting from the 3′ end of the genome and extending for various distances toward the 5′ end. All of these mRNAs have been shown to be associated with polysomes,[57,69] thus representing functional mRNA species in infected cells.

The coding function of each mRNA has been determined by in vitro translation of individual mRNA species.[34,67,71,72] More recently, these gene assignments have been better defined by sequence analysis.[15,73-77a] The coding assignments of various mRNAs are depicted in Figure 2. It should be noted that only the 5′-unique portion of each mRNA is translatable in vitro; thus, each mRNA is monocistronic, although the possibility has not been ruled out that the downstream cistrons may also be translated if the mRNAs are degraded to the extent that the upstream gene of each mRNA is removed. The 5′-most gene encodes a larger-than-200-kdalton protein[34] which, by analogy to other positive-stranded RNA viruses, may represent the RNA polymerases. The gene assignment in different coronaviruses is not colinear; for instance, a nonstructural protein gene is present between the N and E1 genes in IBV, while, in other coronaviruses, N and E1 genes are adjacent to each other (Figure 2). More recently, it has been shown that BCV synthesizes an additional mRNA species between mRNAs 2 and 3 (Keck, J., unpublished data). This may be related to the observation that BCV contains a hemagglutinin protein, gp65,[22,23] which is not found in other coronaviruses.

Recent sequence analysis of some of the mRNAs of coronaviruses suggests that there may be more than one overlapping or nonoverlapping open reading frames (ORF) in some of the mRNAs. For instance, the unique coding region of mRNA5 (gene E) of MHV contains two ORFs which overlap by 5 nucleotides and are in different reading frames.[74] These two

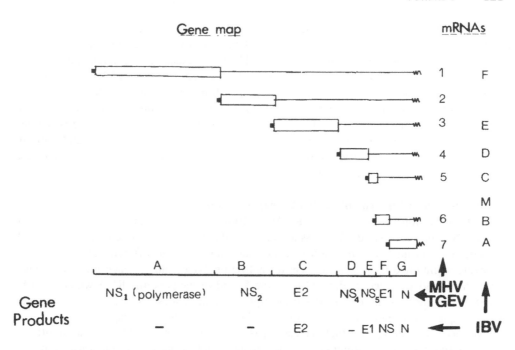

FIGURE 2. The gene map, gene products, and structure of coronavirus mRNAs. The open boxes represent the unique portions of mRNAs, which are translated. The mRNA M of IBV is a minor mRNA species, with a size in between those of mRNAs B and C.[107]

ORFs have a capacity to code for two small proteins, p10 and p12, respectively. Preliminary data suggest that the two proteins may actually be translated in the infected cell.[74] No separate-spliced mRNA has been observed for the second ORF. This is consistent with the fact that the replication of coronaviruses is limited to cytoplasm; thus, the conventional RNA splicing mechanism, which normally takes place in the nucleus, would not operate during coronavirus mRNA synthesis. This is clearly different from influenza viruses which synthesize spliced mRNAs for the downstream ORFs in cases of overlapping ORFs.[78,79] How the downstream ORF of MHV is translated is not clear. One possibility is that these two ORFs could be translated into a single protein by ribosomal frame shifting. This mechanism of translation has been demonstrated in the retrovirus system.[80] Another possibility is that the second ORF is translationally active by itself. Indeed, in vitro translation studies show that the second ORF of MHV mRNA 5 is more favorably translated as a separate protein than the first ORF.[80a] The translational control of these ORFs remains unknown. Another short ORF which has a potential coding capacity is found in the noncoding region of BCV at the 3' end of the genomic RNA;[61] however, no corresponding protein product for this ORF has been detected. Sequence studies of IBV genomic RNA reveal that the 5'-most gene of IBV RNA contains two long ORFs, with capacities to code for two proteins of 441 kdaltons and 300 kdaltons separately.[39a] Preliminary data suggest that these two ORFs are translated into a single protein by a ribosomal frame-shifting mechanism (Boursnell, M., personal communication). Such a large protein product has yet to be detected in IBV-infected cells. The examples cited above indicate that the coding functions of coronavirus mRNAs are more complex than the conventional monocistronic mRNAs.

The abundance of the different mRNAs varies. It appears that the smaller mRNAs are relatively more abundant than the larger mRNA species,[54] but this inverse relationship between the mRNA size and abundance is not universal. The relative ratio of individual mRNAs remains roughly constant throughout the replication cycle.[54] Thus, factor(s) which regulate the synthesis of various mRNAs operate throughout the replication cycle.

Another unique feature of coronavirus mRNAs is the presence of identical leader sequences of 50 to 70 nucleotides at the 5′ ends of all of the mRNAs and genomic RNA.[81-84] The presence of a leader sequence was first suggested by the unusual T1-oligonucleotides present in the mRNAs of MHV,[69,70] and was subsequently confirmed by cDNA cloning and sequencing[82,83] and also by heteroduplex mapping.[82] It has now been established that all of the MHV mRNAs contain a leader RNA sequence of roughly 70 nucleotides, which is also present at the 5′ end of the genomic RNA, suggesting that the leader sequence is derived from the 5′ end of the genomic RNA. A comparable sequence of approximately 50 nucleotides has also been identified in IBV mRNAs.[84,85] No sequence homology between these two leader sequences was detected.

D. Mechanism of Subgenomic mRNA Transcription

The transcription of coronavirus mRNAs probably starts immediately after the negative strand RNA is synthesized. Two issues concern the mechanism of coronavirus mRNA transcription: (1) the different mRNA species are transcribed at different rates which are, however, maintained throughout the entire replication cycle, and (2) each mRNA contains a stretch of leader sequences derived from the 5′ end of the genomic RNA which must be joined to the mRNAs during or after transcription. Both of these issues may be explained by an unique transcriptional mechanism involving a free leader RNA species.

Two pieces of data suggest that the leader RNA sequences present at the 5′ end of mRNAs are not derived by conventional eukaryotic splicing mechanism: (1) coronaviruses replicate exclusively in the cytoplasm of infected cells,[46,47] while the RNA splicing takes place in nucleus; and (2) UV transcriptional mapping studies reveal that the UV target size of each subgenomic mRNA is approximately the same as its physical size, suggesting that each mRNA is transcribed independently.[59,60] In these studies, UV irradiation was administered to the infected cells at 6 hr postinfection, when the negative strand RNA synthesis has probably stopped or decreased to an insignificant level.[52] Because of the small size of leader RNA, the requirement for the synthesis of leader RNA did not affect the UV target size of the mRNAs. Thus, this result suggests that the subgenomic mRNAs of coronaviruses are not derived by cleavage of a big precursor RNA and that the synthesis of leader RNA and mRNAs is independent. Several transcriptional models have been proposed (Figure 3):[86] (1) the "looping out" model in which the negative strand template forms a loop in the "intron" region, thus bringing the leader RNA in close proximity to the initiation sites of mRNAs. The RNA polymerase could thus jump from the leader sequences to the mRNAs in continuous transcription. For smaller mRNAs, the loops would, therefore, be larger; (2) the "leader-primed transcription" model in which the leader RNA is transcribed and becomes dissociated from the RNA template. This "free" leader RNA then rebinds to the template at the initiation sites of various mRNAs and serves as the primer for transcription; (3) the "posttranscriptional processing model" in which the leader RNA and the body sequences are transcribed independently and then joined together by an unknown splicing mechanism posttranscriptionally. The third model is the least likely since the replicative intermediate (RI) RNA structure contains the leader RNA sequences,[86] suggesting that the incomplete nascent RNA chains already contain the leader sequences. The first model has also been considered unlikely since single-stranded RNA loops have not been detected in the RI structure.[86] However, these data can only be considered as circumstantial evidence in support of the "leader-primed transcription" model.

The direct evidence in support of this unusual transcriptional mechanism in MHV are several: (1) several free leader RNA species of 50 to 90 nucleotides have been detected in the cytoplasm of MHV-infected cells.[87] They are discrete RNA species, some of which are dissociated from the membrane-bound transcription complex. It is noteworthy that some of the free leader RNA species detected are larger than the estimated leader RNA sequences

MODEL 1: "LOOP OUT"

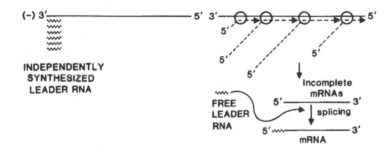

MODEL 2: LEADER-PRIMED TRANSCRIPTION

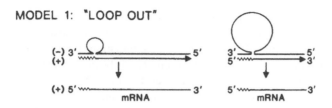

MODEL 3: POST-TRANSCRIPTIONAL PROCESSING

FIGURE 3. Three possible mechanisms for the joining of the leader RNA to mRNA without the use of the conventional RNA splicing mechanism. (Modified from Baric, R. S., Stohlman, S. A., and Lai, M. M. C., *J. Virol.*, 48, 633, 1983. With permission.)

present in the mRNAs, suggesting that the leader RNA will have to be cleaved before transcription of mRNAs; (2) a temperature-sensitive mutant has been isolated which synthesizes only the small leader RNA, but not mRNAs at the nonpermissive temperature,[87] suggesting that the synthesis of the leader RNA and the synthesis of the mRNAs are discontinuous and require different viral proteins; (3) during a mixed infection with two different MHVs, the mRNAs of each virus contains a population with the leader RNA sequences derived from the coinfecting virus.[88] This result strongly suggests that the leader RNA represents a separate transcriptional unit and can be freely exchanged between the mRNAs of two coinfecting viruses. Thus, the free leader-containing RNA species detected in the cytoplasm of MHV-infected cells represent bona fide transcriptional intermediates, rather than abortive transcription products. This piece of data thus strongly eliminates other transcriptional models for coronaviruses; (4) it has recently been shown that an RNA species complementary to the leader RNA ("anti-sense" leader RNA) could be expressed in L-2 cells by a mammalian expression vector using a feline LTR promotor. When the cells expressing this anti-sense leader RNA were infected with MHV, the RNA transcription of the superinfecting virus was substantially inhibited.[88a] This result suggests that the availability of the leader RNA sequences is required for MHV RNA synthesis. These data are most compatible with the interpretation that the leader RNA serves as a primer for transcription, and eliminates the model of posttranscriptional RNA splicing; (5) sequence analysis of the leader RNA and the intergenic regions of various genes in coronavirus genomes indicates that there is sequence homology of roughly 10 nucleotides between the 3' end of the leader RNA and the initiation sites of various mRNAs.[89,90] Therefore, the leader RNA could conceivably bind to the negative strand RNA template at the initiation sites of various mRNA species via these complementary sequences. This homology has been detected in both MHV and IBV;[85,89,90] (6) a recombinant virus has been derived from two MHV strains, A59 and JHM.[91] This recombinant contains 3 kilobases of sequences at the 5' end from JHM while the remaining sequences from A59. This recombinant synthesizes a new mRNA species with a new initiation start site located about 10 kilobases from the 5' end of the genome.[91] This finding suggests that the different leader RNA sequences could recognize different initiation signals in the RNA template. Thus, the initiation of mRNA synthesis is very likely regulated by the leader RNA sequences.

The exact structure and mechanism of synthesis of free leader RNA are not known. At the present time, the precise leader-body junction site for the different subgenomic mRNAs could only be speculated to be the 3'-most point of the intergenic regions where the homology between the 3' end of the leader RNA and the intergenic regions end. This point appears to vary with different mRNA species.[89,90] In fact, any point upstream of this putative leader-body junction sites, but within the homologous stretch, could potentially be the leader-body junction site. It is possible that different leader RNA species are utilized for the transcription of the different mRNAs. In fact, several leader-containing RNA species of 50 to 90 nucleotides have been detected in MHV-infected cells.[87] The understanding of the structure of the leader RNA and its mechanism of synthesis could be aided by the sequencing of the 5' end of the genomic RNA, where the leader RNA was derived. The 5' end sequences have been obtained for IBV and MHV.[85,92] More revealing is the sequence of the 5' end of MHV genomic RNA. Figure 4 shows the partial sequence of the leader region of the JHM strain of MHV. There is a hairpin loop near the possible termination point of the leader RNA synthesis. Indeed, four species of leader RNA species have been localized around this hairpin loop.[92a] This finding is consistent with the observations made in several prokaryotic systems in which RNA transcription has been shown to "pause" or terminate at the hairpin loop regions.[93,94] These leader RNAs are thus likely the species involved in the transcription of mRNAs. It should be noted that these leader RNA species are larger than the presumptive leader sequences present in the subgenomic mRNAs. Thus, these leader RNAs would require processing before being utilized as transcriptional primers. The comparison of 5' end genomic sequences and the intergenic sequences reveals that there is a stretch of 9- to 18-nucleotide homology at the putative leader-body junction sites (Figure 5). Some intergenic regions contain a mismatching nucleotide within this homologous region. It is conceivable that the free leader RNA binds to this site and is cleaved at the 3' end of the homologous region or at the mismatching point (Figure 6). The 3' end of the leader fragment could then serve as the starting point for the transcription of subgenomic mRNAs. This model predicts that the RNA polymerase(s) of coronaviruses possesses endonucleolytic activity and that different mRNAs have different leader-body junction sites, depending on the region of homology and the presence or absence of mismatching nucleotides. It is noteworthy that the degree of homology between the intergenic regions and the 3' end of the leader RNA closely parallels the abundance of individual mRNA species (Figure 5). Thus, the rate of mRNA transcription may be regulated by the degree of homology which, in turn, influences the strength of binding of the leader RNA to the intergenic sites. It is possible that the rate of transcription could also be influenced by other regulatory sequences. It is not clear whether the homology at the initiation sites of mRNAs is enough for the leader RNA to bind. Conceivably, the free leader RNA is associated with a polymerase, which will participate in the leader binding to the intergenic site.

The mechanism of synthesis of the genomic-sized mRNA (RNA1) has not been studied. Whether it requires a free leader RNA species or not is not clear. It is perhaps more likely

FIGURE 4. Partial sequence of the 5' end of the genomic RNA of MHV. The region of the possible leader termination site is shown for the JHM strain of MHV. The three repeats are underlined by dashed lines. The region underlined by a solid line represents an AU-rich sequence which might be the termination point for the leader RNA synthesis. (From Shieh, C.-K., Soe, L. H., Makino, S., Chang, M.-F., Stohlman, S. A., and Lai, M. M. C., *Virology*, 156, 321, 1987. With permission.)

Intergenic site		No. bases of homology	Ratio of RNA amount
3-4	........*UCUU*UUAGAUUUG*UUAAAUAUCG*......	9	1.1
4-5	........*UGAUCA*AGAUUUG*GAGUAGAAUU*......	7	0.9
5-6	...*UACU*AUUAGAUUAGGUUUG*UAAUAC*..........	10 + 4	32.0
6-7	....*ACUC*UUAGAUUAGAUUUGAAAU*UCCUAC*......	18	100.0
Leader	....AUCUAAUCUAAUCUAAACUUUAUAAACG......		

FIGURE 5. Comparison of the leader sequences at the 5' end of the genomic RNA with the intergenic sequences. The sequences at the intergenic sites are presented in (−) sense (template RNA). The regions of complementarity with the 3' end of the leader RNA are underlined by thin lines. The UAAAC and AUUUG underlined by heavy lines are the sequences common to the initiation sites of all of the mRNAs. The numbers of homologous sequences and relative ratio of mRNAs are included for comparison. (From Shieh, C.-K., Soe, L. H., Makino, S., Chang, M.-F., Stohlman, S. A., and Lai, M. M. C., *Virology*, 156, 321, 1987. With permission.)

that the genomic RNA involves continuous transcription, without participation of a free leader RNA species. However, the study of this issue is complicated by the presence of both the mRNA1 and the genomic RNA destined for packaging into virions. These two classes of RNA might be synthesized by different mechanisms.

E. Replicative Intermediate (RI) and Replicative Form (RF) RNA

Since the negative strand RNA is a single species of genome-sized RNA, the mechanism by which the multiple subgenomic and genomic mRNAs are transcribed from this template

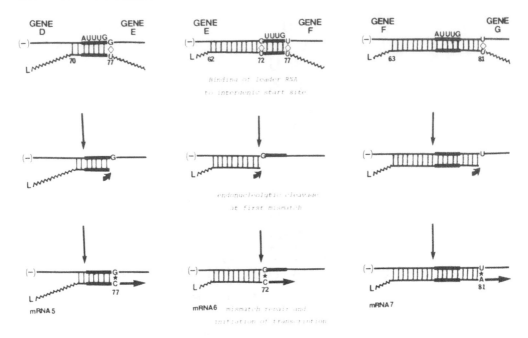

FIGURE 6.   The proposed model of the leader RNA priming. The leader RNA derived from the 5' end of the RNA genome is represented by the wavy line. It binds to the complementary region on the RNA template. The mismatching nucleotides are denoted by open rhombuses, where cleavage of the leader RNA possibly occurs. The transcription starts from the 3' end of the cleaved primer. (From Shieh, C.-K., Soe, L. H., Makino, S., Chang, M.-F., Stohlman, S. A., and Lai, M. M. C., *Virology*, 156, 321, 1987. With permission.)

is of considerable interest. The understanding of this issue has been aided by the study of the structures of the RI and RF. RI is defined as the RNA structure which is actively synthesizing RNA. It is composed of partially double-stranded and partially single-stranded RNA structure, and could be isolated from the intracellular RNA by Sepharose®-column chromatography and separated from the rest of intracellular viral RNAs by virtue of its large size.[86] The RF is defined as the double-stranded RNA obtained from RNase digestion of RI. Both the RI and RF of MHV are single RNA species of roughly genomic size.[53,86] No subgenomic RI or RF RNAs were detected. These data suggest that most of the negative strand RNA template is used for multiple initiation of mRNAs, and practically the entire RNA template is used for transcription simultaneously. The number of nascent RNA chains of each RNA template was estimated to be about six;[86] however, this number could be a considerable underestimate. It should be pointed out that the nascent chains on the RI complex contain the leader RNA sequences,[86] providing further evidence that leader sequences are not added posttranscriptionally.

F. RNA Replication

In most of the coronavirus-infected cells, the relative rate of synthesis of all of the mRNA species is constant throughout the infection; however, late in the infection, there is a gradual accumulation of the genomic RNA species. The most dramatic example of this regulatory shift has been demonstrated in BCV-infected cells (Keck, J., unpublished observation). The change of the relative ratio between the genomic RNA and subgenomic mRNA species late in the infection is probably the result of a switch from RNA transcription to replication. Transcription is defined as the synthesis of mRNAs, while replication is defined as the synthesis of genomic RNA destined to be packaged into virion. It has been determined that late in the infection 90% of the genomic-sized RNA in the infected cells is associated with

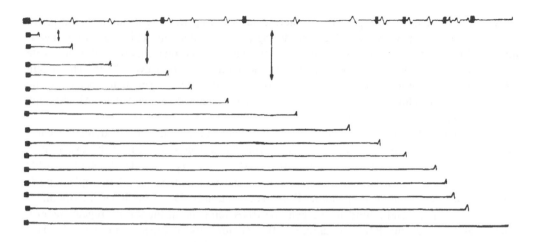

Replication

FIGURE 7. The proposed model of discontinuous coronaviral RNA replication. The RNA replication "pauses" at the regions of hairpin loops marked on the template RNA. Some of the pausing intermediates dissociate from and reassociate with the RNA template and continue transcription. (From Baric. R. S., Shieh, C.-K., Stohlman, S. A., and Lai, M. M. C., *Virology*, 156, 342, 1987. With permission.)

nucleocapsid.[57,68] The remaining 10% is associated with polysomes. Very little is known concerning the difference between the mechanisms of transcription and replication and what is responsible for such a switch. It is quite likely that transcription and replication are carried out by different RNA synthetic machineries. This possibility is strongly suggested by the study of RNA polymerase-membrane complexes isolated from the infected cells late in the infection.[50,58] These studies showed that two separate membrane complexes containing polymerase activity could be detected.[50,58] Brayton et al. further showed that one of these complexes synthesize all seven mRNA species, while the other complex synthesizes only genome-sized RNA,[58] suggesting that the former is a transcription complex, while the latter is a replication complex.

Additional details of the mechanism of RNA replication have been suggested from studies of the leader-containing RNA intermediates in MHV-infected cells and RNA recombination. Baric et al.[87,92a] have shown that discrete species of leader-containing RNA intermediates of more than 150 nucleotides long, which are dissociated from the RNA template, could be detected in MHV-infected cells. These RNA species could represent either abortive transcriptional products or true RNA intermediates. If the latter is the case, it could be expected that, in a mixed infection, these RNA intermediates could rebind to a different template, resulting in RNA recombination via a copy-choice mechanism. Indeed, it has recently been shown that coronaviruses could undergo RNA recombination at a very high frequency.[95] These results suggest that RNA replication of coronaviruses proceed by a discontinuous and nonprocessive manner, i.e., replication "pauses" at many different sites along the RNA template. As a result of these pauses during replication, some of the RNA intermediates may fall off the template (maybe because of the possible nonprocessive nature of the coronavirus RNA polymerase), thus creating a pool of the segmented RNA intermediates which may rejoin the RNA template to continue replication (Figure 7). The pausing sites for RNA replication are probably around the hairpin loops in the RNA products. Transcriptional pausing has been noted in many prokaryotic systems, such as Qβ, MS2, and T7 phages.[93,94] It is probable that coronaviruses possess an additional unusual property, i.e., a nonprocessive RNA polymerase which can cause the dissociation of the pausing intermediates from the

RNA template. The rebinding of these intermediates during RNA replication would result in high frequency of RNA recombination during a mixed infection. This replication mechanism is further supported by the recent finding that many coronavirus recombinants were derived from multiple crossovers during a single viral replication cycle.[95a,104] This result suggests that frequent dissociation and reassociation of RNA intermediates occur during RNA replication. Thus, coronavirus behaves as if it contained segmented RNAs although its genome is clearly nonsegmented.

This mechanism of RNA replication could also lead to the generation of defective-interfering (DI) RNA if rebinding of the RNA intermediates does not occur at the exact sites. This phenomenon has been noted in the MHV system (see Section VII). Another possible consequence of this mechanism of replication is due to the presence of multiple repeats (UCUAA) at the 5' end of the genomic RNA (see Section III.D). If the RNA intermediates derived from the 5' end rebind to the genome RNA template at the wrong repeat, heterogeneity of the genomic RNA containing different numbers of repeats could result. This phenomenon has been noted in a small plaque mutant of JHM strain.[96] The heterogeneity of the 5' end of RNA genome appears to occur very frequently in the JHM strain, as suggested from the oligonucleotide fingerprinting analysis (unpublished observations). Some of the heterogeneity might have resulted from the insertion of deletion or the UCUAA repeat.

The genome-sized RNA synthesized could be bound to the nucleocapsid protein immediately after or during RNA replication. It has been suggested that genomic RNA synthesis and encapsidation are coupled in MHV-infected cells.[60a] The binding of N protein could be one of the mechanisms which allow the RNA intermediates to be dissociated from the RNA template, resulting in discontinuous, nonprocessive RNA replication. The recognition signals for the N protein binding have not been identified. Very likely, these signals are present at the 5' end of the genomic RNA, but not within the leader region since the subgenomic mRNAs are not packaged into the virion.

IV. RNA-DEPENDENT RNA POLYMERASES AND REGULATORY PROTEINS

A. RNA-Dependent RNA Polymerases

Similar to most of the positive-stranded RNA viruses, coronavirus virions do not contain detectable RNA polymerase activity.[49,51] Thus, the RNA polymerase(s) required for the transcription and replication of coronavirus RNAs have to be synthesized *de novo* from the incoming genomic RNA and the subsequently amplified viral RNAs. The synthesis of an RNA-dependent RNA polymerase is, therefore, the first biosynthetic event in the cells after viral infection. Indeed, protein inhibitors, such as cycloheximide, have been shown to inhibit viral RNA transcription when applied to the cells immediately after virus adsorption.[49-52,60a] Furthermore, the requirement of RNA transcription for continuous protein synthesis is evident throughout the replication cycle of the virus. Thus, either the RNA polymerase itself or cofactors have to be synthesized *de novo* continuously. This dependency on protein synthesis is different from rhabdovirus, in which mRNA transcription does not require continuous protein synthesis.[97,98]

The first RNA polymerase activity has been detected 2 to 3 hr postinfection in MHV-infected cells and disappeared by 4 to 5 hr postinfection.[49] This activity appears to be responsible for the synthesis of negative sense RNA.[58] This enzyme is most likely translated from the incoming RNA genome. The second or "late" RNA polymerase activity was detected about 5 to 6 hr postinfection.[49-51] It was the only RNA polymerase activity detected in TGEV.[50] This activity is different from the "early" RNA polymerase in its pH optimum and cationic requirements.[49] It is probably responsible for the transcription of mRNAs and replication of viral RNA genome.[58] These two activities could be further separated into two different membrane-associated complexes probably responsible for transcription and replication, respectively;[50,58] however, their enzymatic properties could not be distinguished.[58] These "late" RNA polymerases have been shown to synthesize all of the mRNAs,[49-51,58] including genomic RNA. Since Sawicki and Sawicki have shown that negative strand RNA synthesis continues late in the infection,[52] the "late" RNA polymerase activities could include several different polymerases responsible for the synthesis of different RNA. It has also been noted that the negative and positive strand RNA syntheses have different sensitivity to inhibition by cycloheximide,[52] further suggesting that they are carried out by different enzymes.

The proteins responsible for these polymerase activities have not been identified, probably because they are present in very small quantity in virus-infected cells. However, in vitro translation in reticulocyte lysate of the MHV virion RNA yielded proteins of up to 250 kdaltons,[34,35] which theoretically should be translated from the 5'-most gene of the RNA genome. Since it is generally assumed that, by analogy to other RNA viruses, the first gene at the 5' end of the RNA genome should code for an RNA polymerase, this 250-kdalton protein might represent the RNA polymerase. No polymerase activity has been associated with this protein. It should be noted that, Leibowitz et al.[34] found three proteins of nearly similar size, which have identical peptide maps, from the in vitro translation products of the MHV RNA genome. Denison and Perlman[35] also noted several proteins, including a 28-kdalton protein, from the same translation products. These proteins were attributed by these authors to degradation of the primary translation products. However, it cannot be ruled out that these proteins might represent primary translation products derived from the usage of different initiation codons or different termination sites in the same template RNA. They might represent different functional polymerases which carry out synthesis of different RNA species such as negative strand RNA, leader RNA, and mRNAs, respectively. The synthesis of different polymerases should presumably be under regulation of viral or cellular proteins. Recent sequence studies indicate that the 28-kdalton protein is indeed the N-terminal protein of the putative RNA polymerase.[98a] This protein has been detected in MHV-infected cells late in the infection.[98b] Thus, it represents the first coronavirus polymerase component identified in vivo.

Since coronavirus RNA polymerase is thought to be at least 700 kdalton in size, based on the translational capacity of the ORFs in the gene,[39a] the 200-kdalton protein made by in vitro translation probably represents only part of the RNA polymerases of coronaviruses. The nature of the rest of the gene products is currently unknown.

Consideration of the RNA replication scheme of coronaviruses suggests that at least three or four different RNA polymerases, or three or four different modified forms of the same polymerase, are required for the complete replication cycle of coronaviruses. These polymerases would carry out the transcription of negative strand RNA, leader RNA and mRNAs, and replication of genomic RNA. The isolation of temperature-sensitive mutants (see Section V.) which are blocked, at the nonpermissive temperature, at the synthesis of leader RNA[87] and mRNAs[99-101] separately, suggests that at least two different viral proteins carry out these different functions. No temperature-sensitive mutants have yet been obtained which are defective in negative strand RNA synthesis or in switching from RNA transcription to replication. Nevertheless, the enzymatic requirement of negative and positive strand RNA synthesis is different,[49] and RNA transcription and replication appear to be carried out in different subcellular compartments.[58] Furthermore, at least in BCV, there is a temporal shift from transcription to replication. Thus, it is probable that different proteins are required for transcription and for replication.

Theoretical considerations place a constraint on the number of polymerase proteins available for carrying out different functions of RNA synthesis early in the infection. Although coronavirus genome contains several genes encoding nonstructural proteins which may fulfill the roles of regulating different RNA synthesis, these nonstructural proteins are not likely translated from the incoming genomic RNA. Instead, these proteins will have to be translated from the subgenomic mRNAs. Thus, these proteins would not be made until all the mRNAs are synthesized. They would not be able to participate in the early regulatory events of RNA synthesis. Therefore, it is more likely that the primary gene products of the 5' end gene (gene A) are multiple, probably as a result of initiation at different AUG sites or termination at different termination codons. The choice of different initiation sites or termination sites could be controlled by host factors. These different translation products might be the different polymerases used for the synthesis of negative strand RNA, leader RNA, and mRNAs, respectively. Alternative possibilities are that the primary polymerase interacts with the host cellular factors to change the specificity of the polymerase, or that small amounts of viral nonstructural proteins are translated from the incoming genomic RNA after it is degraded. Such degraded RNA will allow the internal genes to be translated in eukaryotic cells.

B. RNA Regulatory Proteins

The different RNA polymerase activities described could be due to the presence of different regulatory proteins. None of these putative proteins have been identified. It is conceivable that most of the viral nonstructural proteins are involved in RNA transcription or replication, since six of the complementation groups of temperature-sensitive mutants (see Section V.) are defective in RNA synthesis.[99-101]

The nucleocapsid protein (N) may participate in RNA replication in an uncertain role. In an in vitro transcription system, monoclonal antibody against N inhibited RNA transcription.[101a] Also, N has been shown to bind to virion RNA in an in vitro binding assay.[20] In vesicular stomatitis virus, the N protein has been suggested to be responsible for the switching of RNA transcription to RNA replication.[102] The N protein of coronavirus may play a similar role.

An intriguing finding is that when cycloheximide was added to the infected cells 3 hr postinfection, more viral RNA was synthesized than when it was added later,[49] suggesting that there might be a viral protein which is responsible for shutting off negative strand RNA synthesis.

V. TEMPERATURE-SENSITIVE MUTANTS

Several groups have reported the isolation of temperature-sensitive (*ts*) mutants of MHV.[99-101] Leibowitz et al. have classified these *ts* mutants into seven complementation groups, six of which have RNA(−) phenotypes.[100] Although the number of complementation groups agrees with the number of genes in MHV genome, it seems unlikely that these complementation groups correspond exactly to the seven genes of the virus since three of the viral genes code for structural proteins. Nevertheless, the number of virus-specified proteins involved in RNA replication is high. At the present time, the genetic lesions of these mutants have not been localized.

Another group of *ts* mutants of A59 strain of MHV has been analyzed in greater details (Egbert, J., unpublished observation). These mutants include some which make negative strand, but no positive strand, RNA, a mutant (*LA10*) which makes negative strand and leader RNA, but no positive strand mRNAs,[87] and also mutants which make RNA in normal amounts, but do not produce virus particles. These mutants support the concept that several viral proteins are involved in various steps of RNA synthesis. The genetic lesions of some of these temperature-sensitive mutants have been mapped by use of RNA recombination (Section VI.). For instance, the *ts* lesion of *LA9* of MHV has been mapped within the 3 kilobases from the 5' end of gene A which encodes RNA polymerase.[91]

No *ts* mutants for other coronaviruses have been reported.

VI. RNA RECOMBINATION

Although coronaviruses possess a single piece of RNA genome, it has recently been demonstrated in MHV that coronaviruses could undergo RNA recombination at a very high frequency.[95] The frequency of recombination almost matches the frequency of RNA reassortment in segmented RNA viruses.[103] The evidence presented in Section III.F. suggests that RNA replication of coronaviruses proceeds in a discontinuous, nonprocessive manner, leading to the possibility that coronaviral RNA recombination could be mediated by a copy-choice mechanism involving free RNA intermediates.

Recombinants were obtained by coinfection of two *ts* mutants from two different viruses,[91] or, more remarkably, by randomly examining the progeny viruses derived from coinfection of a *ts* mutant and a wild-type virus.[95] The frequency of recombination was apparently high enough to allow detection of recombinants in such a random population.[95] The recombinants obtained so far contain not only single crossovers, but also double crossovers,[95a,104] suggesting that recombination occurs very readily. In those recombinants which contain single crossovers, the 5' end and 3' end of the genomes were derived from different parental viruses. Therefore, the leader RNA and body sequences of subgenomic mRNAs were derived from different parents. Thus, every mRNA species is a hybrid RNA consisting of the leader RNA of one parent and the body sequences of the second parent.[91] This shows that different leader RNA species can be used interchangeably.

RNA recombination will be discussed more extensively elsewhere in the volume.

VII. DEFECTIVE-INTERFERING (DI) RNA

DI particle has been described for MHV.[105] This class of DI particle has several unusual properties which set it apart from DI particles of other viruses and which suggest additional unusual characteristics of coronavirus RNA synthesis. DI RNA of MHV has a size which is about 95% of the standard virus RNA.[105] It does not contain merely a single deletion, but rather contains multiple base changes or small deletions along the entire genomic RNA.[106] It can interfere with the replication of the standard virus only to a small extent.[105] When the intracellular RNA of the cells infected with DI-containing virus preparation was examined, it was found that all of the standard mRNA species, except the smallest mRNA, which encodes the nucleocapsid N protein, were inhibited. Most surprisingly, several novel poly(A)-containing RNA species were synthesized in the infected cells.[106] These RNA species represent sequences derived from several different regions of the genome. Upon further passage of the DI-containing virus populations, the size of the DI genome remained roughly the same; however, the novel intracellular RNA species synthesized by the DI varied with passage. These DI-specific RNAs represent different parts of the genome being fused together. They are not packaged in the virion. However, recent studies suggest that a small amount of these DI RNAs could be encapsidated nonspecifically in virion particles. These packaged DI RNAs may serve as the template for the synthesis of intracellular DI RNA (Makino, S., unpublished). Since these DI RNAs are predominant RNA species in DI-infected cells, the DI RNA template must have an enormous advantage for RNA replication, when compared with the wild-type RNA genome. The elucidation of the mechanism of generation and replication of these DI RNAs will give significant insights into the mechanism of coronavirus RNA replication.

VIII. PERSPECTIVES

The RNA replication of coronaviruses clearly utilizes very unique mechanisms. These mechanisms include leader-primed RNA transcription and discontinuous, nonprocessive

RNA replication. These unique features of coronavirus replication are very likely to contribute to the complexity of the coronavirus biology: (1) high frequency of RNA recombination would result in divergence of virus strains, (2) discontinuous RNA synthesis could lead to deletion or duplication of RNA genome, (3) the presence of redundant sequences in the leader RNA region creates a possibility that the leader regions of mRNAs would be heterogeneous by using different repeats for initiation. Recent data from my laboratory have already fulfilled some of these predictions. Currently, there are still many unanswered questions in the coronavirus RNA replication scheme. What is the mechanism and kinetics of negative strand RNA synthesis? Is there a shut off of negative strand RNA synthesis? Is there a switch of RNA synthesis from mRNA to genomic RNA? Does the synthesis of the full-length genomic RNA require a free leader RNA? What is the mechanism of leader RNA priming? Most importantly, what is the nature of RNA polymerases? How many polymerases are involved in different phases of RNA replication? What proteins are involved in the regulation of RNA replication?

Clearly, the understanding of the nature of RNA polymerases would be aided by the cloning, sequencing, and expression of the gene encoding these enzymes. The understanding of the detailed mechanism of RNA synthesis has to come from studies using in vitro transcription and replication systems. The understanding of the nature of regulatory proteins would perhaps require analysis of more temperature-sensitive mutants. There is no question that coronavirus RNA replication involves many unique mechanisms, which add to the repertoire of viral RNA synthesis in nature. Future studies should be able to unravel many of the exciting features in this system.

ACKNOWLEDGMENT

I thank my colleagues at the University of Southern California, particularly Drs. Stephen A. Stohlman, Lisa Soe, Susan Baker, and James Keck for helpful comments on the manuscript. I also thank Carol Flores for typing the manuscript. The research contained in this paper was supported by Public Health Service Grants AI 19244 and NS 18146, National Multiple Sclerosis Society Grant RG 1449, and National Science Foundation Grant PCM-4507.

REFERENCES

1. **Tyrrell, D. A., Almedia, J. D., Cunningham, C. H., Dowdle, W. R., Hofsted, M. S., McIntosh, K., Tajima, M., Zakstelskaya, L. Y. A., Easterday, B. C., Kapikian, A., and Bringham, R. W.,** Coronaviridae, *Intervirology,* 5, 76, 1975.
2. **Larson, H. E., Reed, S. E., and Tyrrell, D. A. J.,** Isolation of rhinoviruses and coronaviruses from 38 colds in adults, *J. Med. Virol.,* 5, 221, 1980.
3. **Resta, S., Luby, J. P., Rosenfeld, C. R., and Siegel, J. D.,** Isolation and propagation of a human enteric coronavirus, *Science,* 229, 978, 1985.
4. **Burks, J. S., Devald, B. D., Jankovsky, L. C., and Gerdes, C.,** Two coronaviruses isolated from central nervous system tissue of two multiple sclerosis patients, *Science,* 209, 933, 1980.
5. **Sturman, L. S. and Holmes, K. V.,** The Molecular biology of coronaviruses, *Adv. Virus Res.,* 28, 35, 1983.
6. **Wege, H., Siddell, S., and ter Meulen, V.,** The biology and pathogenesis of coronaviruses, *Curr. Top. Microbiol. Immunol.,* 99, 165, 1982.
7. **Siddell, S., Wege, H., and ter Meulen, V.,** The structure and replication of coronaviruses, *Curr. Top. Microbiol. Immunol.,* 99, 131, 1982.
8. **Sturman, L. S., Holmes, K. V., and Behnke, J.,** Isolation of coronavirus envelope glycoproteins and interaction with the viral nucleocapsid, *J. Virol.,* 33, 449, 1980.

9. **Sturman, L. S. and Holmes, K. V.,** Characterization of a coronavirus. II. Glycoproteins of the viral envelope: tryptic peptide analysis, *Virology*, 77, 650, 1977.

10. **Sturman, L. S., Ricard, C. S., and Holmes, K. V.,** Proteolytic cleavage of the E2 glycoprotein of murine coronaviruses: activation of cell-fusion activity of virions by trypsin and separation of two different 90 K cleavage fragments, *J. Virol.*, 56, 905, 1985.

11. **Frana, M. F., Behnke, J. N., Sturman, L. S., and Holmes, K. V.,** Proteolytic cleavage of the E2 glycoprotein of murine coronaviruses: host-dependent differences in proteolytic cleavage and cell fusion, *J. Virol.*, 56, 912, 1985.

12. **Stern, D. F. and Sefton, B. M.,** Coronavirus proteins: biogenesis of avian infectious bronchitis virus virion proteins, *J. Virol.*, 44, 794, 1982.

13. **Stern, D. F. and Sefton, B. M.,** Coronavirus proteins: structure and function of the oligosaccharides of the avian infectious bronchitis virus, *J. Virol.*, 42, 208, 1982.

14. **Armstrong, J., Niemann, H., Smeekens, S., Rottier, P., and Warren, G.,** Sequence and topology of a model intracellular membrane protein, E-1 glycoprotein, from a coronavirus, *Nature (London)*, 308, 751, 1984.

15. **Boursnell, M. E. G., Brown, T. D. K., and Binns, M. M.,** Sequence of the membrane protein gene from avian coronavirus IBV, *Virus Res.*, 1, 303, 1984.

16. **Rottier, P., Brandenburg, D., Armstrong, J., van der Zeijst, B., and Warren, G.,** Assembly in vitro of a spanning membrane protein of the endoplasmic reticulum: the E1 glycoprotein of coronarvirus mouse hepatitis virus A59, *Proc. Natl. Acad. Sci. U.S.A.*, 81, 1421, 1984.

16a. **Rottier, P., Welling, G. W., Welling-Wester, S., Niesters, H., Lenstra, J. A., and van der Zeijist, B. A. M.,** Predicted membrane topology of the coronavirus protein E1, *Biochemistry*, 25, 1335, 1986.

17. **Niemann, H. and Klenk, H. D.,** Coronavirus glycoprotein E1, a new type of viral glycoprotein, *J. Mol. Biol.*, 153, 993, 1981.

18. **Holmes, K. V., Doller, E. W., and Sturman, L. S.,** Tunicamycin resistant glycosylation of a coronavirus glycoprotein: demonstration of a novel type of viral glycoprotein, *Virology*, 115, 334, 1981.

19. **Stohlman, S. A. and Lai, M. M. C.,** Phosphoproteins of murine hepatitis viruses, *J. Virol.*, 32, 672, 1979.

19a. **Sturman, L. S.,** Characterization of coronavirus. I. Structural proteins: effects of preparation conditions on the migration of protein in polyacrylamide gels, *Virology*, 77, 637, 1977.

20. **Robbins, S. G., Frana, M. F., McGowan, J. J., Boyle, J. F., and Holmes, K. V.,** RNA-binding proteins of coronavirus MHV: detection of monomeric and multimeric N proteins with an RNA overlay-protein blot assay, *Virology*, 150, 402, 1986.

21. **Siddell, S. G., Barthel, A., and ter Meulen, V.,** Coronavirus JHM. A virion-associated protein kinase, *J. Gen. Virol.*, 52, 235, 1981.

22. **King, B. and Brian, D. A.,** Bovine coronavirus structural proteins, *J. Virol.*, 52, 700, 1982.

23. **King, B., Potts, B. J., and Brian, D. A.,** Bovine coronavirus hemagglutinin protein, *Virus Res.*, 2, 53, 1985.

24. **Hogue, B. G., King, B., and Brian, D. A.,** Antigenic relationships among proteins of bovine coronavirus, human respiratory coronavirus OC43 and mouse hepatitis coronavirus A59, *J. Virol.*, 51, 384, 1984.

25. **Macnaughton, M. R.,** The genomes of three coronaviruses, *FEBS Lett.*, 94, 191, 1978.

26. **Stern, D. F. and Kennedy, S. I. T.,** Coronavirus multiplication Strategy. I. Identification and characterization of virus-specified RNA, *J. Virol.*, 34, 665, 1980.

27. **Lai, M. M. C. and Stohlman, S. A.,** The RNA of mouse hepatitis virus, *J. Virol.*, 26, 236, 1978.

28. **Wege, H., Muller, A., and ter Meulen, V.,** Genomic RNA of the murine coronavirus JHM, *J. Gen. Virol.*, 41, 217, 1978.

29. **Brian, D. A., Dennis, D. E., and Guy, J. S.,** Genome of porcine transmissible gastroenteritis virus, *J. Virol.*, 34, 410, 1980.

30. **Lomniczi, B.,** Biological properties of avian coronavirus RNA, *J. Gen. Virol.*, 36, 531, 1977.

31. **Lomniczi, B. and Kennedy, I.,** Genome of infectious bronchitis virus, *J. Virol.*, 24, 99, 1977.

32. **Yogo, Y., Hirano, N., Hino, S., Shibuta, H., and Matumoto, M.,** Polyadenylate in the virion RNA of mouse hepatitis virus, *J. Biochem. (Tokyo)*, 82, 1103, 1977.

33. **Lai, M. M. C. and Stohlman, S. A.,** Comparative analysis of RNA genomes of mouse hepatitis viruses, *J. Virol.*, 38, 661, 1981.

34. **Leibowitz, J. L., Weiss, S. R., Paavola, E., and Bond, C. W.,** Cell-free translation of murine coronavirus RNA, *J. Virol.*, 43, 905, 1982.

35. **Denison, M. R. and Perlman, S.,** Translation and processing of mouse hepatitis virus virion RNA in a cell-free system, *J. Virol.*, 60, 12, 1986.

36. **Stern, D. F. and Kennedy, S. I. T.,** Coronavirus multiplication strategy. II. Mapping the avian infectious bronchitis virus intracellular RNA species to the genome, *J. Virol.*, 36, 440, 1980.

37. **Lai, M. M. C., Brayton, P. R., Armen, R. C., Patton, C. D., Pugh, C., and Stohlman, S. A.,** Mouse hepatitis virus A59: messenger RNA structure and genetic localization of the sequence divergence from the hepatotropic strain MHV 3, *J. Virol.,* 39, 823, 1981.

38. **Oshiro, L. S.,** Coronaviruses, in *Ultrastructure of Animal Viruses and Bacteriophages: An Atlas,* Dalton, A. J. and Haguenau, F., Eds., Academic Press, New York, 1973, 331.

39. **Oshiro, L. S., Schieble, J. H., and Lennette, E. H.,** Electron microscopic studies of a coronavirus, *J. Gen. Virol.,* 12, 161, 1971.

39a. **Boursnell, M. E. G., Brown, T. D. K., Foulds, I. J., Green, P. F., Tomley, F. M., and Binns, M. M.,** Completion of the sequence of the genome of the coronavirus avian infectious bronchitis virus, *J. Gen. Virol.,* 68, 57, 1987.

40. **Pike, B. V. and Gawes, D. J.,** Lipids of transmissible gastroenteritis virus and their relation to those of two different host cells, *J. Gen. Virol.,* 34, 531, 1977.

41. **Lapps, W. and Brian, D. A.,** Oligonucleotide fingerprints of antigenically related bovine coronavirus and human coronavirus OC 43, *Arch. Virol.,* 86, 101, 1985.

41a. **Boyle, J. F., Weismiller, D. G., and Holmes, K. V.,** Genetic resistance to mouse hepatitis virus correlates with absence of virus-binding activity on target tissues, *J. Virol.,* 61, 185, 1987.

42. **Patterson, S. and Bingham, R. W.,** Electron microscope observations on the entry of avian infectious bronchitis virus into susceptible cells, *Arch. Virol.,* 53, 267, 1976.

43. **David-Ferreira, J. F. and Manaker, R. A.,** An electron microscope study of the development of a mouse hepatitis virus in tissue culture cells, *J. Cell Biol.,* 24, 57, 1965.

44. **Doughri, A. M., Storz, J., Hajer, I., and Fernando, H. S.,** Morphology and morphogenesis of a coronavirus infecting intestinal epithelial cells of newborn calves, *Exp. Mol. Pathol.,* 25, 355, 1976.

45. **Krystyniak, K. and Dupuy, J. M.,** Early interaction between mouse hepatitis virus 3 and cells, *J. Gen. Virol.,* 57, 53, 1981.

46. **Brayton, P. R., Ganges, R. G., and Stohlman, S. A.,** Host cell nuclear function and murine hepatitis virus replication, *J. Gen. Virol.,* 56, 457, 1981.

47. **Wilhemsen, K. C., Leibowitz, J. L., Bond, C. W., and Robb, J. A.,** The replication of murine coronaviruses in enucleated cells, *Virology,* 110, 225, 1981.

48. **Evans, M. R. and Simpson, R. W.,** The coronavirus avian infectious bronchitis virus requires the cell nucleus and host transcriptional factors, *Virology,* 105, 582, 1980.

49. **Brayton, P. R., Lai, M. M. C., Patton, C. D., and Stohlman, S. A.,** Characterization of two RNA polymerase activities induced by mouse hepatitis virus, *J. Virol.,* 42, 847, 1982.

50. **Dennis, D. E. and Brian, D. A.,** RNA-dependent RNA polymerase activity in coronavirus-infected cells, *J. Virol.,* 42, 153, 1982.

51. **Mahy, B. W. J., Siddell, S., Wege, H., and ter Meulen, V.,** RNA-dependent RNA polymerase activity in murine coronavirus-infected cells, *J. Gen. Virol.,* 64, 103, 1983.

52. **Sawicki, S. G. and Sawicki, D. L.,** Coronavirus minus-strand RNA synthesis and effect of cycloheximide on coronavirus RNA synthesis, *J. Virol.,* 57, 328, 1986.

53. **Lai, M. M. C., Patton, C. D., and Stohlman, S. A.,** Replications of mouse hepatitis virus: negative-stranded RNA and replicative form RNA are of genome length, *J. Virol.,* 44, 487, 1982.

54. **Leibowitz, J. L., Wilhemsen, K. C., and Bond, C. W.,** The virus-specific intracellular RNA species of two murine coronaviruses: MHV-A59 and MHV-JHM, *Virology,* 114, 29, 1981.

55. **Massalski, A., Coulter-Mackie, M., and Dales, S.,** Assembly of mouse hepatitis virus strain JHM, in *Biochemistry and Biology of Coronaviruses,* ter Meulen, V., Siddell, S., and Wege, H., Eds., Plenum Press, New York, 1981, 111.

56. **Dubois-Dalcq, M. E., Doller, E. W., Haspel, M. V., and Holmes, K. V.,** Cell tropism and expression of mouse hepatitis viruses (MHV) in mouse spinal cord cultures, *Virology,* 119, 317, 1982.

57. **Spaan, W. J. M., Rottier, P. J. M., Horzinek, M. C., and van der Zeijst, B. A. M.,** Isolation and identification of virus-specific mRNAs in cells infected with mouse hepatitis virus (MHV-A59), *Virology,* 108, 424, 1981.

58. **Brayton, P. R., Stohlman, S. A., and Lai, M. M. C.,** Further characterization of mouse hepatitis virus RNA-dependent RNA polymerases, *Virology,* 133, 197, 1984.

59. **Jacobs, L., Spaan, W. J. M., Horzinek, M. C., and van der Zeijst, B. A. M.,** The synthesis of the subgenomic mRNAs of mouse hepatitis virus is initiated independently: evidence from UV transcription mapping, *J. Virol.,* 39, 401, 1981.

60. **Stern, D. F. and Sefton, B. M.,** Synthesis of coronavirus mRNAs: kinetics of inactivation of IBV RNA synthesis by UV light, *J. Virol.,* 42, 755, 1982.

60a. **Perlman, S., Ries, D., Bolger, E., Chang, L-J., and Stoltzfus, C. M.,** *Virus Res.,* 6, 261, 1986.

61. **Kapke, P. A. and Brian, D. A.,** Sequence analysis of the porcine transmissible gastroenteritis coronavirus nucleocapsid protein gene, *Virology,* 151, 41, 1986.

62. **Armstrong, J., Smeekens, S., and Rottier, P.,** Sequence of the nucleocapsid gene from murine coronavirus MHV-A59, *Nucleic Acids Res.,* 11, 833, 1983.

63. **Boursnell, M. E. G., Binns, M. M., Foulds, I. J., and Brown, T. D. K.,** Sequence of the nucleocapsid genes from two strains of avian infectious bronchitis virus, *J. Gen. Virol.,* 66, 573, 1985.
64. **Skinner, M. A. and Siddell, S. G.,** Coronavirus JHM: nucleotide sequence of the mRNA that encodes nucleocapsid protein, *Nucleic Acids Res.,* 11, 5045, 1983.
65. **Wege, H., Siddell, S., Sturm, M., and ter Meulen, V.,** Coronavirus JHM: characterization of intracellular viral RNA, *J. Gen. Virol.,* 54, 213, 1981.
66. **Cheley, S., Anderson, R., Cupples, M. J., Lee Chan, E. C. M., and Morris, V. L.,** Intracellular murine hepatitis virus-specific RNAs contain common sequences, *Virology,* 112, 596, 1981.
67. **Jacobs, L., van der Zeijst, B. A. M., and Horzinek, M. C.,** Characterization and translation of transmissible gastroenteritis virus mRNAs, *J. Virol.,* 57, 1010, 1986.
68. **Robb, J. and Bond, C. W.,** Pathogenic murine coronaviruses. I. Characterization of biological behavior in vitro and virus-specific intracellular RNA of strongly neurotropic JHMV and weakly neurotropic A59 viruses, *Virology,* 94, 352, 1979.
69. **Lai, M. M. C., Patton, C. D., and Stohlman, S. A.,** Further characterization of mouse hepatitis virus: presence of common 5′-end nucleotides, *J. Virol.,* 41, 557, 1982.
70. **Spaan, W. J. M., Rottier, P. J. M., Horzinek, M. C., and van der Zeijst, B. A. M.,** Sequence relationships between the genome and the intracellular RNA species 1, 3, 6 and 7 of mouse hepatitis virus strain A59, *J. Virol.,* 42, 432, 1982.
71. **Rottier, P. J. M., Spaan, W. J. M., Horzinek, M., and van der Zeijst, B. A. M.,** Translation of three mouse hepatitis virus (MHV-A59) subgenomic RNAs in Xenopus laevis oocytes, *J. Virol.,* 38, 20, 1981.
72. **Siddell, S., Wege, H., Barthel, A., and ter Meulen, V.,** Intracellular protein synthesis and the in vitro translation of coronavirus JHM mRNA, in *Biochemistry and Biology of Coronaviruses,* ter Meulen, V., Siddell, S., and Wege, H., Eds., Plenum Press, New York, 1981, 193.
73. **Boursnell, M. E. G. and Brown, T. D. K.,** Sequencing of coronavirus IBV genomic RNA: a 195-base open reading frame encoded by mRNA B, *Gene,* 29, 87, 1984.
74. **Skinner, M., Ebner, D., and Siddell, S. G.,** Coronavirus MHV-JHM mRNA 5 has a sequence arrangement which potentially allows translation of a second, downstream open reading frame, *J. Gen. Virol.,* 66, 581, 1985.
75. **Skinner, M. A. and Siddell, S. G.,** Coding sequences of coronavirus MHV-JHM mRNA 4, *J. Gen. Virol.,* 66, 593, 1985.
76. **Boursnell, M. E. G., Binns, M. M., and Brown, T. D. K.,** Sequencing of coronavirus IBV genomic RNA: three open reading frames in the 5′-"unique" region of mRNA D, *J. Gen. Virol.,* 66, 2253, 1985.
77. **Binns, M. M., Boursnell, M. E. G., Canavagh, D., Pappins, D. J. C., and Brown, T. D. K.,** Cloning and sequencing of the gene encoding the spike protein of the coronavirus IBV, *J. Gen. Virol.,* 66, 719, 1985.
77a. **Schmidt, I., Skinner, M., and Siddell, S.,** Nucleotide sequence of the gene encoding the surface projection glycoprotein of coronavirus MHV-JHM, *J. Gen. Virol.,* 68, 47, 1987.
78. **Lamb, R. A. and Lai, C. J.,** Sequence of interrupted and uninterrupted mRNAs and cloned DNA coding for the two overlapping nonstructural proteins of influenza virus, *Cell,* 21, 475, 1980.
79. **Lamb, R. A., Lai, C. J., and Choppin, P. W.,** Sequences of mRNAs derived from genomic RNA segment of influenza virus colinear and interrupted mRNAs code for overlapping proteins, *Proc. Natl. Acad. Sci. U.S.A.,* 78, 4170, 1981.
80. **Jacks, T. and Varmus, H. E.,** Expression of the Rous sarcoma virus *pol* gene by ribosomal frameshifting, *Science,* 230, 1237, 1985.
80a. **Budzilowicz, C. J. and Weiss, S. R.,** *In vitro* synthesis of two polypeptides from a nonstructural gene of coronavirus mouse hepatitis virus strain A59, *Virology,* 157, 509, 1987.
81. **Lai, M. M. C., Patton, C. D., Baric, R. S., and Stohlman, S. A.,** Presence of leader sequences in the mRNA of mouse hepatitis virus, *J. Virol.,* 46, 1027, 1983.
82. **Spaan, W., Delius, H., Skinner, M., Armstrong, J., Rottier, P., Smeekens, S., van der Zeijst, B. A. M., and Siddell, S. G.,** Coronavirus mRNA synthesis involves fusion of noncontiguous sequences, *EMBO J.,* 2, 1939, 1983.
83. **Lai, M. M. C., Baric, R. S., Brayton, P. R., and Stohlman, S. A.,** Characterization of leader RNA sequences on the virion and mRNAs of mouse hepatitis virus, a cytoplasmic virus, *Proc. Natl. Acad. Sci. U.S.A.,* 81, 3626, 1984.
84. **Brown, T. D. K., Boursnell, M. E. G., and Binns, M. M.,** A leader sequence is present on mRNA A of avian infectious bronchitis virus, *J. Gen. Virol.,* 65, 1437, 1984.
85. **Brown, T. D. K., Boursnell, M. E. G., Binns, M. M., and Tomley, F. M.,** Cloning and sequencing of 5′-terminal sequences from avian infectious bronchitis virus genomic RNA, *J. Gen. Virol.,* 67, 221, 1986.
86. **Baric, R. S., Stohlman, S. A., and Lai, M. M. C.,** Characterization of replicative intermediate RNA of mouse hepatitis virus: presence of leader RNA sequences on nascent chains, *J. Virol.,* 48, 633, 1983.

87. **Baric, R. S., Stohlman, S. A., Razavi, M. K., and Lai, M. M. C.,** Characterization of leader-related small RNAs in coronavirus-infected cells: further evidence for leader-primed mechanism of transcription, *Virus Res.,* 3, 19, 1985.

88. **Makino, S., Stohlman, S. A., and Lai, M. M. C.,** Leader sequences of murine coronavirus mRNAs can be freely reassorted: evidence for the role of free leader RNA in transcription, *Proc. Natl. Acad. Sci. U.S.A.,* 83, 4204, 1986.

88a. **Lai, M. M. C., Makino, S., Baric, R. S., Soe, L., Shieh, C.-K., Keck, J. G., and Stohlman, S. A.,** Leader RNA-primed transcription and RNA recombination of murine coronaviruses, in *Positive Strand RNA Viruses,* Brinton, M. A. and Rueckert, R. R., Eds., Alan R. Liss, New York, 1987, 285.

89. **Budzilowicz, C. J., Wilczynski, S. P., and Weiss, S. R.,** Three intergenic regions of coronavirus mouse hepatitis virus strain A59 genomic RNA contain a common nucleotide sequence that is homologous to the 3'-end of the viral mRNA leader sequence, *J. Virol.,* 53, 834, 1985.

90. **Brown, T. D. K. and Boursnell, M. E. G.,** Avian infectious bronchitis virus genomic RNA contains sequence homologies at the intergenic boundaries, *Virus Res.,* 1, 15, 1984.

91. **Lai, M. M. C., Baric, R. S., Makino, S., Keck, J. G., Egbert, J., Leibowitz, J. L., and Stohlman, S. A.,** Recombination between nonsegmented RNA genomes of murine coronaviruses, *J. Virol.,* 56, 449, 1985.

92. **Shieh, C.-K., Soe, L. H., Makino, S., Chang, M.-F., Stohlman, S. A., and Lai, M. M. C.,** The 5'-end sequence of the murine coronavirus genome: implications for multiple fusion sites in leader-primed transcription, *Virology,* 156, 321, 1987.

92a. **Baric, R. S., Shieh, C.-K., Stohlman, S. A., and Lai, M. M. C.,** Analysis of intracellular small RNAs of mouse hepatitis virus: evidence for discontinuous transcription, *Virology,* 156, 342, 1987.

93. **Mills, D. R., Dabkin, C., and Kramer, F. R.,** Template-determined, variable rate of RNA chain elongation, *Cell,* 15, 541, 1978.

94. **Kassavetis, G. A. and Chamberlin, M. J.,** Pausing and termination of transcription within the early region of bacteriophage T7 DNA in vitro, *J. Biol. Chem.,* 256, 2777, 1981.

95. **Makino, S., Keck, J. G., Stohlman, S. A., and Lai, M. M. C.,** High frequency RNA recombination of murine coronaviruses, *J. Virol.,* 57, 729, 1986.

95a. **Makino, S., Fleming, J. O., Keck, J. G., Stohlman, S. A., and Lai, M. M. C.,** RNA recombination of coronaviruses: localization of neutralizing epitopes and neuropathogenic determinants of the carboxyl terminus of peplomers, *Proc. Natl. Acad. Sci., U.S.A.,* 84, 6567, 1987.

96. **Makino, S., Taguchi, F., Hirano, N., and Fujiwara, K.,** Analysis of genomic and intracellular viral RNAs of small plaque mutants of mouse hepatitis virus, JHM strain, *Virology,* 139, 138, 1984.

97. **Perlman, S. M. and Huang, A. S.,** Synthesis of vesicular stomatitis virus. V. Interactions between transcription and replication, *J. Virol.,* 12, 1395, 1973.

98. **Wertz, G. W. and Levine, M.,** RNA synthesis by vesicular stomatitis virus and small plaque mutant: effects of cycloheximide, *J. Virol.,* 12, 253, 1973.

98a. **Soe, L. H., Shieh, C.-K., Baker, S. C., Chang, M.-F., and Lai, M. M. C.,** Sequence and translation of the murine coronavirus 5'-end genomic RNA reveals the N-terminal structure of the putative RNA polymerase, *J. Virol.,* 1987, in press.

98b. **Denison, M. R. and Perlman, S.,** Identification of putative polymerase gene product in cells infected with murine coronavirus A59, *Virology,* 157, 565, 1987.

99. **Robb, J. A., Bond, C. W., and Leibowitz, J. L.,** Pathogenic murine coronaviruses. III. Biological and biochemical characterization of temperature-sensitive mutants of JHMV, *Virology,* 91, 385, 1979.

100. **Leibowitz, J. L., DeVries, J. R., and Haspel, M. V.,** Genetic analysis of murine hepatitis virus strain JHM, *J. Virol.,* 42, 1080, 1982.

101. **Koolen, M. J. M., Osterhaus, D. M. E., van Steenis, G., Horzinek, M. C., and van der Zeijst, B. A. M.,** Temperature-sensitive mutants of mouse hepatitis virus strain A59: isolation, characterization and neuropathogenic properties, *Virology,* 125, 393, 1983.

101a. **Compton, S. R., Rogers, D. B., Holmes, K. V., Fertsch, D., Remenick, J., and McGowan, J. J.,** *J. Virol.,* 61, 1814, 1987.

102. **Blumberg, B. M., Leppert, M., and Kolakofsky, D.,** Interaction of VSV leader RNA and nucleocapsid protein may control VSV genome replication, *Cell,* 23, 837, 1981.

103. **Fields, B. N.,** Genetics of reovirus, *Curr. Top. Microbiol. Immunol.,* 91, 1, 1981.

104. **Keck, J. G., Stohlman, S. A., Soe, L., Makino, S., and Lai, M. M. C.,** Multiple recombination sites at the 5'-end of murine coronavirus RNA, *Virology,* 156, 331, 1987.

105. **Makino, S., Taguchi, F., and Fujiwara, K.,** Defective interfering particles of mouse hepatitis virus, *Virology,* 133, 9, 1984.

106. **Makino, S., Fujioka, N., and Fujiwara, K.,** Structure of the intracellular defective viral RNAs of defective interfering particles of mouse hepatitis virus, *J. Virol.,* 54, 329, 1985.

107. **Stern, D. F. and Sefton, B. M.,** Coronavirus multiplication: locations of genes for virion proteins on the avian infectious bronchitis virus virion proteins, *J. Virol.,* 50, 22, 1984.

RNA Replication of Negative Strand RNA Viruses

Chapter 7

REPLICATION OF NONSEGMENTED NEGATIVE STRAND RNA VIRUSES

Paul S. Masters and Amiya K. Banerjee

TABLE OF CONTENTS

I.   Introduction 138

II.   Morphology and Protein Composition 138

III. Genetic Organization 140
      A.   Genetic Studies 140
      B.   Physical Studies 141

IV.   Viral Transcription 143
      A.   Basic Phenomena of Transcription 143
            1.   The Transcription Reaction 143
            2.   Roles of Viral and Cellular Proteins in Transcription 144
            3.   Intergenic Junctions 146
      B.   Models of Transcription 148

V.   Viral Replication 150
      A.   Basic Phenomena of Replication 150
      B.   Models of Replication 151

VI.   Conclusion 152

References 153

I. INTRODUCTION

A nonsegmented negative strand RNA virus, by definition, is distinguished by having a unipartite RNA genome that is complementary to and, hence, the template for the (positive polarity) mRNAs it encodes. Viruses of this category fall almost without exception into one of two taxonomic families: the rhabdoviruses and the paramyxoviruses (Table 1). Besides being intrinsically interesting biological entities, the nonsegmented negative strand viruses are of practical concern due to the array of diseases they cause in humans as well as in domestic animals, fish, and plants. The common childhood illnesses mumps and measles are caused by members of two paramyxovirus genera. Two other paramyxoviruses, respiratory syncytial virus (RSV) and human parainfluenza virus (PIV), are serious pathogens of the lower respiratory tract and constitute the principal causes of infantile pneumonia and bronchiolitis. Rabies virus, a rhabdovirus, is infectious in and fatal for most mammalian species. Rapidly lethal forms of hemorrhagic fever have been found to be caused by two unclassified nonsegmented negative strand viruses, Marburg virus and Ebola virus.

It is perhaps, not coincidental that the most intensively studied nonsegmented negative strand viruses, to date, are those which are among the least pathogenic to humans; but other qualities have also influenced this choice. Vesicular stomatitis virus (VSV), a sublethal pathogen of horses and cattle, is the prototype of the rhabdoviruses. The popularity of VSV in experimental investigation is due, in no small measure, to its short generation time and ease of propogation in tissue culture, its stability upon purification, and its amenability to the in vitro study of viral transcription as well as, more recently, viral replication. In addition, well prior to more extensive cytopathic effects, VSV infection produces a shutoff of host cell RNA and protein synthesis, thereby greatly facilitating the analysis of viral macromolecular synthesis. Sendai virus, a murine parainfluenza virus that has become the prototype of the paramyxoviruses, has many of these same features and is therefore the most widely studied member of its family. Recent advances in molecular biology, however, have multiplied the extent of our understanding of other rhabdoviruses and paramyxoviruses. Although this review will focus on VSV and Sendai virus, available information about other viruses will be presented, especially where it further illuminates generalizations drawn from the two prototypes.

II. MORPHOLOGY AND PROTEIN COMPOSITION

The rhabdoviruses appear under negative contrast electron microscopy as highly regular, bullet-shaped particles roughly 75 nm in diameter and 180 nm long.[1,2] This shape is unique, and it allows immediate classification of an unknown virus as a rhabdovirus. The rhabdoviral particle is enveloped in a lipid bilayer acquired from the host cell cytoplasmic membrane, and from this surface protrude some 1200 glycoprotein (G) spikes.[3] The bulk of the G protein, including the amino terminus and its N-linked carbohydrate residues, projects from the exterior of the membrane envelope and is essential for infectivity.[4]

In sharp contrast to the relative uniformity of the rhabdoviruses, the paramyxoviruses are quite pleomorphic. Negative contrast electron microscopy reveals these viruses to be roughly spherical particles usually 100 to 250 nm in diameter, but often much larger.[5] Members of one genus, the morbilliviruses, occasionally have been observed to be filamentous. Like the rhabdoviruses, the paramyxoviruses are surrounded by a host cell-derived membrane, and their surfaces are studded with viral glycoprotein spikes. In this case, though, two types of glycoprotein are found, and both are essential for infectivity.[6] The first is a 65-kdalton fusion protein (F), which contains N-linked oligosaccharides. The second is an O-linked oligosaccharide-containing protein (circa 70 kdaltons), the nature of which varies among the genera within the paramyxovirus family. For members of the paramyxovirus genus, the second

Table 1
NONSEGMENTED NEGATIVE STRAND RNA VIRUSES

Family/genus	Representative viruses/principal hosts
Rhabdoviruses	
Vesiculoviruses	Vesicular stomatitis virus (VSV)
	Cattle, horses, humans, mosquitoes
	Chandipura virus
	Humans, sandflies
Lyssaviruses	Rabies virus
	Most mammals
Unclassified	Spring viremia of carp virus (SVCV)
	Carp
	Infectious hematopoietic necrosis virus (IHNV)
	Salmon, trout
Paramyxoviruses	
Paramyxoviruses	Human parainfluenzaviruses types 1-4
	(PIV 1-4)
	Humans
	Sendai virus
	Mice
	Newcastle disease virus (NDV)
	Poultry
	Mumps virus
	Humans
	Simian virus 5 (SV5)
	Dogs, monkeys
Morbilliviruses	Measles virus
	Humans
	Canine distemper virus (CDV)
	Dogs
Pneumoviruses	Respiratory syncytial virus (RSV)
	Humans
Unclassified	Marburg virus
	Humans
	Ebola virus
	Humans

glycoprotein (HN) has both hemagglutinin and neuraminidase activities. For the morbilli-viruses, the corresponding protein (H) has only the hemagglutinin function, whereas the second glycoprotein (G) of the pneumoviruses has neither of the two activities. Like the rhabdoviral G proteins, the paramyxoviral glycoproteins transverse the membrane, lying predominantly on the external face, although for the HN, H, and RSV G proteins the membrane-anchoring domain is found near the amino terminus rather than the carboxy terminus.[7-9]

Packed within the membrane envelope of each of the nonsegmented negative strand viruses is the viral nucleocapsid. This consists of a single strand genome RNA molecule, 11 to 15 kilobases in length, compactly wrapped in a nucleocapsid protein (circa 50 kdaltons) designated either N or NP. For VSV the stoichiometry of this binding has been determined to be 9 nucleotides of RNA per monomer of N protein.[3] The tightness of the N (NP)-RNA association is such that the nucleocapsids of VSV and Sendai virus are stable to banding in 2.5 *M* CsCl, a condition under which all cellular nucleoprotein complexes are dissociated.[10] Moreover, the encapsidation of the genome by the nucleocapsid protein renders the phosphodiester backbone inaccessible to the action of ribonucleases.

Associated with the nucleocapsid of both families of virus are multiple copies of two proteins which together make up the RNA polymerase that transcribes and replicates the viral genome. These are a large (>200 kdalton) protein, L, together with an acidic phosphoprotein (circa 30 kdaltons in the rhabdoviruses and RSV, 65 kdaltons in the paramyxoviruses) which is variously designated NS (in vesiculoviruses), M1 (in lyssaviruses), or P (in paramyxoviruses). In the VSV virion, the molar amounts of NS and L, relative to N protein, are 0.4 and 0.04, respectively.[3] Although not as rigorously quantitated, the amounts of the RNA polymerase proteins of the other nonsegmented negative strand viruses appear to fall in the same range.

The final major component found in the rhabdoviruses and paramyxoviruses is the matrix or membrane protein, M (denoted M2 for the lyssaviruses). This basic polypeptide (20 to 30 kdaltons) possibly bridges the nucleocapsid and the glycoprotein-containing membrane envelope and may be the most important determinant of virus structure. For VSV, the interaction of the M protein with the nucleocapsid and with the cell membrane, in the presence and absence of G protein, has been well documented.[11-14]

In general, no other proteins are found in the virions of the rhabdoviruses and paramyxoviruses, but a few exceptions are known. The pneumovirus RSV contains a second matrix protein (22K) as well as small amounts of a 7.5-kdalton hydrophobic polypeptide (1A) of unknown function. The virion nucleocapsid of human parainfluenza virus type 3 (PIV3) contains a 22-kdalton basic protein (C) encoded by an internal, overlapping reading frame in the mRNA for the P protein.[15,16] C proteins are also encoded by the P mRNAs of other paramyxoviruses (Sendai virus,[17] measles virus,[18] and canine distemper virus [CDV][19]), but all of these, although present in infected cells, appear not to become incorporated into their respective virions.

III. GENETIC ORGANIZATION

A virus capable of carrying out a productive infection has contained within its genome the information necessary to program those events which culminate in the replication and transmission of that genome. A notion of the makeup of the viral genome, then, is fundamental to an understanding of the process of infection.

A. Genetic Studies

A treatment of the subject of rhabdovirus and paramyxovirus genetics is beyond the scope of this review. The reader is referred to excellent reviews of this topic by Pringle[20,21] and Flamand,[22] in whose laboratories the majority of the work in this field has been done. Genetic studies have been carried out with a number of nonsegmented negative strand viruses: VSV and other vesiculoviruses,[20-22] rabies virus,[22] Newcastle disease virus (NDV),[23] Sendai virus,[24] and RSV.[25] By far, however, the greatest amount of attention has been focused on VSV of the Indiana serotype. Mutants have generally been isolated on the basis of temperature sensitivity, thermolability, or restriction of host range, but plaque morphology variants and heat-resistant viral strains have also been obtained. The major strength of VSV genetics has been the assignment of functions to viral polypeptides. This approach has been indispensible in sorting out the roles of multifunctional proteins and in gaining insight into virus-host interactions.

Despite these successes, the genetic study of rhabdoviruses and paramyxoviruses has been severely hampered by the constraints imposed by a single strand RNA genome. Since the genome is haploid and nonsegmented, it cannot undergo reassortment. Thus, mutational analysis must be largely restricted to varius categories of conditional lethal mutations, a specialized subset of all possible mutations. Moreover, unlike picornaviruses and coronaviruses (both nonsegmented positive strand RNA viruses), the nonsegmented negative strand RNA viruses do not recombine. This limitation has been noted with NDV,[23] VSV,[22,26] and

RSV.[25] Holland and co-workers[27] devised a highly sensitive test for intertypic recombination between the Indiana and New Jersey serotypes of VSV, but such cross-over events were never detected. The unfortunate outcome of the absence of recombination for these viruses has been that genetic experiments fail to yield information about the arrangement of genes on the viral chromosome.

B. Physical Studies

The first picture of the organization of the genomes of rhabdoviruses and paramyxoviruses came from ultraviolet (UV) mapping.[28] This technique makes use of the fact that UV irradiation induces the formation of pyrimidine dimers in nucleic acid templates, and RNA polymerases, apparently without exception, terminate polymerization upon encountering pyrimidine dimers.[29] Exploitation of this can allow determination of the number of transcription units in a viral genome and the ordering of all genes within a transcription unit. Ball and White[30] and Abraham and Banerjee[31] used this approach to derive the gene order of VSV, aided by the fortuity that the five genes of this virus turned out to compose a single transcription unit in the order 3'-N-NS-M-G-L-5'. UV mapping was subsequently employed to analyze Sendai virus,[32] rabies virus,[33] NDV,[34] and RSV,[35] and from these it seems reasonable to generalize that all of the rhabdoviral and paramyxoviral genomes are transcribed by their RNA polymerases as single transcription units.

In some cases it has been possible to establish or confirm gene orders by various other methods: hybrid duplex analysis by electron microscopy;[36,37] kinetics of chemical inhibition of in vitro transcription reactions;[38] and examination of the sequence content of polycistronic mRNAs, which are occasional, aberrant transcription products resulting from polymerase readthrough between adjacent genes.[39-42] However, by far the most comprehensive information about genetic organization has come from the sequencing of viral genomes by way of cDNA clones prepared from viral mRNAs or from randomly copied portions of the viral genomes.[43-51] Besides yielding a wealth of information about gene and protein sequences, this methodology has unequivocally established the order of viral genes and has revealed hitherto undetected genes as well. All of these results are summarized in Table 2.

The most striking aspect of the gene orders tabulated to date, which include at least one representative from each genus of the two viral families, is the degree of uniformity that is exhibited. The canonical pattern observed is 3'-(nucleocapsid protein)-(phosphoprotein)-(matrix protein)-(one or two glycoproteins)-(large polymerase protein)-5'. The only glaring exception to this rule is RSV, which seems to be nearly twice as complex as its relatives.[50] The RSV genome has two nonstructural genes encoding polypeptides of unknown function (1C and 1B; 16 and 15 kdaltons, respectively) situated prior to the N gene at the 3' extreme of the genome. In addition, it has a gene (1A) encoding a small (7.5 kdaltons) hydrophobic protein situated between the M and G genes, and also a second matrix protein gene (22K) positioned between the F and L genes. Moreover, the order of the two glycoprotein genes of RSV, G and F, is the reverse of that found in all other paramyxoviruses sequenced thus far. The only other presently known exception to the canonical gene order occurs in spring viremia of carp virus (SVCV), where a partial sequence 3'-N-M-G-5' has been obtained from overlapping genomic clones.[44] It will be interesting to see where the phosphoprotein (NS) gene is located in this virus.

Besides those already noted for RSV, "extra" genes have been found in a number of rhabdoviruses and paramyxoviruses. The simian virus 5 (SV5) genome contains, between F and HN, a gene encoding a small hydrophobic protein (SH; 5 kdaltons) which has been shown to be present in infected cells, but not in the virions.[48] A gene (NV) encoding a small (12 to 17 kdaltons) nonviral polypeptide has also been found between the G and L genes of the rhabdoviral fish pathogen infectious hematopoietic necrosis virus (IHNV).[37] Perhaps the most interesting of the "extra" genes, however, are the C genes, which are found in a

Table 2
GENE ORDERS OF RHABDOVIRUSES AND PARAMYXOVIRUSES

Virus	Gene order	Method of determination	Ref.
VSV	N NS M G L	UV mapping	30,31
		Sequence analysis	43
		Heteroduplex analysis	36
		Polycistronic mRNAs	39
Rabies virus	N M1 M2 G L	Sequence analysis	51
		UV mapping	33
SVCV	N M G....(partial)	Sequence analysis	44
IHNV	N M1 M2 G NV L	Heteroduplex analysis	37
Sendai virus	NP P/C M F HN L	Sequence analysis	45,46
		UV mapping	32
PIV3	NP P/C M F HN L	Sequence analysis	47
SV5	NP P/V M F SH HN L	Polycistronic mRNAs	40
		Sequence analysis	48
NDV	NP P M F HN L	Polycistronic mRNAs	41
		UV mapping	34
Measles virus	N P/C M F H L	Sequence analysis	18,49,165,166
CDV	N P/C M F H L	Sequence analysis	19,165
RSV	1C 1B N P M 1A G F 22K L	Sequence analysis	50
		UV mapping	35
		Polycistronic mRNAs	42

number of paramyxoviruses and encode relatively small (22 to 23 kdaltons), predominantly basic proteins.[16-19] The inability of investigators to find mRNAs of the appropriate size to code for the various C proteins was explained by the discovery by Giorgi et al.[17] that the Sendai virus C protein is expressed from an overlapping (+ 1) reading frame contained within the mRNA encoding the P protein. Subsequently, this was also found to be the case for measles virus, CDV, and PIV3. These genes, then, appear to be the counterparts of the overlapping reading frames which have been found in the smaller genome segments of bunyaviruses, reoviruses, and influenza viruses. It is curious, as was first noted by Hudson et al.,[52] that each of the phosphoprotein genes (NS or M1) of four different strains of VSV, as well as Chandipura virus[53] and rabies virus[51] (all rhabdoviruses), also internally encode a small, basic polypeptide reminiscent of the C proteins in the + 1 reading frame. However, for most of the examples cited, these are initiated by AUG codons within weak ribosomal recognition sequences,[54] and there is as yet no evidence that any of them are expressed. A further variant of internal coding within the P gene occurs with SV5. In this case, the smaller polypeptide (designated V; 24 kdaltons) is in the same reading frame as P.[40] A similarly in-frame polypeptide of 7 kdaltons has been reported to be encoded by the 3' terminal portion of the mRNA of the NS protein of VSV of the Indiana serotype.[55]

Presently, no functions have been assigned to the products of any of the newly discovered genes of the nonsegmented negative strand RNA viruses. Certain considerations suggest that these genes must play a role in the replication of the viruses in which they occur. The relatively low fidelity of RNA polymerases ensures a high rate of evolution of RNA viruses.[56] Thus, it seems likely that nonessential genes would be rapidly eliminated from these viruses and that the presence of an additional polypeptide encoded within a gene must indicate that it has a role in the viral life cycle. The independent preservation of the C gene reading frame in at least four different viruses supports this notion. Indeed, conservation of homology between the C proteins of Sendai virus and PIV3 appears to have occurred at the expense of homology between the P proteins.[16] On the other hand, many of the newly discovered genes are quite small, and it is possible that some of them may turn out to be remnants of ancestral genes with no present functions. This explanation has been proposed for the 423-base region between the G and L genes on the rabies virus genome, which retains, at each end, vestiges of the intergenic sequences for transcription, but can only encode a polypeptide of 18 amino acids at most.[51] Similarly, the M gene mRNAs of measles virus and CDV have unusually large 3' untranslated regions (>400 bases) which may represent gene remnants.[57]

IV. VIRAL TRANSCRIPTION

The infection that ensues upon entry of a rhabdoviral or paramyxoviral nucleocapsid into the cell cytoplasm begins with the transcription of the viral genome by its associated RNA polymerase. Primary transcripts are translated by the ribosomes of the host, and some of the proteins produced then become available for assembly into progeny nucleocapsids. Progeny nucleocapsids are produced in two stages. Initially, the negative strand genome serves as the template for full-length, positive-polarity RNA molecules. Then these genome-complementary intermediates, in turn, act as templates for the synthesis of more copies of the genome. Both types of genome-length RNA molecules are encapsidated concomitantly with their synthesis. The appearance of progeny nucleocapsids enables a wave of secondary transcription and translation, leading to continued replication, maturation, and eventual budding of virus particles through the surface of the infected cell.

The central feature of the replicative cycle of a nonsegmented negative strand virus, then, is the synthesis of RNA catalyzed by a viral RNA polymerase. This is the major point of control during infection. For VSV it has been shown that viral macromolecular synthesis is governed mainly by the rates of transcription of the various viral genes.[58] There is little or no differential regulation of the translation of viral proteins,[59,60] and there are virtually no differences in the in vivo rates of decay of functional viral mRNAs.[61] Consequently, much effort has been devoted to understanding the mechanisms by which the viral RNA polymerase transcribes and replicates its genome.

A. Basic Phenomena of Transcription

1. The Transcription Reaction

The defining characteristic of negative strand RNA viruses was first demonstrated with a paramyxovirus when it was shown by Kingsbury[62] that more than 80% of the RNA synthesized in NDV-infected chick embryo fibroblasts, in the presence of actinomycin D, could be hybridized to the viral genome. This established the polarity of the NDV genome and, barring more arcane possibilities, made it very likely that the viral RNA of NDV and other similar viruses must be synthesized by a viral RNA-dependent RNA polymerase. The discovery of such a polymerase within purified virions was made by Baltimore et al.[63] with the rhabdovirus VSV. Subsequently, similar enzyme complexes were described for NDV,[64] Sendai virus,[65,66] mumps virus,[67] SV5,[68] rabies virus,[69,70] measles virus,[71] and RSV.[72] However, the VSV RNA polymerase has proven to be by far the most amenable to in vitro study, and almost the entirety of our present knowledge of nonsegmented negative strand RNA virus transcription and replication comes from the intensive study of VSV that has taken place over the past decade.

The VSV transcriptase catalyzes the sequential synthesis of six major RNA species. Five of these are the viral mRNAs which encode the five structural proteins of the virus. Like most eukaryotic messages, those of VSV contain capped, methylated 5' termini and 3' polyadenylate tails,[73] although both of these modifications appear to be added by viral, not cellular, enzymic activities. The mechanisms of capping and methylation are both unique. In the VSV capping reaction, the triphosphate bridge of the cap is formed from the α- and β-phosphates of GTP and the α-phosphate of the mRNA 5' terminus.[74] This stands in contradistinction to all other eukaryotic caps, in which the GTP contributes only its α-

phosphate to the bridge, and the mRNA contributes the α- and β-phosphates of its 5′ terminus.[75] In the VSV methylation reaction, two methyl groups are transferred from *S*-adenosylmethionine molecules to the capped 5′ terminus of the mRNA, the first occuring at the 2′-hydroxyl position of the penultimate base (A) and the second at the 7-position of the guanosine cap residue.[76] This sequence of methyl transfer is the reverse of that which occurs with all other capped eukaryotic mRNAs.[75] It has not been possible thus far to decouple either the guanylyltransferase or the two methyltransferase activities from ongoing VSV transcription in vitro, so the enzymology of these unusual reactions remains largely unexamined. It is also not clear why VSV (and by extension, the rhabdoviruses and paramyxoviruses) find it necessary to employ a unique strategy to construct the 5′ termini of their mRNAs. The same question may be asked of the orthomyxoviruses (Chapter 8), bunyaviruses (Chapter 9), coronaviruses (Chapter 6), and picornaviruses (Chapter 2). Perhaps this reflects some intrinsic biochemical constraint of RNA-templated transcription.

The sixth major RNA species synthesized during VSV transcription is a 47-base leader RNA, which was discovered and first sequenced by Colonno and Banerjee.[77,78] The leader RNA, which is neither capped nor polyadenylated, is complementary to the exact 3′ terminus of the VSV genome. Thus, the full sequential order of VSV transcription is: leader RNA-N mRNA-NS mRNA-M mRNA-G mRNA-L mRNA. Since it is the first transcribed species, the leader RNA is made in substantial amounts, but the role of this molecule remained obscure for a number of years. It is now known that in vivo the nascent leader is encapsidated by the viral N protein[79,80] and is also bound by a host cell factor, the lupus antigen, La.[81] The leader ribonucleoprotein complex is translocated to the nucleus shortly following its synthesis[82] and is thought to inhibit cellular transcription by the host RNA polymerases II and III.[83] An expanded treatment of the role of the leader RNA in VSV cytopathology may be found in a recent review by Wagner et al.[84]

In addition to the six major transcription products, the VSV transcriptase also synthesizes a number of minor RNA species. Some of these appear to be abortive or aberrant products which may provide clues to the mechanism of transcription, whereas others may possibly have roles in the course of infection. Singh et al.[85] have described a 9-S RNA, synthesized in vitro and in vivo, which corresponds to the 5′ terminus of the N mRNA and is produced in amounts roughly equimolar with the N mRNA. This species, approximately 400 bases in length, is capped, but not polyadenylated and has variable 3′ termini. A short, 40-base, capped RNA corresponding to the 5′ terminus of the N mRNA has been identified by Schubert et al.[86] These workers have also detected roughly equimolar quantities of a second short RNA species, 24 to 28 bases long, which is internally initiated 41 bases within the N gene — at precisely the next base following the termination of the 40-base-long RNA. The remarkable feature of this uncapped intracistronic RNA is that it is initiated with GTP, in contrast to all of the VSV messages, which start with an A residue. Finally, a number of workers have described various sets of small RNAs initiated at the starts of the N and NS genes.[87-89] The debated significance of these is discussed below.

2. Roles of Viral and Cellular Proteins in Transcription

The transcribing nucleocapsid of VSV is composed of the viral genome and the viral proteins N, L, and NS. As discussed in Section II., the genomic RNA is tightly encapsidated by the N protein, and this complex, not free genomic RNA, constitutes the template for transcription and replication. Thus, the N protein protects the phosphodiester backbone of the genome from cytoplasmic nucleases and yet somehow allows the viral polymerase sufficient access to the genome to form the nucleotide base-pairs necessary for templated RNA synthesis.

The viral RNA-dependent RNA polymerase requires both the L and NS proteins, as was shown by Emerson and Yu.[90] Earlier studies indicated that the polymerase had a 1:1 stoichiometry of its two constituent proteins,[90,91] but kinetic studies of reconstituted in vitro transcription have demonstrated that a large stoichiometric excess of NS protein to L protein is required for the optimal rate of transcript elongation.[92] This may account for the observation that in infected cells, NS protein is synthesized well in excess of the amount required for assembly of progeny virions, and a considerable fraction of NS protein remains free in the cytoplasm.[93] This cytoplasmic NS has been found to be highly active in supporting VSV transcription.[94]

The assignment of the various activities of the RNA polymerase to one or the other of its constituent polypeptides is presently incomplete. Based on reconstitution experiments, it is thought that the actual catalysis of RNA synthesis resides in the comparatively huge L protein,[92,95,96] while the NS protein is proposed to play an auxilliary role in RNA chain elongation, perhaps by unwinding the N-RNA template[92,95] or by directly facilitating the interaction of the L protein and the template by binding to each of them.[97] One of the enzymic activities of mRNA modification, polyadenylation, has been shown by Hunt et al.[98] to be due to the L protein, since a particular mutation in the L gene causes unusually large polyadenylate tails to be added to viral transcripts. Horikami and Moyer[99] have analyzed a class of host range mutants of VSV defective in one or both of the 5′ terminal methyltransferase activities, clearly demonstrating that these activities are virally encoded. The mutations in these viruses have recently been shown to reside in the L gene.[162] The unusual nature of the VSV guanylyltransferase strongly suggests that this activity, too, is virally encoded, but assignment of it to either of the two polymerase proteins is not possible at this time. The finding of a GDP binding site on the NS protein,[100] however, raises the possibility that NS directly participates in the capping reaction.

One activity that can unambiguously be attributed to the NS protein is that of phosphate acceptor. NS of the Indiana serotype of VSV has been shown by Hsu et al.[101] to be phosphorylated on various combinations of some 21 serine and threonine residues. These numerous phosphorylated subspecies can be resolved by gel electrophoresis into two discrete sets of phosphorylated states, NS1 and NS2, which probably represent distinct conformations of the molecule.[94,102,103] Kingsford and Emerson[104] have also shown that DEAE-cellulose chromatography separates NS into less phosphorylated and more highly phosphorylated forms, NSI and NSII, respectively; each of these are then further resolvable into NS1 and NS2 subspecies. In vitro, NS1 has been shown to convert into NS2 by phosphorylation with ATP, and conversely, phosphatase treatment of NS2 has been shown to convert it to NS1.[101,102] Although this extremely acidic phosphoprotein has proved quite intractable to standard methods of proteolytic digestion and peptide analysis, several groups have been able to determine that both NS1 and NS2 contain a major cluster of phophoserine and phosphothreonine residues within the amino terminal third of the molecule.[105-107] NS2 differs from NS1 in that it contains additional phosphorylated residues elsewhere in the molecule.[101,106]

The NS phosphoprotein is the target of multiple protein kinases. It has been demonstrated that VSV virions, as well as the cell cytoplasm, contain one or more protein kinases of cellular origin which can phosphorylate NS.[108-110] In addition, Sanchez et al.[111] have shown that purified preparations of L protein possess a protein kinase for which NS is a highly preferred substrate. The L-associated protein kinase, in the presence of the N-RNA template, shows a marked preference for the NS2 form of the molecule, whereas protein kinase(s) of the cellular cytoplasm preferentially phosphorylate NS1.[94]

The existence of multiple phosphorylated states of NS has prompted speculation that phosphorylation-dephosphorylation may govern transcriptional efficiency or modulate the switch from transcription to replication. A number of observations appear to bear on these issues: (1) NSII exhibits much greater activity than NSI in reconstitution of transcription;[104] (2) the complete, irreversible dephosphorylation of NS results in a threefold diminution of the transcriptive activity of virion nucleocapsids;[101] (3) chemical inhibitors of NS phosphorylation cause a parallel inhibition of in vitro VSV RNA synthesis;[111-114] and (4) NS forms multiple types of complexes with free N protein in vitro and in vivo, and these directly participate in genome replication.[115-117] These results portend a critical role for various phosphorylated states of the NS protein in both types of viral RNA synthesis, but further information will be required in order to clearly delineate this role. A promising approach to dissect the functions of the NS molecule has been initiated by Gill et al.,[118] who have inserted the NS gene of the New Jersey serotype of VSV into a DNA construct which allows it to be transcribed by the bacteriophage SP6 polymerase. Full length or progressively truncated forms of the NS mRNA produced by this system were translated in vitro, and the resulting proteins were then used to reconstitute in vitro transcription with purified L protein and N-RNA template. This deletion mapping technique has led to the definition of domains within the NS molecule essential for transcription and for the binding of NS to the N-RNA template. The same system has allowed the site-specific mutagenesis of individual serine residues to define phosphorylation targets on the NS protein that are essential for transcription.[163]

A further intriguing characteristic of the NS protein is that the gene which encodes it is highly mutable within the vesiculovirus genus. The NS proteins of the closely related Indiana and New Jersey serotypes of VSV have only 32% amino acid sequence homology, whereas the N, M, and G proteins of the two viruses are nearly twice as homologous to each other.[119,120] The NS protein of a more distant vesiculovirus, Chandipura virus, shows even poorer homology, 21 and 23%, respectively, to its VSV Indiana and New Jersey counterparts.[53] By contrast, the Chandipura N protein is 50% homologous to the N proteins of each of the two VSV serotypes. Despite this divergence among the primary sequences of the NS proteins, there is a great deal of similarity in their amino acid compositions and in plots of the relative hydropathicities of their polypeptide chains, suggesting that the overall structures of the NS molecules have been conserved throughout evolution. Moreover, the only region that is relatively highly conserved among the three NS proteins is a 20-amino acid carboxy-terminal domain which has been shown to be essential for binding of NS to the N-RNA template.[118]

Although very active VSV transcription can be demonstrated in reactions reconstituted from highly purified viral components, the presence of trace amounts of host proteins in these reactions cannot be ruled out. The participation of one or more host factors in VSV transcription has long been suggested both by the existence of host range mutants affecting viral polymerase function[121] and by the observation that cell extracts could stimulate viral RNA synthesis in vitro. Recently, two groups have independently reported an obligatory role for tubulin or microtubule-associated proteins (MAPs) in VSV RNA synthesis. Moyer et al.[122] found that a monoclonal antibody directed against β-tubulin almost completely inhibited in vitro transcription by purified VSV virions or by extracts from VSV- or Sendai virus-infected cells, suggesting that the ability of purified virions to catalyze viral RNA synthesis in vitro may depend on the degree to which host cell tubulin is incorporated into assembled virions. Evidence for an intracellular interaction between tubulin and the VSV L protein was shown by the nearly quantitative immunoprecipitation of L protein from VSV-infected cell extracts by the anti-β-tubulin monoclonal antibody. However, the presence of tubulin in purified virions has not been clearly demonstrated. Using nucleocapsids prepared from virions, Hill et al.[123] observed a dose-dependent stimulation of VSV RNA synthesis by microtubules to an extent as great as 16-fold. When the microtubule preparation was further fractionated into tubulin and MAPs, virtually all of the RNA polymerase-stimulatory activity was found in the MAPs fraction. The role of tubulin or MAPs in rhabdoviral and paramyxoviral RNA synthesis remains to be explored.

3. Intergenic Junctions

One of the results accruing from the many recent rhabdovirus and paramyxovirus sequencing studies has been the definition of the viral genome sequences at the junctions

Table 3
INTERGENIC JUNCTION SEQUENCES OF
RHABDOVIRUSES AND PARAMYXOVIRUSES

Virus	End sequence[a]	Intergenic sequence[a]	Start sequence[a]
VSV[43] and SVCV[b,44]	UAUG(A)$_7$	CU	AACAGNNNUC
Rabies virus[51]	UG(A)$_{7\text{-}8}$	C(N)$_{1\text{-}422}$	AACANNNCU
Sendai virus[124]	ANUAAG(A)$_5$	CUU	AGGGUNAAAG
PIV3[47,125]	AANUANN(A)$_5$	CUU	AGGANNAAAG
Measles virus[b,49]	UUAU(A)$_6$	CUU	AGGANCNANGU
RSV[50]	AGUNAUNU (A)$_4$ or AGUUANNNN(A)$_4$	(N)$_{0\text{-}51}$U	GGGGCAAAU

[a] All consensus sequences are given in the positive strand (message) sense and are in the 5' to 3' direction. N = any base.
[b] Based on partial set of intergenic sequences.

between adjacent genes. These regions have received much attention since it is a reasonable assumption that they contain at least part of the signals for many of the events that occur in the transcription reaction. The consensus intergenic junction sequences known to date are listed in Table 3 (in the positive strand sense). Using the nomenclature of Gupta and Kingsbury,[124] each of these has been divided into three portions: (1) the end sequence corresponding to the 3' extreme of the preceding mRNA; (2) the intergenic sequence which does not appear in mature mRNA; and (3) the start sequence corresponding to the 5' extreme of the succeeding mRNA. It should be noted that although the end and start sequences are often referred to as termination and initiation signals, there is as yet no evidence that they constitute the entirety of such signals or, conversely, that all parts of the sequences are necessary to ensure proper termination or initiation of transcription. Similarly, the intergenic sequences are generally held to be nontranscribed, but this also remains to be proven.

A striking feature of the tabulated intergenic junctions is that they are extremely well conserved within a given virus. Even more noteworthy, however, are the degrees of similarity among these sequences in different viruses across genus and family lines. All rhabdoviral and paramyxoviral messages begin with a consensus sequence consisting of a block of 9 to 11 residues at their 5' termini of which at least 6 are identical within a given virus. With the exception of RSV, the messages of all viruses listed in Table 3, as well as those of SV5 and NDV, initiate with an A residue. A similar conservation is seen at the 3' termini, where a consensus block of at least a dinucleotide precedes a run of 4 to 8 A residues encoded by the viral template. These occur immediately prior to the polyadenylate tails of the messages, which are not template-encoded. For VSV, it has been suggested that the viral polymerase synthesizes the polyadenylate tails of the messages by repeated slippage or "chattering" when it encounters the (U)$_7$ tract on the genome template.[126] The intergenic sequences of rhabdoviruses and paramyxoviruses, in most cases, are strictly conserved di- or trinucleotides (CU or CUU). The two known exceptions to this rule are rabies virus and RSV, which have intergenic sequences that are highly variable, both in size and base composition.[50,51] Even for each of these, an element of similarity can be seen with the CU or CUU consensus sequences: all rabies virus intergenic sequences begin with C; all RSV intergenic sequences end with U.

Some clues as to the significance of the junction sequences derive from the occurrence of mutant junctions within a particular virus which happen to be at variance with the consensus sequences. The intergenic sequence between the NS and M genes of the Indiana serotype of VSV is GU (in the positive strand sense).[43,52] This single-base deviation from the consensus dinucleotide, CU, appears to cause an increased probability of polymerase readthrough at this junction, resulting in a threefold higher frequency of NS-M dicistronic transcripts compared to other polycistronic transcripts.[39] Similarly, the Sendai virus HN-L intergenic sequence, CCC instead of the consensus CUU, results in increased readthrough at that junction.[127] These observations argue that the intergenic di- or trinucleotides are not merely spacers between the end and start sequences, and that their base content partially affects proper termination at the junctions. Another anomalous junction region occurs in the end sequence of the M gene of PIV3, where there is an 8-base insertion situated immediately prior to the $(A)_5$ tract which starts polyadenylation.[47,125] This disruption of the consensus sequence causes a large increase in polymerase readthrough from the M gene to the F gene, indicating that an intact end sequence is also important for proper transcript termination. It will be interesting to see the intergenic sequences of NDV, due to the fact that NDV polycistronic messages account for 25% of the total transcription by that virus.[41] An insertational mutation also occurs in the G-L junction of VSV of the New Jersey stereotype, where an anomalous block of 19 bases occurs between the intergenic CU and the consensus start sequence.[164] Suprisingly, this leads to two alternative start sites for the L mRNA in vitro.

It should be noted that the first intergenic junction in rhabdoviruses and paramyxoviruses, that between the leader and N (NP) genes, differs from those between the other genes. The sequence of this junction is known for a number of vesiculoviruses as well as rabies virus and Sendai virus.[51,124,128,129] Although the start sequence of the N (NP) gene conforms to the other start sequences for each virus, the end sequence of the leader contains neither the oligo(A) tract nor the consensus block which precedes it in all the mRNAs. Moreover, the intergenic sequence between the leader gene and the N (NP) gene is UUU, UGU, or UUUU (for vesiculoviruses,[128,129] rabies virus,[51] and Sendai virus,[124] respectively, in the positive strand sense). The uniqueness of the leader-N (NP) junction presumably reflects the fact that the leader RNA is not polyadenylated.

Each intergenic junction, in the course of multiple rounds of transcription, is potentially the site of a number of processes: (1) termination of the preceding mRNA, usually, but not always, including polyadenylation of that mRNA; (2) initiation and capping of the succeeding mRNA; (3) readthrough from the preceding to the succeeding gene; and (4) exit of the viral polymerase from the template. This latter phenomenon, attenuation, was first suggested from a study by Villarreal et al.[58] of the in vivo molar ratios of the VSV messages, which were seen to decrease as a function of distance from the 3′ end of the genome. Iverson and Rose[130] examined this more closely by hybridizing in vivo synthesized VSV mRNA to cDNA clones representing the 3′ proximal or 5′ proximal portions of each of the first four genes of VSV of the Indiana serotype. In this manner they established that there is a uniform 30% decrease of transcription between each successive gene on the viral genome, and that this attenuation occurs at the junctions between adjacent genes. Furthermore, analysis of in vitro transcription kinetics showed that significant pauses occur between the transcription of adjacent genes. Thus, some event at the intergenic junctions is slow with respect to the rate of mRNA elongation. Attenuation appears to be independent of whatever causes occasional readthrough at intergenic junctions, since the frequency of attenuation at the anomalous NS-M junction is the same as that at the N-NS and M-G junctions. Curiously, the frequency of attenuation is very similar to the frequency of synthesis of poly(A)-minus transcripts,[131] suggesting a possible correlation between failure to polyadenylate and detachment of the polymerase from the template.

B. Models of Transcription

Any model of VSV transcription must offer an explanation for the sequential nature of VSV mRNA synthesis as well as some description of the events which occur at the junctions between the viral genes. At various times, three different models of VSV transcription have been proposed. The first of these suggested that transcription proceeds by a single initiation of the polymerase at the 3' end of the genome resulting in synthesis of a genome-length precursor which is then processed into the leader RNA and five mature mRNA species.[73] The alternative to this notion of a single initiation is the proposal that the polymerase initiates at the start and terminates at the end of each of the genes encoding the leader and the five mRNAs. This further subdivides into two models: one in which the polymerase can enter the genome only at the 3' end,[132] and the other in which the polymerase binds to independent promoters at the start of each of the six cistrons.[87]

The single initiation/processing model, as originally stated,[73] is not tenable, since it is now clear that a genome-length precursor to the VSV messages does not exist. The original attractiveness of this model was that it made obligate the unique aspect of VSV cap formation, since a nucleolytically processed RNA precursor would have only a single phosphate at its 5' terminus to donate to the triphosphate bridge of the cap. However, the single initiation/processing model is generally regarded as unsatisfactory because of the precise nucleolytic cleavages it invokes prior to polyadenylation.[124] These become even less feasible upon consideration of the recently determined heterogeneous intergenic sequences of rabies virus and RSV.[50,51] Nevertheless, no experiment has been described to date which unequivocally rules out this mechanism.

The single entry/multiple initiation model,[132] which is formally analogous to the ribosome scanning model for translation of eukaryotic mRNAs,[54] proposes that the viral polymerase gains access to the genome only at the 3' terminus and proceeds toward the 5' terminus synthesizing the leader and five mRNAs sequentially. At each intergenic junction the polymerase terminates, with or without polyadenylation, and either attenuates or reinitiates at the start of the next gene. By this model, polycistronic transcripts would arise from occasional failure to recognize termination and initiation signals at particular junctions.

The multiple entry/multiple initiation model proposes that viral polymerase molecules gain access to the genome at the 3' terminus and also internally at the start of each of the five genes.[87] Sequential RNA synthesis is maintained by the stipulation that mRNA elongation at each promoter is arrested some short distance after initiation (<50 bases) and cannot proceed until the polymerase transcribing the previous gene has completed its transcript. By this model, either attenuation or readthrough events generating polycistronic transcripts would be alternatives open to a transcribing polymerase approaching an unoccupied promoter. This cascade model was invoked by Testa et al.[87] to explain the existence of short, uncapped RNA species produced during in vitro transcription reactions. Two of these are 5' triphosphate initiated RNAs, 42 and 28 bases in length, corresponding to the 5' termini of the N and NS mRNAs, respectively. These species have UV target sizes consistent with their small molecular weights and have been shown by kinetic studies to be synthesized within the first minute of the transcription reaction, prior to the sequential appearance of their corresponding mRNAs.

The efforts by a number of groups to distinguish between the single and multiple entry models of VSV transcription have focussed on two questions: (1) can the viral polymerase enter and initiate internally within the genome independent of entry at the 3' end?; and (2) can the short, triphosphate initiated N- and NS-start RNA species be shown to be precursors to full-length mRNAs?[133]

A number of observations bear on the question of internal initiation. Pinney and Emerson[88] have described a set of four triphosphate-initiated oligonucleotides which are produced during in vitro transcription and correspond to the first 11 to 14 bases of the N gene. Since these are synthesized in as much as a tenfold molar excess over the leader RNA, they cannot depend on prior synthesis of leader and must be internally initiated. Talib and Hearst[134] examined the effects on VSV transcription of aurintricarboxylic acid and vanadyl ribonucleoside, and they characterized the single species synthesized in the presence of either of these inhibitors as a capped RNA corresponding to the first 68 bases of the N gene. Since this was produced in the absence of any detectable leader RNA synthesis, it, too, must have been due to internal initiation. It has been pointed out, though, that all of these internally initiated species may arise from viral polymerases arrested at or near intergenic junctions during virion assembly.[135] Synthesis of the various N- and NS-start products thus may have no bearing on whether the polymerase can gain access to an internal promoter without entering by way of the 3′ end of the genome. This interpretation is consistent with a study by Emerson[132] in which purified N-RNA template (free of polymerase) was reconstituted with L and NS proteins in the presence of a partial or full complement of ribonucleoside triphosphates. Reconstituted nucleocapsids provided with just ATP and CTP synthesized only the dinucleotide pppAC, which was taken to be the initiated leader RNA. Short oligonucleotides corresponding to the 5′ ends of the viral mRNAs, pppAAC and pppAACA, could be detected only after the reconstituted nucleocapsids were permitted to synthesize full-length leader RNA in the presence of all four NTPs, suggesting that these were absolutely dependent upon prior passage of the polymerase through the leader gene. At variance with this study, however, Thornton et al.[136] observed synthesis of both leader and mRNA initiating nucleotides in partial (ATP + CTP) transcription reactions using UV-irradiated N-RNA template reconstituted with L and NS proteins. In total transcription reactions in this system, all N- and NS-start short RNA products, as well as leader RNA, were produced in undiminished quantities, whereas full-length mRNA synthesis was 90% inhibited.

The second question, whether the N- and NS-start species are true precursors to full-length mRNAs, is similarly unresolved. Attempts to pulse chase the small triphosphate initiated RNAs or a possibly related set of small capped RNAs into mature messages have all had negative results.[89,135,137,138] However, if the short RNA species result from reiterative initiation events in the promoter region, then most would be released from the template unelongated, and only a small fraction would be expected to end up in completed mRNAs and might be exceedingly difficult to detect.

V. VIRAL REPLICATION

A. Basic Phenomena of Replication

The most fundamental and motivational observation in the study of nonsegmented negative strand RNA viral genome replication was the finding that in vivo synthesis of genome-length Sendai virus[139] or VSV[140,141] RNA is dependent upon continued protein synthesis within the host cell. This suggested a requirement for some viral or cellular protein which is either needed in stoichiometric amounts or is quite labile and must be produced continuously. It also accounted for the consistently noted fact that, with the exceptions cited below, purified VSV virions, although capable of very active transcription, do not carry out replication in vitro. Consequently, much subsequent effort by a number of investigators went toward developing in vitro systems to assay positive and negative strand genome-length RNA synthesis by either VSV or Sendai virus. These have involved the coupling of RNA-synthesizing viral nucleocapsids, obtained either from purified virions or from infected cells, with in vitro protein synthesizing cell extracts or with pools of proteins from infected cells.[142-149] With varying degrees of definition, these systems have been successful at demonstrating synthesis of N (NP) protein-encapsidated, genome-length RNA of both polarities. These in vitro-generated ribonucleoproteins have been shown to be ribonuclease-resistant and to band in CsCl at the same density as authentic viral N-RNA (NP-RNA) complexes. Their RNA is genome length, as assayed by gel electrophoresis, and in the case of the negative strand of VSV, it hybridizes to all five viral mRNAs. Moreover, in some of the systems, inhibition of in vitro protein synthesis by cycloheximide or pactamycin eliminates viral N-RNA synthesis, albeit at various rates, dependent upon the sizes of the pools of viral proteins which have accumulated.[144,145,147]

Wertz and colleagues[145,150,151] have developed a system consisting of a rabbit reticulocyte lysate programmed with viral mRNAs, to which is added nucleocapsids derived either from virions or from viral defective interfering particles (DIPs). The use of DIPs (which transcribe only a 46-base-long leader RNA[152,153]) eliminates the high background of viral mRNA synthesis in in vitro replication studies, allowing the clear examination of leader and genome-length RNA synthesis. By programming this system with individual hybrid-selected mRNAs, Patton et al.[151] have been able to show that, of the three smaller VSV proteins, N protein alone fulfills the protein synthesis requirement of replication. The extent of production of positive and negative full-length DIP RNA was seen to be directly proportional to the amount of N protein concurrently synthesized, and this N protein was demonstrated to encapsidate the RNA.

In contrast to the pivotal role established for N protein in replication, the function of the NS protein is less clear. Hill and Summers[147] found that polyclonal antibodies to NS inhibited VSV replication in a coupled in vitro system, but it is not clear whether this indicates a requirement for free NS protein or an effect on the RNA polymerase since the same antibodies also inhibited transcription. It has been shown that in extracts from infected cells,[115] as well as in an in vitro coupled system programmed with purified N and NS mRNAs,[117] most of the free N and NS proteins are associated with each other in multiple types of complexes. Since free N protein is known to form insoluble aggregates both in vivo[154] and in vitro,[155] it has been hypothesized that the function of NS protein in replication is to maintain N protein in a soluble form available for binding to nascent RNA.

By contrast with the above studies, two groups have demonstrated conditions under which the viral RNA polymerase can circumvent the requirement for free N protein in order to read through some or all of the intergenic junctions on the VSV genome. Testa et al.[156] found that although the β,γ-imido analog of ATP could not directly substitute for ATP in in vitro transcription reactions, it could do so if viral nucleocapsids were first preinitiated with ATP and CTP and then recovered. Under these conditions, synthesis of a full-length positive strand copy of the VSV genome was observed. This suggested a role for the β,γ-bond of ATP and, hence, for some phosphorylation reaction, in the switch from transcriptive to replicative RNA synthesis. Similarly, Chanda et al.[157] found synthesis of genome-length positive strand RNA in in vitro transcription reactions in which ITP was substituted for GTP. This result may indicate that RNA:RNA base-pairing is also important in the proper recognition of transcriptional termination signals on the genome template.

Perrault et al.[158,159] have generated a series of VSV mutants, designated pol R, which are specifically repressed in the ability to terminate transcription at the junction between the leader and N genes. Thus, these viruses, assayed for transcription in vitro, produce linked leader RNA-N mRNA transcripts at a frequency of >80%, compared to 10% for their wild-type parent. In addition, the pol R mutants have a tenfold greater ability than wild-type virus to use β,γ-imido ATP as a substrate (in the absence of preinitiation with ATP and CTP), indicating that they have a markedly reduced requirement for ATP in the initiation of RNA synthesis.[160] Surprisingly, reconstitution experiments and results with pol R DIPs have shown that the pol R mutations reside in the VSV N protein and not in the template RNA or in the polymerase proteins. Thus, an alteration in the template N protein can at least partially obviate the need for free N protein in junction readthrough by the viral polymerase.

B. Models of Replication

The basic problem which a model of VSV replication must address is how the viral polymerase reads through the intergenic junctions, ignoring signals for termination, polyadenylation, and reinitiation, which it faithfully recognizes during transcription. The two models which have been put forward to explain VSV replication differ in whether they envisage either a passive or an active role for the genome template in the readthrough process.

Blumberg et al.[161] have proposed that VSV replication is governed by the availability of free N protein, which acts as an antiterminator at an attenuation signal occurring in the neighborhood of the leader-N gene junction. In this model, the leader RNA species results from termination at this site, while genome-length RNA synthesis requires suppression of termination due to the binding of N protein to the nascent leader RNA, which is thought to initiate nucleocapsid assembly. Two predictions of this model have been confirmed by in vivo results: (1) the cycloheximide-induced inhibition of protein synthesis in cells infected with a mixture of wild-type VSV and DIPs brings about a reduction of genome RNA synthesis accompanied by an increase of DIP leader RNA, suggesting an inverse correlation between production of these two types of RNA species; and (2) leader RNAs have been found to be encapsidated by N protein in vivo, indicating that the nucleation point for nucleocapsid formation occurs within the leader RNA.[161] The in vitro VSV replication systems are consistent with the most salient aspect of this antitermination model, i.e., that replication is governed by the amount of free N protein available. It should be noted, however, that although the in vitro DIP-replicating system of Wertz shows a modest increase of leader synthesis in the presence of cycloheximide, the encapsidation of leader RNA is not seen.[150]

A significantly different model of VSV replication has been proposed by Perrault et al.[159] on the basis of the pol R mutants. These workers hypothesize the existence of separate populations of "transcription nucleocapsids" and "replication nucleocapsids" that are predetermined by the state of modification of N protein in the N-RNA template. The mutations in pol R mutants are thus seen to bias N toward the confirmation or modification state occurring in replication nucleocapsids. The notion that there occurs posttranslational modification of N protein is supported by the finding that N protein from wild-type virus can be resolved into four differently charged species by isoelectric focussing and that the positions of these are shifted for N proteins from the pol R mutants.[159] However, the proposal that the template alone can determine the switch from transcription to replication appears to be at variance with results obtained by Arnheiter et al.[116] with two different monoclonal antibodies against the N protein. The first of these binds both free N protein and N protein complexed with the genome template, and it inhibits both transcription and replication. The second antibody binds only to free N protein and not to template N protein, and it inhibits only replication and not transcription. This finding strongly supports an essential role for free N protein in replication.

It deserves mention that each of the models of replication has difficulty in addressing the data that supports the other. Also, neither explicitly deals with events at intergenic junctions other than that between the leader and N genes. Finally, neither model clearly explains either the readthrough events promoted by β,γ-imido ATP[156] and ITP[157] or the paradoxical result of Hill and Summers that polyclonal antibodies to L protein were seen to inhibit transcription, but stimulated replication in vitro.[147] Thus, as with VSV transcription, further work remains in order to elucidate mechanistic details.

VI. CONCLUSION

The study of nonsegmented negative strand RNA viruses has delineated much about the unique scheme by which these viruses replicate, although the details of many processes remain to be discovered, especially as investigation widens from the prototype viruses to their less-studied relatives. Future work in this field might be expected to yield additional exciting insights into the unique mechanisms of transcription and replication of these viruses and into the complexities of virus-host interactions.

REFERENCES

1. **Nakai, T. and Howatson, A. F.**, The fine structure of vesicular stomatitis virus, *Virology*, 35, 268, 1968.
2. **Murphy, F. A. and Harrison, A. K.**, Electron microscopy of the rhabdoviruses of animals, in *Rhabdoviruses*, Vol. 1, Bishop, D. H. L., Ed., CRC Press, Boca Raton, Florida, 1979, chap. 4.
3. **Thomas, D., Newcomb, W. W., Brown, J. C., Wall, J. S., Hainfeld, J. F., Trus, B. L., and Steven, A. C.**, Mass and molecular composition of vesicular stomatitis virus: a scanning transmission electron microscopy analysis, *J. Virol.*, 54, 598, 1985.
4. **Cartwright, B., Smale, C. J., and Brown, F.**, Surface structure of vesicular stomatitis virus, *J. Gen. Virol.*, 5, 1, 1969.
5. **Choppin, P. W. and Compans, R. W.**, Reproduction of paramyxoviruses, in *Comprehensive Virology*, Vol. 4, Fraenkel-Conrat, H. and Wagner, R. R., Eds., Plenum Press, New York, 1975, 95.
6. **Choppin, P. W. and Scheid, A.**, The role of viral glycoproteins in adsorption, penetration and pathogenicity of viruses, *Rev. Infect. Dis.*, 2, 40, 1980.
7. **Hiebert, S. W., Paterson, R. G., and Lamb, R. A.**, Hemagglutinin-neuraminidase protein of the paramyxovirus simian virus 5: nucleotide sequence of the mRNA predicts an N-terminal membrane anchor, *J. Virol.*, 54, 1, 1985.
8. **Wertz, G. W., Collins, P. L., Huang, Y., Gruber, C., Levine, S., and Ball, L. A.**, Nucleotide sequence of the G protein gene of respiratory syncytial virus reveals an unusual type of viral membrane protein, *Proc. Natl. Acad. Sci. U.S.A.*, 82, 4075, 1985.
9. **Elango, N., Coligan, J. E., Jambou, R. C., and Venkatesan, S.**, Human parainfluenza type 3 virus hemagglutinin-neuraminidase glycoprotein: nucleotide sequence of mRNA and limited amino acid sequence of the purified protein, *J. Virol.*, 57, 481, 1986.
10. **Leppart, M., Rittenhouse, L., Perrault, J., Summers, D. F., and Kolakofsky, D.**, Plus and minus strand leader RNAs in negative strand virus-infected cells, *Cell*, 18, 735, 1979.
11. **Wilson, T. and Lenard, J.**, Interaction of wild-type and mutant M protein of vesicular stomatitis virus with nucleocapsids in vitro, *Biochemistry*, 20, 1349, 1981.
12. **De, B. P., Thornton, G. B., Luk, D., and Banerjee, A. K.**, Purified matrix protein of vesicular stomatitis virus blocks viral transcription *in vitro*, *Proc. Natl. Acad. Sci. U.S.A.*, 79, 7137, 1982.
13. **Moller, J. R., Rager-Zisman, B., Quan, P.-C., Schattner, A., Panush, D., Rose, J. K., and Bloom, B. R.**, Natural killer cell recognition of target cells expressing different antigens of vesicular stomatitis virus, *Proc. Natl. Acad. Sci. U.S.A.*, 82, 2456, 1985.
14. **Pal, R., Grinnell, B. W., Snyder, R. M., and Wagner, R. R.**, Regulation of viral transcription by the matrix protein of vesicular stomatitis virus probed by monoclonal antibodies and temperature-sensitive mutants, *J. Virol.*, 56, 386, 1985.
15. **Sanchez, A. and Banerjee, A. K.**, Studies on human parainfluenza virus type 3: characterization of the structural proteins and in vitro synthesized proteins coded by mRNAs isolated from infected cells, *Virology*, 143, 45, 1985.
16. **Luk, D., Sanchez, A., and Banerjee, A. K.**, Messenger RNA encoding the phosphoprotein (P) gene of human parainfluenza virus type 3 is bicistronic, *Virology*, 153, 318, 1986.
17. **Giorgi, C., Blumberg, B. M., and Kolakofsky, D.**, Sendai virus contains overlapping genes expressed from a single mRNA, *Cell*, 35, 829, 1983.
18. **Bellini, W. J., Englund, G., Rozenblatt, S., Arnheiter, H., and Richardson, C. D.**, Measles virus P gene codes for two proteins, *J. Virol.*, 53, 908, 1985.
19. **Barrett, T., Shrimpton, S. B., and Russell, S. E. H.**, Nucleotide sequence of the entire protein coding region of canine distemper virus polymerase-associated (P) protein mRNA, *Virus Res.*, 3, 367, 1985.
20. **Pringle, C. R.**, Genetics of rhabdoviruses, in *Comprehensive Virology*, Vol. 9, Fraenkel-Conrat, H. and Wagner, R. R., Eds., Plenum Press, New York, 1977, chap. 6.
21. **Pringle, C. R. and Szilagyi, J. F.**, Gene assignment and complementation group, in *Rhabdoviruses*, Vol. 2, Bishop, D. H. L., Ed., CRC Press, Boca Raton, Florida, 1979, chap. 8.
22. **Flamand, A.**, Rhabdovirus genetics, in *Rhabdoviruses*, Vol. 2, Bishop, D. H. L., Ed., CRC Press, Boca Raton, Florida, 1979, chap. 7.
23. **Granoff, A.**, Studies on mixed infection with Newcastle disease virus: isolation of Newcastle disease virus mutants and tests for genetic recombination between them, *Virology*, 9, 636, 1959.
24. **Portner, A., Marx, P. A., and Kingsbury, D. W.**, Isolation and characterization of Sendai virus temperature-sensitive mutants, *J. Virol.*, 13, 298, 1974.
25. **Gimenez, H. B. and Pringle, C. R.**, Seven complementation groups of respiratory syncytial virus temperature-sensitive mutants, *J. Virol.*, 27, 459, 1978.
26. **Wong, P. K. Y., Holloway, A. F., and Cormack, D. V.**, Search for recombination between temperature-sensitive mutants of vesicular stomatitis virus, *J. Gen. Virol.*, 13, 477, 1971.
27. **Holland, J., Spindler, K., Horodyski, F., Grabau, E., Nichol, S., and Vande Pol, S.**, Rapid evolution of RNA genomes, *Science*, 215, 1577, 1982.

28. **Sauerbier, W. and Hercules, K.,** Gene and transcription unit mapping by radiation effects, *Annu. Rev. Genet.,* 12, 329, 1978.

29. **Michalke, H. and Bremer, H.,** RNA synthesis in *Escherichia coli* after irradiation with ultraviolet light, *J. Mol. Biol.,* 41, 1, 1969.

30. **Ball, L. A. and White, C. N.,** Order of transcription of genes of vesicular stomatitis virus, *Proc. Natl. Acad. Sci. U.S.A.,* 73, 442, 1976.

31. **Abraham, G. and Banerjee, A. K.,** Sequential transcription of the genes of vesicular stomatitis virus, *Proc. Natl. Acad. Sci. U.S.A.,* 73, 1504, 1976.

32. **Glazier, K., Raghow, R., and Kingsbury, D. W.,** Regulation of Sendai virus transcription: evidence for a single promoter in vivo, *J. Virol.,* 21, 863, 1977.

33. **Flamand, A. and Delagneau, J. F.,** Transcriptional mapping of rabies virus in vivo, *J. Virol.,* 28, 518, 1978.

34. **Collins, P. L., Hightower, L. E., and Ball, L. A.,** Transcriptional map for Newcastle disease virus, *J. Virol.,* 35, 682, 1980.

35. **Dickens, L. E., Collins, P. L., and Wertz, G. W.,** Transcriptional mapping of human respiratory syncytial virus, *J. Virol.,* 52, 364, 1984.

36. **Herman, R., Adler, S., Lazzarini, R. A., Colonno, R. J., Banerjee, A. K., and Westphal, H.,** Intervening polyadenylate sequences in RNA transcripts of vesicular stomatitis virus, *Cell,* 15, 587, 1978.

37. **Kurath, G., Ahern, K. G., Pearson, G. D., and Leong, J. C.,** Molecular cloning of the six mRNA species of infectious hematopoietic necrosis virus, a fish rhabdovirus, and gene order determination by R-loop mapping, *J. Virol.,* 53, 469, 1985.

38. **Chanda, P. K. and Banerjee, A. K.,** Inhibition of vesicular stomatitis virus transcriptase *in vitro* by phosphonoformate and Ara-ATP, *Virology,* 107, 562, 1980.

39. **Masters, P. S. and Samuel, C. E.,** Detection of *in vivo* synthesis of polycistronic mRNAs of vesicular stomatitis virus, *Virology,* 134, 277, 1984.

40. **Paterson, R. G., Harris, T. J. R., and Lamb, R. A.,** Analysis and gene assignment of a paramyxovirus, simian virus 5, *Virology,* 138, 310, 1984.

41. **Wilde, A., McQuain, C., and Morrison, T.,** Identification of the sequence content of four polycistronic transcripts synthesized in Newcastle disease virus infected cells, *Virus Res.,* 5, 77, 1986.

42. **Collins, P. L. and Wertz, G. W.,** cDNA cloning and transcriptional mapping of nine polyadenylated RNAs encoded by the genome of human respiratory syncytial virus, *Proc. Natl. Acad. Sci. U.S.A.,* 80, 3208, 1983.

43. **Rose, J. K.,** Complete intergenic and flanking sequences from the genome of vesicular stomatitis virus, *Cell,* 19, 415, 1980.

44. **Wu, J.-G., Kiuchi, A., and Roy, P.,** Intergenic sequences and gene order (N-M-G) of spring viremia of carp virus, personal communication.

45. **Dowling, P. C., Giorgi, C., Roux, L., Dethlefsen, L. A., Galantowicz, M., Blumberg, B. M., and Kolakofsky, D.,** Molecular cloning of the 3'-proximal third of Sendai virus genome, *Proc. Natl. Acad. Sci. U.S.A.,* 80, 5213, 1983.

46. **Shioda, T., Iwasaki, K., and Shibuta, H.,** Determination of the complete nucleotide sequence of the Sendai virus genome RNA and the predicted amino acid sequences of the F, HN and L proteins, *Nucleic Acids Res.,* 14, 1545, 1986.

47. **Spriggs, M. K. and Collins, P. L.,** Human parainfluenza virus type 3: messenger RNAs, polypeptide coding assignments, intergenic sequences, and genetic map, *J. Virol.,* 59, 646, 1986.

48. **Paterson, R. G., Harris, T. J. R., and Lamb, R. A.,** Fusion protein of the paramyxovirus simian virus 5: nucleotide sequence of mRNA predicts a highly hydrophobic glycoprotein, *Proc. Natl. Acad. Sci. U.S.A.,* 81, 6706, 1984.

49. **Richardson, C. D., Berkovich, A., Rozenblatt, S., and Bellini, W. J.,** Use of antibodies directed against synthetic peptides for identifying cDNA clones, establishing reading frames, and deducing the gene order of measles virus, *J. Virol.,* 54, 186, 1985.

50. **Collins, P. L., Dickens, L. E., Buckler-White, A., Olmsted, R. A., Spriggs, M. K., Camargo, E., and Coelingh, K. V. W.,** Nucleotide sequences for the gene junctions of human respiratory syncytial virus reveal distinctive features of intergenic structure and gene order, *Proc. Natl. Acad. Sci. U.S.A.,* 83, 4594, 1986.

51. **Tordo, N., Poch, O., Ermine, A., Keith, G., and Rougeon, F.,** Walking along the rabies genome: is the large G-L intergenic region a remnant gene?, *Proc. Natl. Acad. Sci. U.S.A.,* 83, 3914, 1986.

52. **Hudson, L. D., Condra, C., and Lazzarini, R. A.,** Cloning and expression of a viral phosphoprotein: structure suggests vesicular stomatitis virus NS may function by mimicking an RNA template, *J. Gen. Virol.,* 67, 1571, 1986.

53. **Masters, P. S. and Banerjee, A. K.,** Sequences of chandipura virus N and NS genes: evidence for high mutability of the NS gene within vesiculoviruses, *Virology,* 157, 298, 1987.

54. **Kozak, M.,** Point mutations define a sequence flanking the AUG initiator codon that modulates translation by eukaryotic ribosomes, *Cell*, 44, 283, 1986.
55. **Herman, R. C.,** Internal initiation of translation on the vesicular stomatitis virus phosphoprotein mRNA yields a second protein, *J. Virol.*, 58, 797, 1986.
56. **Reanney, D. C.,** The evolution of RNA viruses, *Annu. Rev. Microbiol.*, 36, 47, 1982.
57. **Bellini, W. J., Englund, G., Richardson, C. D., Rozenblatt, S., and Lazzarini, R. A.,** Matrix genes of measles virus and canine distemper virus: cloning, nucleotide sequences, and deduced amino acid sequences, *J. Virol.*, 58, 408, 1986.
58. **Villarreal, L. P., Breindl, M., and Holland, J. J.,** Determination of molar ratios of vesicular stomatitis virus induced RNA species in BHK21 cells, *Biochemistry*, 15, 1663, 1976.
59. **Wagner, R. R., Snyder, R. M., and Yamazaki, S.,** Proteins of vesicular stomatitis virus: kinetics and cellular sites of synthesis, *J. Virol.*, 5, 548, 1970.
60. **Kang, C. Y. and Prevec, L.,** Proteins of vesicular stomatitis virus: III. Intracellular synthesis and extracellular appearance of virus-specified proteins, *Virology*, 46, 678, 1971.
61. **Pennica, D., Lynch, K. R., Cohen, P. S., and Ennis, H. L.,** Decay of vesicular stomatitis virus mRNAs *in vivo*, *Virology*, 94, 484, 1979.
62. **Kingsbury, D. W.,** Newcastle disease virus RNA: II. Preferential synthesis of RNA complementary to parental viral RNA by chick embryo cells, *J. Mol. Biol.*, 18, 204, 1966.
63. **Baltimore, D., Huang, A. S., and Stampfer, M.,** Ribonucleic acid synthesis of vesicular stomatitis virus, II. An RNA polymerase in the virion, *Proc. Natl. Acad. Sci. U.S.A.*, 66, 572, 1970.
64. **Huang, A. S., Baltimore, D., and Bratt, M. A.,** Ribonucleic acid polymerase in virions of Newcastle disease virus: comparison with the vesicular stomatitis virus polymerase, *J. Virol.*, 7, 389, 1971.
65. **Robinson, W. S.,** Ribonucleic acid polymerase activity in Sendai virus virions and nucleocapsids, *J. Virol.*, 8, 81, 1971.
66. **Stone, H. O., Portner, A., and Kingsbury, D. W.,** Ribonucleic acid transcriptases in Sendai virions and infected cells, *J. Virol.*, 8, 174, 1971.
67. **Bernard, J. P. and Northrop, R. L.,** RNA polymerase in mumps virions, *J. Virol.*, 14, 183, 1974.
68. **Buetti, E. and Collins, P. L.,** The transcriptase complex of the paramyxovirus SV5, *Virology*, 82, 493, 1977.
69. **Kawai, A.,** Transcriptase activity associated with rabies virion, *J. Virol.*, 24, 826, 1977.
70. **Flamand, A., Delagneau, J. F., and Bussereau, F.,** An RNA polymerase activity in purified rabies virion, *J. Gen. Virol.*, 40, 233, 1978.
71. **Seifried, A. S., Albrecht, P., and Milstien, J. B.,** Characterization of an RNA-dependent RNA polymerase activity associated with measles virus, *J. Virol.*, 25, 781, 1978.
72. **Mbuy, G. N. and Rochovansky, O. M.,** RNA-dependent RNA polymerase associated with respiratory syncytial virus, in *Nonsegmented Negative Strand Viruses*, Bishop, D. H. L. and Compans, R., Eds., Academic Press, San Francisco, 1984, 175.
73. **Banerjee, A. K., Abraham, G., and Colonno, R. J.,** Vesicular stomatitis virus: mode of transcription, *J. Gen. Virol.*, 34, 1, 1977.
74. **Abraham, G., Rhodes, D. P., and Banerjee, A. K.,** The 5'-terminal structure of the methylated messenger RNA synthesized *in vitro* by vesicular stomatitis virus, *Cell*, 5, 51, 1975.
75. **Baneerjee, A. K.,** 5'-Terminal cap structure in eucaryotic messenger ribonucleic acids, *Microbiol. Rev.*, 44, 175, 1980.
76. **Testa, D. and Banerjee, A. K.,** Two methyltransferase activities in the purified virions of vesicular stomatitis virus, *J. Virol.*, 24, 786, 1977.
77. **Colonno, R. J. and Banerjee, A. K.,** A unique RNA species involved in initiation of vesicular stomatitis virus RNA transcription *in vitro*, *Cell*, 8, 197, 1976.
78. **Colonno, R. J. and Banerjee, A. K.,** Complete nucleotide sequence of the leader RNA synthesized *in vitro* by vesicular stomatitis virus, *Cell*, 15, 93, 1978.
79. **Blumberg, B. M. and Kolakofsky, D.,** Intracellular vesicular stomatitis virus leader RNAs are found in nucleocapsid structures, *J. Virol.*, 40, 568, 1981.
80. **Blumberg, B. M., Giorgi, C., and Kolakofsky, D.,** N protein of vesicular stomatitis virus selectively encapsidates leader RNA *in vitro*, *Cell*, 32, 559, 1983.
81. **Wilusz, J., Kurilla, M. G., and Keene, J. D.,** A host protein (La) binds to a unique species of minus-sense leader RNA during replication of vesicular stomatitis virus, *Proc. Natl. Acad. Sci. U.S.A.*, 80, 5827, 1983.
82. **Kurilla, M. G., Piwnica-Worms, H., and Keene, J. D.,** Rapid and transient localization of the leader RNA of vesicular stomatitis virus in the nuclei of infected cells, *Proc. Natl. Acad. Sci. U.S.A.*, 79, 5240, 1982.
83. **McGowan, J. J., Emerson, S. U., and Wagner, R. R.,** The plus-strand leader RNA of VSV inhibits DNA-dependent transcription of adenovirus and SV40 genes in a soluble whole cell extract, *Cell*, 28, 325, 1982.

84. **Wagner, R. R., Thomas, J. R., and McGowan, J. J.,** Rhabdovirus cytopathology: effects on cellular macromolecular synthesis, in *Comprehensive Virology,* Vol. 19, Fraenkel-Conrat, H. and Wagner, R. R., Eds., Plenum Press, New York, 1984, 223.

85. **Singh, B., Thornton, G. B., Roy, J., Talib, S., De B. P., and Banerjee, A. K.,** Frequent intragenic transcription termination within the N gene of vesicular stomatitis virus, *Virology,* 122, 239, 1982.

86. **Schubert, M., Harmison, G. G., Sprague, J., Condra, C. S., and Lazzarini, R. A.,** In vitro transcription of vesicular stomatitis virus: initiation with GTP at a specific site within the N cistron, *J. Virol.,* 43, 166, 1982.

87. **Testa, D., Chanda, P. K., and Banerjee, A. K.,** Unique mode of transcription *in vitro* by vesicular stomatitis virus, *Cell,* 21, 267, 1980.

88. **Pinney, D. F. and Emerson, S. U.,** Identification and characterization of a group of discrete initiated oligonucleotides transcribed *in vitro* from the 3' terminus of the N-gene of vesicular stomatitis virus, *J. Virol.,* 42, 889, 1982.

89. **Piwnica-Worms, H. and Keene, J. D.,** Sequential synthesis of small capped RNA transcripts *in vitro* by vesicular stomatitis virus, *Virology,* 125, 206, 1983.

90. **Emerson, S. U. and Yu, Y.-H.,** Both NS and L proteins are required for *in vitro* RNA synthesis by VSV, *J. Virol.,* 15, 1348, 1975.

91. **Naito, S. and Ishihama, A.,** Function and structure of RNA polymerase from vesicular stomatitis virus, *J. Biol. Chem.,* 251, 4307, 1976.

92. **De, B. P. and Banerjee, A. K.,** Requirements and functions of vesicular stomatitis virus L and NS proteins in the transcription process *in vitro, Biochem. Biophys. Res. Commun.,* 126, 40, 1985.

93. **Hsu, C.-H., Kingsbury, D. W., and Murti, K. G.,** Assembly of vesicular stomatitis virus nucleocapsids in vivo: a kinetic analysis, *J. Virol.,* 32, 304, 1979.

94. **Masters, P. S. and Banerjee, A. K.,** Phosphoprotein NS of vesicular stomatitis virus: phosphorylated states and transcriptional activities of intracellular and virion forms, *Virology,* 154, 259, 1986.

95. **De, B. P. and Banerjee, A. K.,** Specific interactions of L and NS proteins of vesicular stomatitis virus with heterologous genome ribonucleoprotein template lead to mRNA synthesis *in vitro, J. Virol.,* 51, 628, 1984.

96. **Ongradi, J., Cunningham, C., and Szilagyi, J. F.,** The role of polypeptides L and NS in the transcription process of vesicular stomatitis virus New Jersey using the temperature-sensitive mutant tsE1, *J. Gen. Virol.,* 66, 1011, 1985.

97. **Keene, J. D., Thornton, B. J., and Emerson, S. U.,** Sequence-specific contacts between the RNA polymerase of vesicular stomatitis virus and the leader RNA gene, *Proc. Natl. Acad. Sci. U.S.A.,* 78, 6191, 1981.

98. **Hunt, D. M., Smith, E. G., and Buckley, D. W.,** Aberrant polyadenylation by a vesicular stomatitis virus mutant is due to an altered L protein, *J. Virol.,* 52, 515, 1984.

99. **Horikami, S. M. and Moyer, S. A.,** Host range mutants of vesicular stomatitis virus defective in *in vitro* RNA methylation, *Proc. Natl. Acad. Sci. U.S.A.,* 79, 7694, 1982.

100. **De, B. P. and Banerjee, A. K.,** Specific binding of guanosine 5'-diphosphate with the NS protein of vesicular stomatitis virus, *Biochem. Biophys. Res. Commun.,* 114, 138, 1983.

101. **Hsu, C.-H., Morgan, E. M., and Kingsbury, D. W.,** Site-specific phosphorylation regulates the transcriptive activity of vesicular stomatitis virus NS protein, *J. Virol.,* 43, 104, 1982.

102. **Clinton, G. M., Burge, B. W., and Huang, A. S.,** Phosphoproteins of vesicular stomatitis virus: identity and interconversion of phosphorylated forms, *Virology,* 99, 84, 1979.

103. **Hsu, C.-H. and Kingsbury, D. W.,** NS phosphoprotein of vesicular stomatitis virus: subspecies separated by electrophoresis and isoelectric focusing, *J. Virol.,* 42, 342, 1982.

104. **Kingsford, L. and Emerson, S. U.,** Transcriptional activities of different phosphorylated species of NS protein purified from vesicular stomatitis virions and cytoplasm of infected cells, *J. Virol.,* 33, 1097, 1980.

105. **Marnell, L. L. and Summers, D. F.,** Characterization of the phosphorylated small enzyme subunit, NS, of the vesicular stomatitis virus RNA polymerase, *J. Biol. Chem.,* 259, 13518, 1984.

106. **Hsu, C.-H. and Kingsbury, D. W.,** Constitutively phosphorylated residues in the NS protein of vesicular stomatitis virus, *J. Biol. Chem.,* 260, 8990, 1985.

107. **Bell, J. C. and Prevec, L.,** Phosphorylation sites on phosphoprotein NS of vesicular stomatitis virus, *J. Virol.,* 54, 697, 1985.

108. **Imblum, R. L. and Wagner, R. R.,** Protein kinase and phosphorylation of vesicular stomatitis virus, *J. Virol.,* 13, 113, 1974.

109. **Moyer, S. A. and Summers, D. F.,** Phosphorylation of vesicular stomatitis virus *in vivo* and *in vitro, J. Virol.,* 13, 455, 1974.

110. **Harmon, S. A., Marnell, L. L., and Summers, D. F.,** The major ribonucleoprotein-associated protein kinase of vesicular stomatitis virus is a host cell protein, *J. Biol. Chem.,* 258, 15283, 1983.

111. **Sanchez, A., De, B. P., and Banerjee, A. K.,** *In vitro* phosphorylation of NS protein by the L protein of vesicular stomatitis virus, *J. Gen. Virol.,* 66, 1025, 1985.

112. **Watanabe, Y., Sakuma, S., and Tanaka, S.,** A possible biological function of the protein kinase associated with vaccinia and vesicular stomatitis virus, *FEBS Lett.*, 41, 331, 1974.

113. **Witt, D. J. and Summers, D. F.,** Relationship between virion-associated kinase-effected phosphorylation and transcription activity of vesicular stomatitis virus, *Virology*, 107, 34, 1980.

114. **Talib, S. and Banerjee, A. K.,** Protamine: a potent inhibitor of vesicular stomatitis virus transcriptase *in vitro, Biochem. Biophys. Res. Commun.*, 98, 875, 1981.

115. **Peluso, R. W. and Moyer, S. A.,** Vesicular stomatitis virus proteins required for the *in vitro* replication of defective interfering particle genome RNA, in *Nonsegmented Negative Strand Viruses*, Bishop, D. H. L. and Compans, R., Eds., Academic Press, San Francisco, 1984, 153.

116. **Arnheiter, H., Davis, N. L., Wertz, G. W., Schubert, M., and Lazzarini, R. A.,** Role of the nucleocapsid protein in regulating vesicular stomatitis virus RNA synthesis, *Cell*, 41, 259, 1985.

117. **Davis, N. L., Arnheiter, H., and Wertz, G. W.,** Vesicular stomatitis virus N and NS proteins form multiple complexes, *J. Virol.*, 59, 751, 1986.

118. **Gill, D. S., Chattopadhyay, D., and Banerjee, A. K.,** Identification of a domain within the phosphoprotein of vesicular stomatitis virus that is essential for transcription *in vitro, Proc. Natl. Acad. Sci. U.S.A.*, 83, 8873, 1986.

119. **Gill, D. S. and Banerjee, A. K.,** Vesicular stomatitis virus NS proteins: structural similarity without extensive sequence homology, *J. Virol.*, 55, 60, 1985.

120. **Gill, D. S. and Banerjee, A. K.,** Complete nucleotide sequence of the matrix protein mRNA of vesicular stomatitis virus (New Jersey serotype), *Virology*, 150, 308, 1986.

121. **Pringle, C. R.,** The tdCE and hrCE phenotypes: host range mutants of vesicular stomatitis virus in which polymerase function is affected, *Cell*, 15, 597, 1978.

122. **Moyer, S. A., Baker, S. C., and Lessard, J. L.,** Tubulin: a factor necessary for the synthesis of both Sendai virus and vesicular stomatitis virus RNAs, *Proc. Natl. Acad. Sci. U.S.A.*, 83, 5405, 1986.

123. **Hill, V. M., Harmon, S. A., and Summers, D. F.,** Stimulation of vesicular stomatitis virus *in vitro* RNA synthesis by microtubule-associated proteins, *Proc. Natl. Acad. Sci. U.S.A.*, 83, 5410, 1986.

124. **Gupta, K. C. and Kingsbury, D. W.,** Complete sequences of the intergenic and mRNA start signals in the Sendai virus genome: homologies with the genome of vesicular stomatitis virus, *Nucleic Acids Res.*, 12, 3829, 1984.

125. **Luk, D., Masters, P. S., Sanchez, A., and Banerjee, A. K.,** Complete nucleotide sequence of the matrix protein mRNA of human parainfluenza virus type 3 and three intergenic sequences, *Virology*, 156, 189, 1987.

126. **Schubert, M., Keene, J. D., Herman, R. C., and Lazzarini, R. A.,** Site on the vesicular stomatitis virus genome specifying polyadenylation and the end of the L gene mRNA, *J. Virol.*, 34, 550, 1980.

127. **Gupta, K. C. and Kingsbury, D. W.,** Polytranscripts of Sendai virus do not contain intervening polyadenylate sequences, *Virology*, 141, 102, 1985.

128. **McGeoch, D. J. and Dolar, A.,** Sequence of 200 nucleotides at the 3' terminus of the genome RNA of vesicular stomatitis virus, *Nucleic Acids Res.*, 6, 3199, 1979.

129. **Giorgi, C., Blumberg, B. M., and Kolakofsky, D.,** Sequence termination of the (+) leader RNA regions of the vesicular stomatitis virus Chandipura, Cocal and Piry serotype genomes, *J. Virol.*, 46, 125, 1983.

130. **Iverson, L. E. and Rose, J. K.,** Localized attenuation and discontinuous synthesis during vesicular stomatitis virus transcription, *Cell*, 23, 477, 1981.

131. **Ball, L. A. and White, C. N.,** Coupled transcription and translation in mammalian and avian cell-free systems, *Virology*, 84, 479, 1978.

132. **Emerson, S. U.,** Reconstitution studies detect a single polymerase entry site on the vesicular stomatitis virus genome, *Cell*, 31, 635, 1982.

133. **Ball, L. A. and Wertz, G. W.,** VSV RNA synthesis: how can you be positive?, *Cell*, 26, 143, 1981.

134. **Talib, S. and Hearst, J. E.,** Initiation of RNA synthesis *in vitro* by vesicular stomatitis virus: single internal initiation in the presence of aurintricarboxylic acid and vandyl ribonucleoside complexes, *Nucleic Acids Res.*, 11, 7031, 1983.

135. **Iverson, L. E. and Rose, J. K.,** Sequential synthesis of 5'-proximal vesicular stomatitis virus mRNA sequences, *J. Virol.*, 44, 356, 1982.

136. **Thornton, G. B., De, B. P., and Banerjee, A. K.,** Interaction of L and NS proteins of vesicular stomatitis virus with its template ribonucleoprotein during RNA synthesis *in vitro, J. Gen. Virol.*, 65, 663, 1984.

137. **Lazzarini, R. A., Chien, I., Yang, F., and Keene, J. D.,** The metabolic fate of independently initiated VSV mRNA transcripts, *J. Gen. Virol.*, 58, 429, 1982.

138. **Chanda, P. K. and Banerjee, A. K.,** Identification of promoter proximal oligonucleotides and a unique dinucleotide, pppGpC, from *in vitro* transcription products of vesicular stomatitis virus, *J. Virol.*, 39, 93, 1981.

139. **Robinson, W. S.,** Sendai virus RNA synthesis and nucleocapsid formation in the presence of cycloheximide, *Virology*, 44, 494, 1971.

140. **Wertz, G. W. and Levine, M.,** RNA synthesis by vesicular stomatitis virus and a small plaque mutant: effects of cycloheximide, *J. Virol.,* 12, 253, 1973.

141. **Perlman, S. M. and Huang, A. S.,** RNA synthesis of vesicular stomatitis virus. V. Interactions between transcription and replication, *J. Virol.,* 12, 1395, 1973.

142. **Breindl, M. and Holland, J. J.,** Studies on the *in vitro* transcription and translation of vesicular stomatitis virus mRNA, *Virology,* 73, 106, 1976.

143. **Condra, J. H. and Lazzarini, R. A.,** Replicative RNA synthesis and nucleocapsid assembly in vesicular stomatitis virus-infected permeable cells, *J. Virol.,* 36, 796, 1980.

144. **Hill, V. M., Marnell, L., and Summers, D. F.,** *In vitro* replication and assembly of vesicular stomatitis virus nucleocapsids, *Virology,* 113, 109, 1981.

145. **Davis, N. L. and Wertz, G. W.,** Synthesis of vesicular stomatitis virus negative strand RNA in vitro: dependence on viral protein synthesis, *J. Virol.,* 41, 821, 1982.

146. **Ghosh, K. and Ghosh, H. P.,** Synthesis *in vitro* of full-length genomic RNA and assembly of the nucleocapsid of vesicular stomatitis virus in a coupled transcription-translation system, *Nucleic Acids Res.,* 10, 6341, 1982.

147. **Hill, V. M. and Summers, D. F.,** Synthesis of VSV RNPs *in vitro* by cellular VSV RNPs added to HeLa cell extracts: VSV protein requirements for replication *in vitro, Virology,* 123, 407, 1982.

148. **Peluso, R. W. and Moyer, S. A.,** Initiation and replication of vesicular stomatitis virus genome RNA in a cell-free system, *Proc. Natl. Acad. Sci. U.S.A.,* 80, 3198, 1983.

149. **Carlsen, S. R., Peluso, R. W., and Moyer, S. A.,** In vitro replication of Sendai virus wild-type and defective interfering particle genome RNAs, *J. Virol.,* 54, 493, 1985.

150. **Wertz, G. W.,** Replication of vesicular stomatitis virus defective interfering particle RNA in vitro: transition from synthesis of defective interfering leader RNA to synthesis of full-length defective interfering RNA, *J. Virol.,* 46, 513, 1983.

151. **Patton, J. T., Davis, N. L., and Wertz, G. W.,** N protein alone satisfies the requirement for protein synthesis during replication of vesicular stomatitis virus, *J. Virol.,* 49, 303, 1984.

152. **Semler, B. L., Perrault, J., Abelson, J., and Holland, J. J.,** Sequence of a RNA templated by the 3'-OH RNA terminus of defective interfering particles of vesicular stomatitis virus, *Proc. Natl. Acad. Sci. U.S.A.,* 75, 4704, 1978.

153. **Schubert, M., Keene, J. D., Lazzarini, R. A., and Emerson, S. U.,** The complete sequence of a unique RNA species synthesized by a DI particle of VSV, *Cell,* 15, 103, 1978.

154. **Sprague, J., Condra, J. H., Arnheiter, H., and Lazzarini, R. A.,** Expression of a recombinant DNA gene coding for the vesicular stomatitis virus nucleocapsid protein, *J. Virol.,* 45, 773, 1983.

155. **Blumberg, B. M., Giorgi, C., Rose, K., and Kolakofsky, D.,** Preparation and analysis of the nucleocapsid proteins of vesicular stomatitis virus and Sendai virus, and the analysis of the Sendai virus leader-NP gene region, *J. Gen. Virol.,* 65, 769, 1984.

156. **Testa, D., Chanda, P. K., and Banerjee, A. K.,** *In vitro* synthesis of the full-length complement of the negative-strand genome RNA of vesicular stomatitis virus, *Proc. Natl. Acad. Sci. U.S.A.,* 77, 294, 1980.

157. **Chanda, P. K., Roy, J., and Banerjee, A. K.,** *In vitro* synthesis of genome length complementary RNA of vesicular stomatitis virus in the presence of inosine 5'-triphosphate, *Virology,* 129, 225, 1983.

158. **Perrault, J., Lane, J. L., and McClure, M. A.,** *In vitro* transcription alterations in a vesicular stomatitis virus variant, in *The Replication of Negative Strand Viruses,* Bishop, D. H. L. and Compans, R. W., Eds., Elsevier, New York, 1981, 829.

159. **Perrault, J., Clinton, G. M., and McClure, M. A.,** RNP template of vesicular stomatitis virus regulates transcription and replication functions, *Cell,* 35, 175, 1983.

160. **Perrault, J. and McLear, P. W.,** ATP dependence of vesicular stomatitis virus transcription initiation and modulation by mutation in the nucleocapsid protein, *J. Virol.,* 51, 635, 1984.

161. **Blumberg, B. M., Leppert, M., and Kolakofsky, D.,** Interaction of VSV leader RNA and nucleocapsid protein may control VSV genome replication, *Cell,* 23, 837, 1981.

162. **Hercyk, N., Horikami, S. M., and Moyer, S. A.,** The vesicular stomatitis virus L protein possesses the mRNA methyltransferase activities, *Virology,* 1987, in press.

163. **Chattopadhyay, D. and Banerjee, A. K.,** Phosphorylation within a specific domain of the phosphoprotein of vesicular stomatitis virus regulates transcription in vitro, *Cell,* 49, 407, 1987.

164. **Luk, D., Masters, P. S., Gill, D. S., and Banerjee, A. K.,** Intergenic sequences of the vesicular stomatitis genome (New Jersey serotype): evidence for two transcription initiation sites within the L gene, *Virology,* 160, 88, 1987.

165. **Rima, B. K., Baczko, K., Clarke, D. K., Curran, M. D., Martin, S. J., Billeter, M. A., and Ter Meulen, V.,** Characterization of clones for the sixth (L) gene and a transcriptional map for morbilliviruses, *J. Gen. Virol.,* 67, 1971, 1986.

166. **Dowling, P. C., Blumberg, B. M., Menonna, J., Adamus, J. E., Cook, P., Crowley, J. C., Kolakofsky, D., and Cook, S. D.,** Transcriptional map of the measles virus genome, *J. Gen. Virol.,* 67, 1987, 1986.

Chapter 8

INFLUENZA VIRAL RNA TRANSCRIPTION AND REPLICATION

Robert M. Krug

TABLE OF CONTENTS

I.   Introduction 160

II.   Viral Messenger RNA Synthesis 160

III.   Template RNA Synthesis 163

IV.   Virion RNA Synthesis and the Regulation of Viral Gene Expression 165

V.   Concluding Remarks 167

References 167

I. INTRODUCTION

Influenza virus is a negative strand RNA virus with a segmented genome. Essentially all the studies of the transcription and replication of influenza virus have been carried out with A strain viruses, which contain eight virion RNA (vRNA) segments. Three types of virus-specific RNAs are synthesized in infected cells: (1) viral messenger RNAs (mRNAs); (2) full-length copies of the vRNAs that serve as templates for vRNA replication; and (3) vRNAs. Much more information has been obtained about the mechanism of synthesis of the viral mRNAs than about the mechanism of synthesis of the template RNAs and vRNAs. We will discus the current state of knowledge about the syntheses of these three types of virus-specific RNAs.

II. VIRAL MESSENGER RNA SYNTHESIS

The synthesis of influenza viral mRNA requires initiation by host cell primers, specifically capped (m^7GpppNm-containing) RNA fragments derived from host cell RNA polymerase II transcripts.[1-5] This occurs in the nucleus of the infected cell.[6] As a consequence, viral mRNA synthesis requires the continuous functioning of the cellular RNA polymerase II and is inhibited by α-amanitin.[7] The host cell primers are generated by a viral cap-dependent endonuclease that cleaves the capped cellular RNAs 10 to 13 nucleotides from their 5′ ends, preferentially at a purine residue.[4] Transcription is initiated by the incorporation of a G residue onto the 3′ end of the resulting fragments, directed by the penultimate C residue of the vRNAs.[4] Viral mRNA chains are then elongated until a stretch of 4 to 7 uridine residues is reached 17 to 22 nucleotides before the 5′ ends of the vRNAs, where transcription terminates and polyadenylate (poly[A]) is added to the mRNAs.[8-10]

Viral mRNA synthesis is catalyzed by viral nucleocapsids[4,11] which consist of the individual vRNAs associated with four viral proteins, the nucleocapsid (NP) protein, and the three P (PB1, PB2, and PA) proteins.[11,12] The P proteins are responsible for viral mRNA synthesis, and some of their roles have been determined by analyses of the in vitro reaction catalyzed by virion nucleocapsids.[12-14] Ultraviolet light-induced cross-linking experiments showed that the three P proteins are in the form of a complex that starts at the 3′ ends of the vRNA templates and moves down the templates in association with the elongating mRNAs during transcription.[13] The PB2 protein in this complex recognizes and binds to the cap of the primer RNA.[12,13] Experiments with two virus temperature-sensitive mutants which have a defect in the vRNA segment coding for the PB2 protein remains associated with the cap during the first 11 to 15 nucleotides of chain growth.[13] The PB1 protein, which was initially found at the first residue (a G residue) added onto the primer, moves as part of the P protein complex to the 3′ ends of the growing viral mRNA chains, indicating that it most likely catalyzes each nucleotide addition.[13] Based on the relative positions of PB1 and PB2 on the nascent chains, it was concluded that the P protein complex most likely has the PB1 protein at its leading edge and the PB2 protein at its trailing edge. Figure 1 shows the model of the functions and movements of the P proteins during viral mRNA synthesis.[13]

The analysis of the transcription reaction catalyzed by virion nucleocapsids left unresolved many important questions about the mechanism of viral mRNA synthesis and the role of the P proteins. Thus, it was not determined how the complex of the three P proteins is assembled at the 3′ ends of the vRNA templates where transcription initiates. Because reinitiation of viral mRNA synthesis was extremely limited, or did not occur at all, with virion nucleocapsids,[13] it is not known whether reinitiation occurs in the infected cell and, if so, by what mechanism. In addition, it is not known which P protein(s) is the endonuclease and whether the PA protein has a definite function in viral mRNA synthesis. Two recent results may provide approaches to answer these questions.

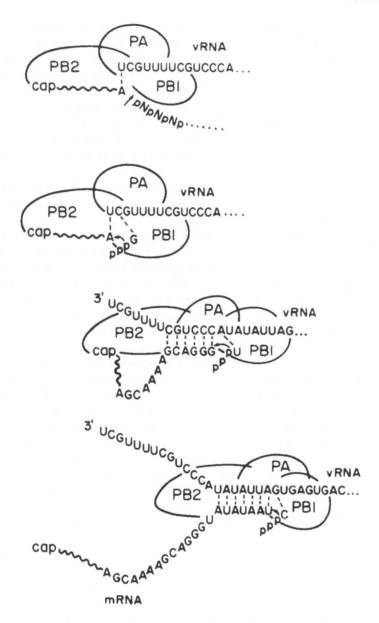

FIGURE 1. Model of the functions and movements of the three P proteins during capped, RNA-primed viral mRNA synthesis. The sequence shown is that of the vRNA and mRNA coding for the NP protein. (From Braam, J., Ulmanen, I., and Krug, R. M., *Cell*, 34, 609, 1983. With permission.)

First, it had been known that some of the P proteins in the infected cell are not associated with viral nucleocapsids,[15,16] and recently it has been shown that this pool of P proteins is large and that the protein in this pool are predominantly, if not entirely, in the form of complexes of the three P proteins.[17] When nucleocapsid-depleted cytoplasmic and nuclear extracts were subjected to immunoprecipitation using either an anti-PB1 or an anti-PB2 antiserum, all three P proteins were precipitated, indicating that the P proteins were in a complex that was largely resistant to disruption by the detergents present in the innumo-precipitation buffer.[17] One of the detergents was sodium dodecyl sulfate (at a concentration of 0.1%), attesting to the strength of the association of the P proteins in the complex. Sucrose density gradient analysis showed that the P protein complexes had sedimentation values ranging from about 11 to 22 S and that almost all of the PB1 and PB2 proteins molecules synthesized during a one hour period (2.5 to 3.5 hr postinfection) were in these complexes.[17] Little or no free PB1 or PB2 protein was detected. These P protein complexes were predominantly in the nuclear fraction of infected cells, where viral mRNA synthesis occurs.[17] Consequently, it is likely that it is these complexes, rather than a particular one of the P proteins, that recognizes and binds to the 3′ ends of the vRNAs to initiate mRNA synthesis. In addition, these P protein complexes may be involved in the reinitiation of viral mRNA synthesis. Future studies on the activities of these P protein complexes, both by themselves (e.g., cap-dependent endonuclease) and in combination with viral nucleocapsids, may be expected to provide new information about the mechanism of viral mRNA synthesis. Some of the P protein complexes in this nonnucleocapsid pool may also be involved in the initiation of template RNA and vRNA synthesis (see below).

The other potential breakthrough is the successful production of large amounts of the individual P proteins using baculovirus vectors.[18] Earlier experiments in which the influenza viral P genes were expressed in eukaryotic cells employed bovine papillomavirus vectors.[19,20] The P proteins synthesized using these vectors were functional in that they complemented the temperature-sensitive defects of viral mutants. However, the level of production of the P protein was insufficient for use in biochemical experiments.[19] In contrast, when the influenza viral P genes were inserted into baculovirus vectors under the control of the extremely strong polyhedrin promoter, large amounts of each of the three P proteins were synthesized in insect cells.[18] This should allow reconstitution of P protein complexes and the determination of their activity both by themselves and in combination with viral nucleocapsid templates. The experiments with the baculovirus vectors have already shown that the PB1 and PB2 proteins have an intrinsic ability to form a complex with each other. Thus, in cells infected simultaneously with both the PB1 and PB2 baculovirus recombinants, a PB1-PB2 complex was formed that was immunoprecipitated by either an anti-PB1 or an anti-PB2 antiserum.[18] In contrast, the PA protein was not incorporated into an immunoprecipitable P protein complex in baculovirus infected cells.[18] In cells infected simultaneously with all three P baculovirus recombinants, a PB1-PB2 complex lacking the PA protein was precipitated with an anti-PB1 or an anti-PB2 antiserum. This suggests the possibility that the incorporation of PA into an immunoprecipitable P protein complex in influenza virus-infected cells requires the participation of other influenza viral gene product(s). Finally, recent experiments indicate that it may also be possible to reconstitute influenza viral ribonucleoprotein templates. The addition of a bacterially expressed NP protein (containing 32 heterologous amino terminal amino acids) to various RNAs generated by SP6 polymerase transcription resulted in the formation of complexes which were similar to authentic influenza virus ribonucleoproteins in ultrastructure and in the ratio of nucleotide residues to NP molecules.[21] Interestingly, the binding of NP to form these ribonucleoproteins was not specific for influenza virus nucleotide sequences. Perhaps specificity is imparted by the P protein complexes.

The novel mechanism of viral mRNA synthesis, which requires initiation by cellular-capped RNA primer in the nucleus of the infected cell, is the apparent target of the interferon (IFN)-induced Mx protein,[22] a 75,000-molecular-weight protein that accumulates in the nucleus.[23-25] In mouse cells, the antiviral state induced by IFN α/β against influenza virus is controlled by the host Mx gene. Only cells that possess this gene develop an efficient antiviral state against influenza virus after exposure to IFN α/β, whereas the antiviral state against other viruses is independent of the Mx gene. Initially, it was reported that in Mx-containing (Mx$^+$) cells IFN α/β caused the inhibition of the translation of apparently functional viral mRNAs in the cytoplasm, but did not affect the synthesis of viral mRNAs in the nucleus.[26] However, this was not confirmed. Rather, it was shown that in Mx$^+$ mouse embryo cells treated with IFN α/β viral mRNA synthesis in the nucleus was severely inhibited.[22] In Mx[+] cells treated with IFN, the amount of viral mRNA synthesized as a result of primary transcription, i.e., transcription in the presence of the protein synthesis inhibitor, anisomycin, was drastically reduced.[22] Only two viral mRNAs could be detected by Northern analysis and by translating the poly A-containing (poly A[+]) RNA from infected cells in wheat germ extract: a reduced amount of the mRNA for nonstructural protein 1 (NS1) and an even lower amount of the mRNA for the matrix (M1) protein. The other viral mRNAs were not made in detectable amounts. In addition, the rate of viral mRNA synthesis catalyzed by the inoculum transcriptase, measured by in vitro RNA synthesis catalyzed by permeabilized cells, was severely inhibited.[22] In contrast, IFN treatment of cells lacking the Mx gene (Mx−) had little or no effect on either the steady-state level or the rate of synthesis of viral mRNAs made by the inoculum transcriptase. These results suggested that the IFN-induced Mx gene product, which accumulates in the nucleus, inhibits viral mRNA synthesis in the nucleus. No Mx-specific effect acting directly on viral protein synthesis in the cytoplasm was detected.[22] Thus, the NS1 and M1 viral mRNAs that continued to be synthesized in IFN-treated Mx[+] cells were translated in vivo, and the amount of this translation was similar to the amount of translation observed when the poly A(+) RNA from these cells was assayed in wheat germ extracts. Subsequently, it was shown that transfection of the cloned DNA encoding the Mx protein into Mx-negative cells caused the establishment of a specific antiviral state against influenza virus,[27] strongly suggesting that the Mx protein by itself causes the inhibition of influenza viral mRNA synthesis.

III. TEMPLATE RNA SYNTHESIS

The first step in the replication of influenza vRNA is the switch from viral mRNA synthesis to the synthesis of template RNAs, i.e., the full-length copies of vRNA that then serve as templates for vRNA replication. This switch requires (1) a change from the capped RNA-primed initiation of transcription used during mRNA synthesis to unprimed initiation; and (2) antitermination at the poly(A) site, 17 to 22 nucleotides from the 5′ ends of the vRNAs, that is used during mRNA synthesis.[8,28] In vivo the switch from mRNA template RNA synthesis requires the synthesis of one of more virus-specific proteins.[28,29]

Progress in identifying these proteins and in determining their roles has come from the establishment of an in vitro system that catalyzes the synthesis of template RNA as well as of viral mRNA.[30,31] Nuclear extracts prepared from infected cells were shown to be active in at least one of the steps involved in the switch from mRNA to template RNA synthesis, the antitermination step.[31] In these experiments, M13 single-stranded DNA, specific for transcripts copied off the NS vRNA, the smallest vRNA, was used to measure the NS1 mRNA and NS template RNA synthesized by the nuclear extract. In the absence of an added primer, these extracts synthesized only low levels of NS1 mRNA and NS template RNA. Addition of a high concentration (0.4 m*M*) of the dinucleotide ApG, which had been shown to act as a primer for viral mRNA synthesis catalyzed by virion nucleocapsids,[32,35] greatly stimulated the synthesis of both NS1 mRNA and NS template RNA catalyzed by these nuclear extracts.[31] Consequently, these nuclear extracts contained the factor(s) that cause antitermination at the poly(A) site used during viral mRNA synthesis, but were deficient in unprimed initiation of template RNA synthesis and in the capped primers needed for viral mRNA synthesis. The addition of ApG circumvented the inefficient, unprimed initiation and, hence, allowed the analysis of the mechanism of antitermination. In contrast to ApG, the addition of a capped RNA primer stimulated the synthesis of only NS1 mRNA; little or no NS template RNA was synthesized.[31] Consequently, viral RNA transcripts that initiated with a capped primer were not antiterminated by the nuclear factor(s) that antiterminated the ApG-initiated viral transcripts.

The antitermination factor(s) could be separated from the viral nucleocapsid templates by ultracentrifugation of the nuclear extract.[31] This ultracentrifugation yielded a pellet fraction that contained the viral nucleocapsids active in viral mRNA synthesis, but not in template RNA synthesis, and a supernatant fraction which contained the antitermination factor. When the supernatant, which had essentially no activity by itself, was added to the pellet in the presence of ApG, template RNA synthesis was restored.[31] Depletion experiments in which this supernatant was incubated with protein A Sepharose® containing antibodies to individual viral proteins demonstrated that the viral NP protein was the viral protein that was required for antitermination.[31] In contrast, depletion of the NS1 protein did not eliminate antitermination activity, suggesting that the NS1 protein does not participate directly in antitermination. Consequently, in addition to the population of NP molecules associated with viral nucleocapsids, there is a population of NP molecules free of nucleocapsids that is required for antitermination during template RNA synthesis.

The mechanism by which the NP protein causes antitermination has not been determined. The most likely possibility is that NP acts by binding to the viral RNA transcript rather than by binding to the P protein complex catalyzing transcription. It has been shown that template RNAs in infected cells are in the form of nucleocapsids containing NP,[8,34] so that it can be presumed that the templates synthesized in vitro also become coated with NP to form nucleocapsids. Indeed, with another negative strand RNA virus, vesicular stomatitis virus, the NP protein is also required for antitermination in vitro,[35] and the resulting RNAs are in the form of nucleocapsids.[35-37] In this case, there is only a single vRNA template that has a termination signal near its 3' end. In the absence of NP protein, RNA synthesis terminates, yielding a small 47-nucleotide-long RNA, and the transcriptase then reinitiates at the cap site of the first downstream mRNA sequence. However, most likely as a consequence of the binding of NP to a sequence in the nascent 47-nucleotide-long RNA, antitermination occurs and a full-length template RNA is made. This binding site is also apparently the site for the initiation of nucleocapsid assembly.[38] With influenza virus, there are eight vRNA templates, all of which have termination signals at their 5' ends rather than their 3' ends.[8,28] One possibility is that, as with vesicular stomatitis virus, the NP initially binds to the nascent transcripts at a sequence close to the site of termination, both causing antitermination and initiating nucleocapsid assembly. However, the eight viral RNA transcripts do not have a common sequence in this region. An alternative possibility is that NP binds at, or close to, the common 12-nucleotide-long sequence at the 5' ends of the nascent transcripts. Subsequent addition of NP molecules to the growing chains would allow readthrough when the termination site is reached. The latter hypothesis would provide an explanation for the observation that influenza viral RNA transcripts initiated with a capped primer were not antiterminated in the presence of the NP molecules that were active in the antitermination of ApG-initiated transcripts.[31] Perhaps the 5'-terminal cap structure and/or the primer-donated sequence preceding the common 5' sequence of the viral transcripts blocks the binding of the NP protein. An alternative explanation would be that a P transcription complex that initiates with a capped RNA primer might be different from the complex found after unprimed or ApG-primed initiation, and that this different structure might not allow recognition of the antitermination signal. In any case, it is clear that the type of initiation used by the transcriptase largely determines whether termination of transcription occurs, which is not the case for vesicular stomatitis virus. The termination of all capped RNA-primed transcripts at the poly(A) site likely ensures that these transcripts, which contain host sequences at their 5' ends, are used only as mRNAs and not as templates for vRNA replication. It is conceivable that if these transcripts had copies of the 5' ends, as well as of the 3' ends, of the vRNAs, they might inadvertently be recognized as templates by the replicating enzymes.

IV. VIRION RNA SYNTHESIS AND THE REGULATION OF VIRAL GENE EXPRESSION

Little is known about the second step in vRNA replication, i.e., the copying of template RNA into vRNA. This synthesis almost certainly occurs without a primer, since the vRNAs contain a triphosphorylated 5' end.[39] It might then be anticipated that the P protein complexes involved in vRNA synthesis would differ from those involved in capped, RNA-primed viral mRNA synthesis. Recent experiments using nonaqueous fractionation of cells showed that vRNA synthesis occurs in the nucleus,[40] thereby establishing that all virus-specific RNA synthesis is nuclear.[6,40,41] The vRNAs, like the viral mRNAs, are efficiently transported to the cytoplasm.[40] In contrast, the template RNAs are sequestered in the nucleus, where they direct vRNA synthesis throughout infection.[40]

The copying of template RNA into vRNA is an important point of regulation during the early phase of virus infection. During the early phase (prior to 2.5 hr in BHK-21 cells), the synthesis of specific vRNAs, viral mRNAs, and viral proteins were coupled.[8,40,42] The first event detected after primary transcription was the synthesis of template RNAs, presumably copied off the parental vRNAs. Approximately equimolar amounts of each of the template RNAs were made. The peak rate of template RNA synthesis occurred early (1.5 hr postinfection in BHK-21 cells), and then sharply declined. Specific template RNAs were selectively transcribed into vRNAs. Specifically, the NS and NP vRNAs were preferentially synthesized early, whereas M vRNA synthesis was delayed. The rate of synthesis of a particular vRNA correlated with, and therefore most likely determined, the rate of synthesis of the corresponding mRNA and of its encoded protein. Thus, the NS1 and NP mRNAs and proteins were preferentially synthesized at early times, whereas the synthesis of the M1 mRNA and protein were delayed. Hence, the control of viral protein synthesis during the early phase is predominantly a direct consequence of the regulation of vRNA synthesis. It will be important to determine the mechanism by which the synthesis of specific vRNAs is turned on during the early phase. This will require the establishment of in vitro systems that initiate and elongate vRNA chains.

Early reports suggested that the relationships between the syntheses of vRNAs, viral mRNAs, and viral proteins that occurred during the first phase of infection continued at later times.[8,42,43] However, a recent study has shown that these relationships change dramatically during the second phase of infection.[40] This study employed single-stranded M13 DNAs specific for various influenza viral genomic segments to analyze the synthesis of virus-specific RNAs in infected cells. It was shown that the rate of synthesis of all the vRNAs remained at, or near, maximum during the second phase, whereas the rate of synthesis of all the viral mRNAs dramatically decreased.[40] All the viral mRNAs behaved similarly. They had a peak rate of synthesis at the same time, 2.5 hr postinfection in BHK-21 cells, and the subsequent reduction in their rates of synthesis were identical. By 4.5 hr in BHK-21 cells, the rate of synthesis of all the viral mRNAs was 5% the maximum rate. Thus, vRNA and viral mRNA syntheses were not coupled during this second phase. In addition, viral mRNA and protein synthesis were not coupled, as the synthesis of all the viral proteins continued at maximum levels during the second phase.[40] Previously synthesized viral mRNAs were undoubtedly used to direct viral protein synthesis. Figure 2 diagrams the relationships between the syntheses of template RNAs, vRNAs, viral mRNAs, and viral proteins during the two phases of virus infection.[40]

A significant part of the control mechanisms in influenza virus-infected cells is directed at the preferential synthesis of the NP and NS1 proteins early and at delaying the synthesis of the M1 protein. The NP and NS1 proteins are synthesized early presumably because they are needed for template RNA and/or vRNA synthesis. As noted above, NP molecules not associated with nucleocapsids have been shown to be required for the antitermination step

A

EARLY PHASE

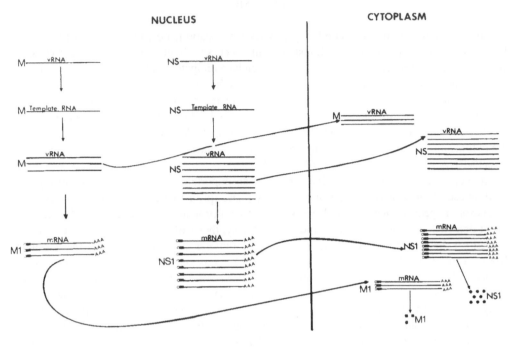

B

LATE PHASE

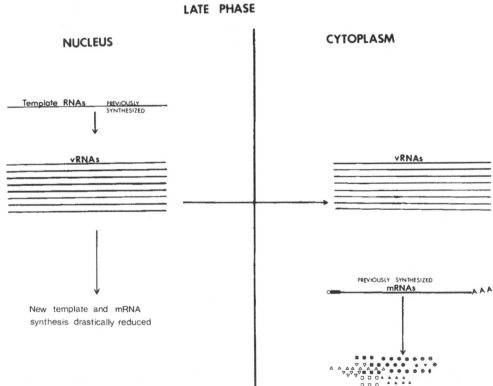

FIGURE 2. The relationships between the synthesis of template RNAs, vRNAs, viral mRNAs, and viral proteins during the early (A) and late (B) phases of infection. (From Shapiro, G. I., Gurney, T., Jr., and Krug, R. M., *J. Virol.*, 61, 764, 1987. With permission.)

that occurs as part of the switch from viral mRNA to template RNA synthesis. However, it has not yet been established that the NS1 protein is involved in template RNA and/or vRNA synthesis. It is conceivable that the synthesis of the M1 protein is delayed because this protein may be involved in the transition between the early and late phases of viral infection, i.e., in stopping the transcription of vRNA into viral mRNA. The matrix (M) protein of another negative strand RNA virus, vesicular stomatitis virus, has been implicated in the shutdown of viral RNA transcription,[44-47] and the influenza viral M1 protein has been shown to inhibit viral RNA transcription in vitro.[48] Perhaps the influenza viral M1 protein in the infected cell selectively interacts with the nucleocapsids containing vRNAs to inhibit the transcription of vRNA into mRNA, but does not interact with the nucleocapsids containing template RNAs since the transcription of template RNA into vRNA continues. In addition, such a selective association of the M1 protein could be involved in the selective transport of vRNAs, but not of template RNAs, from the nucleus. This hypothesis would predict that some M1 protein would be in the nucleus, which has been observed by some investigators.[15,49,50]

V. CONCLUDING REMARKS

Current knowledge about influenza viral RNA transcription and replication is sketchy. The basic mechanism of capped RNA-primed viral mRNA synthesis and some of the roles of the viral P proteins in this process are known. Future studies will need to resolve the remaining unanswered questions about the roles of the P proteins in viral mRNA synthesis and to establish the mechanism by which the IFN-induced Mx protein specifically inhibits influenza viral mRNA synthesis. With regard to the switch from viral mRNA to template RNA synthesis, it is known that the NP protein is required for antitermination and that this protein can only act on transcripts that are initiated without a capped RNA primer. However, the mechanism by which the NP protein causes antitermination has not been established; nor has it been ruled out that other viral proteins are involved. In addition, the proteins involved in the unprimed initiation of template RNAs have not been identified. Finally, the viral proteins involved in the unprimed initiation and elongation of vRNA chains have not been identified, and the mechanism of the selective initiation of specific mRNAs has not been determined.

REFERENCES

1. **Bouloy, M., Plotch, S. J., and Krug, R. M.,** Globin mRNAs are primers for the transcription of influenza viral RNA *in vitro, Proc. Natl. Acad. Sci. U.S.A.,* 75, 4886, 1978.
2. **Plotch, S., Bouloy, M., and Krug, R. M.,** Transfer of 5' terminal cap of globin mRNA to influenza viral complementary RNA during transcription *in vitro, Proc. Natl. Acad. Sci. U.S.A.,* 76, 1618, 1979.
3. **Krug, R. M., Broni, B. A., and Bouloy, M.,** Are the 5' ends of influenza viral mRNAs synthesized *in vivo* donated by host mRNAs?, *Cell,* 18, 329, 1979.
4. **Plotch, S. J., Bouloy, M., Ulmanen, I., and Krug, R. M.,** A unique cap (mGpppXm)-dependent influenza virion endonuclease cleaves capped RNAs to generate the primers that initiate viral RNA transcription, *Cell,* 23, 847, 1981.
5. **Krug, R. M.,** Priming of influenza viral RNA transcription by capped heterologous RNAs, *Curr. Top. Microbiol. Immunol.,* 93, 125, 1981.
6. **Herz, C., Stavnezer, E., Krug, R. M., and Gurney, T., Jr.,** Influenza virus, and RNA virus, synthesizes its messenger RNA in the nucleus of infected cells, *Cell,* 26, 391, 1981.
7. **Mark, G. E., Taylor, J. M., Broni, B., and Krug, R. M.,** Nuclear accumulation of influenza viral RNA and the effects of cyclohexamide actinomycin D and alpha amanitin, *J. Virol.,* 29, 744, 1979.
8. **Hay, A. J., Lomniczi, B., Bellamy, A. H., and Skehel, J. J.,** Transcription of the influenza virus genome, *Virology,* 83, 337, 1977.

9. **Hay, A. J., Abraham, G., Skehel, J. J., Smith, J. C., and Fellner, P.,** Influenza virus messenger RNAs are incomplete transcripts of the genome RNAs, *Nucleic Acids Res.*, 4, 4197, 1977.

10. **Robertson, J. S., Schubert, M., and Lazzarini, R. A.,** Polyadenylation sites for influenza virus mRNA, *J. Virol.*, 38, 157, 1981.

11. **Inglis, S. C., Carroll, A. R., Lamb, R. A., and Mahy, B. W. J.,** Polypeptides specified by the influenza virus genome. I. Evidence for eight distinct gene products specified by fowl plague virus, *Virology*, 74, 489, 1976.

12. **Ulmanen, I., Broni, B. A., and Krug, R. M.,** Role of two of the influenza virus core P proteins in recognizing cap 1 structures (m^7 GpppNm) on RNAs and in initiating viral RNA transcription, *Proc. Natl. Acad. Sci. U.S.A.*, 78, 7355, 1981.

13. **Braam, J., Ulmanen, I., and Krug, R. M.,** Molecular model of a eucaryotic transcription complex: functions and movements of influenza P proteins during capped RNA primed transcription, *Cell*, 34, 609, 1983.

14. **Ulmanen, I., Broni, B. A., and Krug, R. M.,** Influenza virus temperature-sensitive cap (m^7GpppNm)-dependent endonuclease, *J. Virol.*, 45, 27, 1983.

15. **Hay, A. J. and Skehel, J. J.,** Studies on the synthesis of influenza virus proteins, in *Negative Strand Viruses*, Vol. 2, Mahy, B. W. J. and Barry, R. D., Eds., Academic Press, London, 1975, 635.

16. **Krug, R. M. and Etkind, P. R.,** Cytoplasmic and nuclear virus-specific proteins in influenza virus-infected MDCK cells, *Virology*, 56, 334, 1973.

17. **Detjen, B. M., St Angelo, C., Katze, M. G., and Krug, R. M.,** The three influenza virus polymerase (P) proteins not associated with viral nucleocapsids in the infected cell are in the form of a complex, *J. Virol.*, 61, 16, 1987.

18. **St. Angelo, C., Smith, G. E., Summers, M. D., and Krug, R. M.,** Two of the three influenza viral polymerase (P) proteins expressed using baculovirus vectors form a complex in insect cells, *J. Virol.*, 61, 361, 1987.

19. **Braam-Markson, J., Jaudon, C., and Krug, R. M.,** Expression of a functional influenza viral cap-recognizing protein by using a bovine papilloma virus vector, *Proc. Natl. Acad. Sci. U.S.A.*, 82, 4326, 1985.

20. **Krystal, M., Ruan, L., Lyles, D., Paulakis, G., and Palese, P.,** Expression of the three influenza virus polymerase proteins in a single cell allows growth complementation of viral mutants, *Proc. Natl. Acad. Sci. U.S.A.*, 83, 2709, 1986.

21. **Kingsbury, D. W., Jones, I. M., and Murti, K. G.,** Assembly of influenza ribonucleoprotein *in vitro* using recombinant nucleoprotein, *Virology*, 156, 396, 1987.

22. **Krug, R. M., Shaw, M., Broni, B., Shapiro, G., and Haller, O.,** Inhibition of influenza viral mRNA synthesis in cells expressing the interferon-induced Mx gene product, *J. Virol.*, 56, 201, 1985.

23. **Haller, O., Arnheiter, H., Lindenmann, J., and Gresser, I.,** Host gene influences sensitivity to interferon action selectively for influenza virus, *Nature (London)*, 283, 660, 1980.

24. **Horisberger, M. A., Staeheli, P., and Haller, O.,** Interferon induces a unique protein in mouse cells bearing a gene for resistance to influenza virus, *Proc. Natl. Acad. Sci. U.S.A.*, 80, 1910, 1983.

25. **Dreiding, P., Staehli, P., and Haller, O.,** Interferon-induced protein Mx accumulates in nuclei of mouse cells expressing resistance to influenza viruses, *Virology*, 141, 192, 1985.

26. **Meyer, T. and Horisberger, M. A.,** Combined action of mouse α and β interferons in influenza virus-infected macrophages containing the resistance gene Mx, *J. Virol.*, 49, 709, 1984.

27. **Staeheli, P., Haller, O., Boll, W., Lindenmann, J., and Weissmann, C.,** Mx protein: constitutive expression in 3T3 cells transformed with cloned Mx cDNA confers selective resistance to influenza virus, *Cell*, 44, 147, 1986.

28. **Hay, A. J., Skehel, J. J., and McCauley, J.,** Characterization of influenza virus RNA complete transcripts, *Virology*, 116, 517, 1982.

29. **Barrett, T., Wolstenholme, A. J., and Mahy, B. W. J.,** Transcription and replication of influenza virus RNA, *Virology*, 98, 211, 1979.

30. **Beaton, A. R. and Krug, R. M.,** Synthesis of the templates for influenza virion RNA replication *in vitro*, *Proc. Natl. Acad. Sci. U.S.A.*, 81, 4682, 1984.

31. **Beaton, A. R. and Krug, R. M.,** Transcription antitermination during influenza viral template RNA synthesis requires the nucleocapsid protein and the absence of a 5' capped end, *Proc. Natl. Acad. Sci. U.S.A.*, 83, 6282, 1986.

32. **McGeoch, D. and Kitron, N.,** Influenza virion RNA-dependent RNA polymerase: stimulation by guanosine and related compounds, *J. Virol.*, 15, 686, 1975.

33. **Plotch, S. J. and Krug, R. M.,** Influenza virion transcriptase: synthesis *in vitro* of large, polyadenylic acid-containing complementary RNA, *J. Virol.*, 21, 24, 1977.

34. **Krug, R. M.,** Cytoplasmic and nucleoplasmic viral RNPs in influenza virus-infected MDCK cells, *Virology*, 50, 103, 1972.

35. **Patton, J. T., Davis, N. L., and Wertz, G.,** N protein alone satisfies the requirement for protein synthesis during RNA replication of vesicular stomatitis virus, *J. Virol.,* 49, 303, 1984.

36. **Patton, J. T., Davis, N. L. E., and Wertz, G.,** Cell-free synthesis and assembly of vesicular stomatitis virus nucleocapsids, *J. Virol.,* 45, 155, 1983.

37. **Peluso, R. W. and Moyer, S. A.,** Initiation and replication of vesicular stomatitis virus genome RNA in a cell-free system, *Proc. Natl. Acad. Sci. U.S.A.,* 80, 3198, 1983.

38. **Blumberg, B. M., Leppert, M., and Kolakofsky, D.,** Interaction of VSV leader RNA and nucleocapsid protein may control VSV genome replication, *Cell,* 23, 837, 1981.

39. **Young, R. J. and Content, J.,** 5'-terminus of influenza virus RNA, *Nature (London),* 230, 140, 1971.

40. **Shapiro, G. I., Gurney, T., Jr., and Krug, R. M.,** Influenza viral gene expression: control mechanisms at early and late times of infection and nuclear-cytoplasmic transport of virus-specific RNAs, *J. Virol.,* 61, 764, 1987.

41. **Hay, A. J., Lomniczi, B. A., Bellamy, H., and Skehel, J. J.,** Transcription of the influenza virus genome, *Virology,* 83, 337, 1977.

42. **Smith, G. L. and Hay, A. J.,** Replication of the influenza virus genome, *Virology,* 118, 96, 1982.

43. **Inglis, S. C. and Mahy, B. W. J.,** Polypeptides specified by the influenza virus genome, 3: control of synthesis in infected cells, *Virology,* 95, 154, 1979.

44. **Carroll, A. R. and Wagner, R. R.,** Role of the membrane (M) protein in endogenous inhibition of *in vitro* transcription by vesicular stomatitis virus, *J. Virol.,* 29, 134, 1979.

45. **Clinton, G. M., Little, S. P., Hagen, F. S., and Huang, A. S.,** The matrix (M) protein of vesicular stomatitis protein regulates transcription, *Cell,* 15, 1455, 1978.

46. **De, B. P., Thornton, G. B., Luk, D., and Banerjee, A. K.,** Purified matrix protein of vesicular stomatitis virus blocks viral transcription, *in vitro, Pro. Natl. Acad. Sci. U.S.A.,* 79, 7137, 1982.

47. **Pal, R., Grinnell, B. W., Snyder, R. M., and Wagner, R. R.,** Regulation of viral transcription by the matrix protein of vesicular stomatitis virus probed by monoclonal antibodies and temperature-sensitive mutants, *J. Virol.,* 56, 386, 1985.

48. **Zvonarjev, A. Y. and Ghendon, Y. Z.,** Influence of a membrane (M) protein on influenza A virus virion transcriptase *in vitro* and its susceptibility to rimantadine, *J. Virol.,* 33, 583, 1980.

49. **Oxford, J. S. and Schild, G. C.,** Immunological studies with influenza virus matrix protein: negative strand viruses, 2, 611, 1975.

50. **Gregoriades, A.,** Influenza virus-induced proteins in nuclei and cytoplasm of infected cells, *Virology,* 79, 449, 1977.

RNA Replication of Double-Stranded RNA Viruses

Chapter 9

REPLICATION OF THE REOVIRIDAE: INFORMATION DERIVED FROM GENE CLONING AND EXPRESSION

G. W. Both

TABLE OF CONTENTS

I.   Introduction 172

II.   Strategies for Gene Cloning and Identification 172

III.   Reovirus Genes and Proteins 173
       A.   Segment S1 174
             1.   Protein σ_1 174
             2.   The 14-kdalton Protein 176
       B.   Segment S2; Protein σ_2 176
       C.   Segment S3; Protein σ NS 177
       D.   Segment S4; Protein σ_3 177

IV.   Rotavirus Genes and Proteins 177
       A.   Morphogenesis of Rotaviruses: Proteins to Study Cellular
             Processes 178
       B.   Segment 4; Protein VP3 178
       C.   Segment 5; Protein NS53 180
       D.   Segment 6; Protein VP6 180
       E.   Segment 7 180
       F.   Segment 8 181
       G.   Segment 9 181
             1.   VP7 Gene Structure and Immunogenicity 181
             2.   Immunogenicity of Expressed VP7 182
             3.   Expression and Processing of VP7 183
             4.   Cellular Location of VP7 184
       H.   Segment 10 184
       I.   Segment 11 185

V.   Other Reoviridae Genes 186
       A.   Bluetongue Virus Genes and Proteins 186
             1.   Segment L2; Protein VP2 186
             2.   Segment L3; Protein VP3 186
             3.   Segment M5; Protein VP5 186
       B.   Gene Segments of Wound Tumor Virus 186

VI.   Conclusion 187

Acknowledgment 187

References 187

I. INTRODUCTION

The Reoviridae are a family of viruses comprising six genera with a host range extending from insects and plants to vertebrates and mammals. The viruses are characterized primarily by their double-stranded RNA (dsRNA) genomes which vary from 10 to 12 segments in number. Mammalian reovirus, the prototype of the orthoreoviruses, is the best studied member of the family, despite the fact that reoviruses are not associated with any major human disease. The ease of propagation and purification of this virus has greatly facilitated its characterization.[1] Rotaviruses, characterized in animals in the late 1960s and in man as recently as 1973, have both veterinary and medical importance, being the major etiologic agents in acute gastroenteritis in the young in many parts of the world.[2-5] At first, only some bovine strains and the simian strain SA11 could be readily propagated in the laboratory, but the more fastidious human isolates can now also be adapted to grow in tissue culture. The orbiviruses are less well characterized, but viruses such as bluetongue (BTV) are of economic importance in the livestock industry.[6] In comparison with the above genera, even less is known about the cypo-, phytoreo-, and fijiviruses, the prototype viruses for which are cytoplasmic polyhedrosis virus (CPV), wound tumour virus (WTV), and Fijivirus, respectively. The latter genera in particular are more difficult to work with experimentally, and little molecular biology has been done for many of them.[7] Most of the work covered in this chapter will therefore deal with the orthoreo- and rotavirus genera.

It is not intended that all aspects of Reovirdiae replication will be covered in this chapter. Other comprehensive reviews are available.[2-7] Rather, since recent work has concentrated on gene cloning, characterization, and expression, the intention is to relate new information from this work to other data concerning viral replication. There were three main reasons for this interest. The first related to the possible existence of uncharacterized Reoviridae gene products, an analogy drawn from work on RNA viruses, such as influenza, where gene cloning and sequencing exposed cryptic reading frames in gene segments and confirmed the existence of suspected protein products.[8,9] Second, structure/function studies of RNA genes may be facilitated by the availability of DNA copies. For example, expression of a normally minor viral polypeptide in a suitable host/vector system may allow the determination of its biochemical properties. Specific mutations in the gene, introduced easily at the DNA, but not the RNA level, can be useful in identifying functionally important regions of the protein. Third, there are more practical implications for gene cloning and expression. For viruses with medical and veterinary importance, cloned genes or sequences derived from them may be useful diagnostic reagents, and expression of genes encoding the important antigens of the virus might also permit the eventual development of molecular vaccines. A necessary prerequisite for these studies was to devise a way to synthesize DNA copies of the dsRNA segments of these genomes.

II. STRATEGIES FOR GENE CLONING AND IDENTIFICATION

One principal feature classifying viruses as members of the family Reoviridae is a genome containing between 10 and 12 dsRNA segments.[1] Segment sizes have been accurately estimated for reo- and rotaviruses as varying between 1200 to 3825 and 660 to 3400 base-pairs, respectively.[10-12] By comparison with reovirus, genome segments for orbiviruses (BTV) range between 850 and 3800 base-pairs, the total size of the genome being considerably smaller than that of reoviruses. The other genera have similar fragments whose sizes have not been as accurately estimated.[7] A further useful property of the Reoviridae is that virus particles contain the enzymes necessary for mRNA transcription and modification.[1] The inherent stability of these activities in viral cores has made it possible to synthesize large amounts of mRNA and to study transcription in vitro. Such studies led to the finding that

CPV, reovirus, rotavirus, and WTV contained 5' terminal m[7]G cap structures at the 5' end of their plus-stranded mRNAs.[13-17] This structure is also present on the plus strand of the dsRNAs of reovirus,[18] CPV,[13] and rotavirus,[19] but the 5' end of the minus strand in these viruses is not capped. Genomic dsRNA segments and viral mRNAs also lack 3' terminal polyadenylation found on most other eukaryotic mRNAs,[20] an important consideration in the design of cloning strategies involving these RNA species.

Protocols for cloning Reoviridae RNAs were devised considering the terminal structures elucidated for the segments and assuming that the dsRNAs were flush ended, the plus strand being identical to the single-stranded (ss) mRNAs transcribed by viral cores.[18] dsRNA was generally used as the starting material, the rationale, at least for rotaviruses, being that genes from noncultivable isolates might be cloned using dsRNA isolated from fecal specimens. The strategies are similar in principle, but differ in detail.[11,21-30] The mixture of dsRNA segments was modified enzymatically by 3' terminal polyadenylation or the addition of oligo(C)$_{15}$. The dsRNAs were then denatured with heat, DMSO, or methylmercury hydroxide and both strands transcribed simultaneously into ss cDNA using AMV reverse transcriptase primed by oligo(dT)$_{12-18}$ or oligo(dG)$_{10}$. RNA templates were destroyed using alkali, pancreatic ribonuclease, or ribonuclease H and the ss cDNAs annealed to yield a mixture of dsDNAs which were partial and complete gene copies. To maximize the yields of flush-ended dsDNAs, these were enzymatically repaired. Flush-ended DNAs were generally dC-tailed and cloned into a dG-tailed plasmid vector. In one case, cDNA/RNA hybrids were dC-tailed and cloned directly.[31] To improve the chances of recovering full-length clones of the large gene segments, short cDNA transcripts were eliminated before the annealing step.[10,25] In alternative strategies,[32,33] mRNA transcripts were polyadenylated, reverse copied into cDNA, tailed with dC residues, treated with alkali, and copied into dsDNA using oligo(dG) to prime synthesis by reverse transcriptase. Flush-ended dsDNA copies were then cloned as described above.

Cloned genes were identified by their ability to hybridize with cDNA copies of specific RNA segments.[11,21-30,32,33] Radiolabeled cDNAs were synthesized on denatured templates by reverse transcriptase using random or oligo(dT) primers, and dsRNAs or adenylated RNAs, respectively. In some cases, cloned DNAs were radiolabeled by nick translation and used to probe denatured dsRNA segments by Northern hybridization. Alternatively, radiolabeled RNA segments were used to probe unlabeled cloned DNA. Clones selected as full length on the basis of size were confirmed as complete copies by comparison of their terminal sequences with those for the dsRNA segments.[10,22,34] Most cloning strategies were designed such that full-length clones would have oligo(dG)/oligo(dT) and oligo(dA)/oligo(dC) homopolymer tracts at the 5' and 3' ends, respectively.[11,29] A remaining problem was to determine the orientation of a cloned gene in the absence of any sequence information for the dsRNA segment or the protein it encoded. For the first rotavirus clones, the negative sense sequence was determined by using DNA fragments from the clone to prime cDNA synthesis by reverse transcriptase on mRNA template in the presence of dideoxy nucleoside triphosphates.[11,35] For WTV, terminally labeled, plus-sense mRNA transcripts were hybridized with ssDNAs derived by subcloning the gene copy into the bacteriophage M13; only clones carrying the negative-sense strand hybridized.[29]

III. REOVIRUS GENES AND PROTEINS

All ten genes of the Dearing strain of reovirus type 3 have been cloned as full-length copies, as determined by a comparison between their terminal sequences and those of the original dsRNA segments.[10] However, with the exception of the four S segments discussed below, they have not been further characterized. Gene products of the L and M segments and their properties have been discussed.[14,15,18,36,37]

A. Segment S1

1. Protein σ_1

One of the most interesting and important reovirus proteins, with respect to replication and immunity, is the minor outer capsid polypeptide σ_1, which was identified as the product of the S1 segment by direct translation of denatured dsRNA and studies of recombinant viruses.[18,36] The role of σ_1 as the serotype-specific protein and the polypeptide responsible for tissue localization was defined using genetic reassortants. The pathogenicity of viruses carrying different combinations of genes from reovirus serotypes 1, 2, and 3 varied in mice with the type of σ_1 protein, showing that this polypeptide was solely responsible for tissue localization.[36,37] σ_1 also determines the capacity of reovirus to spread from the intestine in mice,[38] causes binding of virus particles to cellular microtubules, and is the major target of cytotoxic T lymphocytes and other types of T cells.[37] It may also play a role in the maintenance of persistent infections.[39] A product of segment S1 also causes inhibition of cellular DNA synthesis (see below).[37]

Monoclonal antibodies directed against σ_1 have been used to define at least three domains on the molecule. One of these is involved in cytotoxic T cell recognition and type-specific neutralization, another has unknown function, and the third is involved in erythrocyte binding.[37] Since σ_1 is the protein responsible for cell attachment and antibodies directed against it prevent binding of the virus to cells,[36,40] presumably the latter domain interacts with a receptor molecule on the surface of the cell. A monoclonal antibody directed against the proposed receptor domain of σ_1 was used to generate an anti-idiotype antibody[41,42] capable of recognizing the σ_1 binding site on the receptor itself. This antibody immunoprecipitated a monomeric glycoprotein of 67 kdaltons present on mouse, monkey, rat, and human cells and diverse tissue types including lymphoid and neuronal cells, suggesting the receptor may be associated with normal cellular functions.[42] In contrast, the reovirus type 3 receptor on human erythrocytes has been characterized as glycophorin (type A), a glycoprotein of 31 kdaltons which is present in 4.5 to 5 x 10^5 copies per cell.[42a] From studies on the kinetics of binding of ^{125}I-labeled virus or σ_1 anti-idiotype antibodies to various cell types, it has been estimated that there are 20,000 to 80,000 virus binding sites per cell.[42,43]

The S1 segment has now been cloned and sequenced for all three reovirus serotypes.[28,30,44] The segments for types 1 and 2 are more similar to each other than to the serotype 3 gene, a relationship also deduced for the respective proteins from serological data; types 1 and 2 cross react, while type 3 is specific. The greater homology between the type 1 and 2 S1 segments contrasts with that seen for the other nine type 3 genes which are more closely related to serotype 1.[44] The S1 segment for serotype 3 is 1416 bases long, has 5' and 3' noncoding regions of 12 and 39 nucleotides, respectively, and an open reading frame which codes for a protein of 455 amino acids (molecular weight 49,071). The type 1 and 2 segments, actually longer than that for type 3, encode shorter proteins of 418 and 399 residues, respectively, due to extensive 3' noncoding regions.[44] There are three sites for attachment of N-linked carbohydrate in σ_1 type 3 and numerous sites in types 1 and 2, but no evidence that these are used.[44] ^{3}H-mannose is not incorporated into virus[45] and there is no convincing signal peptide at the N terminus of any σ_1 protein.[46] These factors argue against the attachment of mannose-rich carbohydrate such as that found on rotavirus glycoproteins.[47,48] However, low levels of ^{3}H-glucosamine and galactose apparently can be incorporated into σ_1 and other reovirus proteins and, being sensitive to 2-deoxyglucose, these may be O-linked carbohydrates.[45]

The most interesting feature of σ_1 to emerge from the characterization of its gene relates to the probable structure of the protein.[28] A heptapeptide repeat pattern occurs between amino acids 28 and 158 and the region is flanked by proline residues. Proline and aromatic amino acids are absent from the repeat, but within it, the first and fourth residues are generally hydrophobic. These features are characteristic of a α-helical coiled-coil structure,

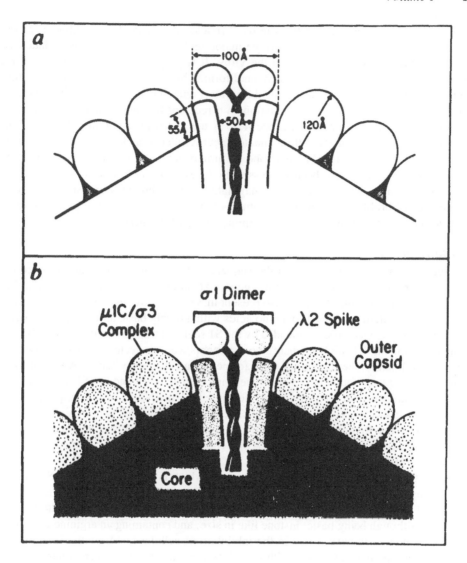

FIGURE 1. Schematic representation of the morphology of the outer capsid of reovirus type 3. (a) Dimensions of the virion. (b) Orientation of the σ_1 protein in the virus. The α-helical region is shown to extend through the λ_2 channel into the viral core. The globular-like structure sits on top of the λ_2 channel and interacts with host cell receptors. (From Bassel-Duby, R., Jayasuriya, A., Chatterjee, D., Sonenberg, N., Maizel, J. V., and Fields, B. N., *Nature (London)*, 315, 421, 1985. With permission. Copyright 1985 Macmillan Journals, Ltd.)

and computer analyses of σ_1 secondary structure support this contention. To form a coiled-coil, σ_1 must at least exist as a dimer and it was estimated that there are 24 σ_1 molecules per virion located on the 12 spikes of the outer capsid.[18] However, recent work suggests that in its undisrupted form the σ1 multimer may be composed of four subunits.[28a] Computer analysis of the remainder of σ_1 indicates that a variety of structural forms exist, suggesting that the C terminal region has an overall globular-like structure.[28] Broadly speaking, there appears to be a structural role for the N terminal and a functional role for the C terminal region. σ_1 also exhibits structural relatedness to nematode myosin and rabbit tropomyosin, however, the significance of this is not yet clear.[28] A representation of the reovirus outer capsid, shown in Figure 1, is based on the predicted structure of σ_1 and information derived from other work. For example, the phenotype of mutations in the spike protein λ_2 can be suppressed by accompanying mutations in σ_1.[49] In addition, monoclonal antibodies against λ_2 also neutralize reovirus, although σ_1 is the major neutralizing antigen.[37] These data argue for the close proximity of σ_1 and λ_2 in the virus. It is proposed that σ_1 exists as a dimer with the coiled-coil contained within a channel formed by λ_2. Only the globular C terminal domain of σ_1 protrudes at the surface of the virus (Figure 1). In support of the functional importance of this domain, a point mutation in a receptor-binding mutant selected with a monoclonal antibody[50] mapped to residue 419 in the C terminal region.[28] Of four other monoclonally selected variants with attenuated neurovirulence and restricted cell tropism, three also mapped to residue 419 and one to amino acid 340 (Fields, B. N., personal communication). σ_1 has now been expressed in *Escherichia coli* as a 47-kdalton fusion polypeptide which appears capable of hemagglutinating erythrocytes and specific binding to mouse L cell fibroblasts.[51] With these domains functionally expressed, it may be possible to analyze them further using site-directed mutagenesis of the gene.

2. The 14-kdalton Protein

The second feature to emerge from the sequence of segment S1 was the presence of a second, shorter, open reading frame in the gene, extending in type 3 from base 71 to base 430.[28,30,44] A similar reading frame is also present in S1 from reovirus types 1 and 2.[44] These each code for a protein of about 120 amino acids with a highly basic amino terminal region. The existence of a second product encoded by the S1 segment was foreshadowed by ribosome binding studies using S1 mRNA where two fragments were protected by 80-S ribosomes[52] and two formylmethionine dipeptides were formed in a cell-free system.[53] A protein of 14 kdaltons has since been found in reovirus type 3-infected cells[54,55] and in cell-free translation systems programmed with S1 mRNA,[54,56] denatured S1 dsRNA,[55] or transcripts produced in vitro from a truncated S1 segment.[57] Thus, reovirus S1 codes for two proteins and is the first segment of the Reoviridae demonstrated to be bicistronic. This is achieved by virtue of the sequences flanking the first two AUG codons. The consensus sequence favoring initiation of protein synthesis[58,59] is present for the second (14 kdalton) AUG codon, but absent from the first one (for σ_1). Thus, ribosomes which bind at the 5' end of S1 mRNA, "scan" along the message, sometimes by-passing the "weak" σ_1 initiation site to begin translation at the second, "strong" AUG codon.[52,59] The function of the 14-kdalton protein is unknown, although being basic, histone like in size, and containing an arginine-rich amino terminal region, it may have a role in the inhibition of cellular DNA synthesis in infected cells,[60] a phenotype that segregates with the S1 gene segment.[37] However, it has also been observed that the inhibition of DNA synthesis is reversible by the removal of the replication complex (nuclei) from infected cells, suggesting that structural or metabolic cellular integrity is also required for inhibition.[61] As a step towards determining its function, the 14-kdalton protein has been expressed in a prokaryotic host/vector system under the control of the temperature-inducible λ P_L promoter. A 14-kdalton product expressed after induction reacts specifically with rabbit antisera made against synthetic peptides, the sequence of which was predicted from the S1 cDNA sequence between bases 71 and 430. Immunofluorescence studies indicate that the 14-kdalton protein accumulates in the cytoplasm of reovirus-infected L cells,[60] a result which appears inconsistent with the histone-like properties suggested for this protein.

B. Segment S2; Protein σ_2

The S2 segment codes for σ_2, an inner core protein of reovirus.[18] The gene is 1329 base-pairs in length with an open reading frame of 331 codons (molecular weight 37,204). A second reading frame of 85 codons begins at base 1020, but it is not known if it is used.[21] The inferred amino acid sequence of σ_2 offers no clue, nor is there presently available other data concerning the function of the protein.

C. Segment S3; Protein σ NS

Segment S3 encodes a major nonstructural protein of reovirus, σ NS.[18] The gene is 1198 base-pairs in length with a single open reading frame extending from base 28 to 1124 (366 codons; molecular weight 41,061).[62] Bases 28 to 30 are used for initiation, as shown by ribosome binding studies using S3 mRNA.[63]

σ NS is known to bind to ssRNA, but not to dsRNAs, and may possibly play a role in assembling the ten reovirus ssRNA segments into subviral particles prior to dsRNA synthesis during infection. Analysis of the amino acid sequence inferred for σ NS shows that there are distinct clusters of charged amino acids within predicted helical regions, some of which could conceivably interact with RNA.[62] A poly(C)-dependent RNA polymerase activity associated with particles containing the protein has also been reported.[18]

In an attempt to obtain more of the protein for study, the S3 gene was incorporated into a plasmid vector and expressed under the control of the λ P_L promoter in *E. coli*. Upon derepression of the promoter at the permissive temperature, synthesis of a 41-kdalton protein (r σ NS) was observed.[64] This protein was stable over a 6-hr chase period and constituted ≈ 6 to 7% of the total cellular protein at 3 hr postinduction. The authenticity of r σ NS was confirmed by its size, its V8 protease pattern, compared with that for σ NS from reovirus infected cells, and its reactivity in immunoblots with antiserum against σ NS. The recombinant protein was purified to virtual homogeneity from *E. coli* cell lysates by a combination of centrifugation, salt extraction, and polyA-agarose column chromatography, and its ability to bind various RNAs in vitro was tested. Like authentic σ NS, r σ NS did not require a 5′ m⁷G cap structure on the RNA for binding, and both proteins showed a distinct preference for binding ssRNA over dsRNA from reovirus and CPV. r σ NS also bound to rRNAs and, to some extent, tRNAs. Binding to ssRNAs was also inhibited by GTP. The lack of apparent specificity of the σ NS proteins for reovirus ssRNAs is surprising if this polypeptide is involved in selection and condensation of ssRNAs during infection, and implies the involvement of some other host or virus factor.[64]

D. Segment S4; Protein σ₃

The S4 segment codes for the virion protein σ_3, a major component of the outer capsid and the first polypeptide removed by treatment of virions with chymotrypsin.[18] σ_3 may be responsible for inhibiting cellular RNA and protein synthesis following infection,[37] and in vitro can stimulate translation of late viral mRNA, suggesting it has a role as a lateviral-mRNA-specific initiation factor.[64a] Mutations in the gene are required for the initiation of persistent infections in cultured cells.[37] The protein also binds dsRNA, suggestive of a role in viral morphogenesis. The gene segment is 1196 bases in length with a single open reading frame between nucleotides 33 and 1127.[65] The 5′ proximal AUG codon is functional; it is protected by ribosomes in vitro and occurs within the consensus sequence considered favorable for initiation.[14] Thus, σ_3 consists of 365 amino acids (molecular weight 41,164). Brief exposure of the protein to chymotrypsin yields polypeptide fragments of 14 and 11.5 kdaltons, but the cleavage sites cannot be deduced from the inferred amino acid sequence due to the number of Phe, Tyr, Trp, and other residues which could be involved.[65]

IV. ROTAVIRUS GENES AND PROTEINS

In contrast with reovirus, not all the segments of the rotavirus genome have been reported cloned, largely due to the considered importance of the genes encoding viral antigens and the interest in studying them first for the purposes of vaccine development. A second reason for the interest in particular rotavirus genes relates to the distinctive morphogenetic pathway of rotaviruses which has been reviewed recently[2-5,66] but is summarized here for the purposes of later discussion.

A. Morphogenesis of Rotaviruses: Proteins to Study Cellular Processes

The complete morphogenetic pathway of rotaviruses is by no means elucidated, but it seems that virus assembly begins in the cytoplasm in inclusion bodies referred to as viroplasms.[2] The inner viral capsid proteins, VP1, VP2, and VP6, probably condense with RNA segments to form core-like structures. Particles visible at the periphery of the viroplasm bud through adjacent endoplasmic reticulum (ER) membrane into the lumenal space, becoming transiently enveloped in the process. Evidence from immuno-electron microscopy[67] and analysis of the viral carbohydrate[47,48,68] suggests that the nonstructural glycoprotein NS29 is located in the membrane; it may act as a receptor for budding particles.[67] The major outer capsid protein, VP7, is translocated across the ER membrane of the infected cell[2-5], but remains amino-terminally anchored.[68,69] The protein is acquired by maturing particles in a calcium-dependent process,[70,70a] but it is not clear whether this occurs during budding or by condensation of VP7 onto the surface of particles after the removal of the temporary envelope. Neither is it clear when VP3, the other major outer capsid protein, is acquired. The mechanism by which the envelope is removed in the ER is also undefined, but appears to involve NS29.[66] Virus particles do not seem to be found elsewhere in the cell,[2] particularly at early times postinfection[69] and only double-shelled virus is released from the cell following lysis; other subviral structures, including single-shelled particles remain in the cell and may be associated with the cytoskeleton.[71]

The type of carbohydrate attached to a glycoprotein is a sensitive indicator of where the protein has been in the cell. Carbohydrate attached to VP7 and NS29 is sensitive to endoglycosidase H and therefore N linked.[2-5,68,69,72] In N-linked glycosylation reactions, a glucose$_3$-mannose$_9$-N-acetylglucosamine$_2$ (Glc$_3$Man$_9$GlcNac$_2$) core is transferred from a dolichol pyrophosphate carrier to an asparagine residue on the nascent polypeptide chain. This core glycosylation is then trimmed and eventually modified in the Golgi apparatus when more complex sugars, e.g., sialic acid are added.[73] Examination of the ^{3}H-mannose-labeled VP7 glycopeptides prepared from purified, double-shelled particles showed that the original Glc$_3$Man$_9$GlcNac$_2$ oligosaccharide had been processed, the major species being Man$_8$GlcNac$_2$; significant amounts of the Man$_7$, Man$_6$, and Man$_5$ forms were also present. NS29 carried mostly the Man$_9$ form, with traces of Man$_8$. Complex glycopeptides, such as those found on Sindbis virus glycoproteins, were absent, indicating that VP7 and NS29 never reached the Golgi apparatus, but were retained in the ER.[47,68,69] The availability of cloned genes for glycoproteins directed to the ER is comparatively rare; they therefore provide tools to probe the cellular location of some carbohydrate processing enzymes and to identify the signal(s) involved in directing proteins to the ER.[68,69]

B. Segment 4; Protein VP3

The fourth largest segment of SA11 codes for the 88-kdalton outer capsid polypeptide VP3[2,4,5,74] which is the hemagglutinin of the virus and the one responsible for limiting the growth of some strains in tissue culture.[5] VP3 also has a major role in determining gastrointestinal tract virulence of viruses in a mouse model system.[75] Proteolytic cleavage of VP3 enhances viral infectivity,[76] perhaps by enhancing uncoating of the virus in the cell;[3] the protein is cleaved by trypsin into VP5* (60 kdaltons) and VP8* (28 kdaltons) polypeptides.[2-5] This arrangement is similar to that seen for hemagglutinin, the major antigen of influenza virus,[8,9] but contrasts with the situation for reovirus where the protease-sensitive site is located on an outer capsid protein, μ1C (encoded by segment M2), and the viral hemagglutinin σ$_1$ is coded for by another segment, S1.[18,37]

The nucleotide sequence of gene segment 4 has now been determined for SA11,[77,77a] the human strains RV-5[77b] and Wa (Y. Furuichi, personal communication) and rhesus rotavirus.[77c] For SA11, the gene is 2362 nucleotides long with 5' and 3' noncoding regions of 9 and 25 bases, respectively. The protein is predicted to be 776 amino acids in length with a molecular weight of 86,751. The N terminal amino acid sequence predicted for VP3 does not appear to contain a signal sequence,[46] and the protein is not glycosylated,[2-5] despite the presence of many potential sites for N-linked glycosylation.[77] It would seem, by these criteria, that VP3 is not translocated across the ER membrane and virus particles must acquire the protein prior to budding and before they acquire VP7.

It was reported that rotavirus structural proteins have blocked amino termini making it difficult to determine their N terminal sequences.[26,77] However, VP5*, the trypsin cleavage product of VP3, was amenable to sequencing by Edman degradation; VP8* was not.[77] Based on their respective molecular sizes, VP8* was therefore derived from the amino terminal one third of VP3; VP5* comprised the C terminal two thirds of the molecule. N terminal sequencing revealed two species of VP5* which differed in size by six amino acids. The major species corresponded to the sequence predicted from the gene 4 clone, beginning at amino acid 247. The minor species indicated cleavage between residues 241/242.[77] These cleavage sites are also conserved in the other sequences of VP3 which have been determined.[77c,d] The intervening six amino acids may be removed by cleavage at both sites, as proposed for fowl plague influenza hemagglutinin.[8,9] The amino acid sequence through the cleavage site(s) was determined for VP3 in several other human and animal rotaviruses. There were significant amino acid differences among the animal strains, e.g., between porcine, simian, and bovine strains and these, in turn, differed from the human isolates examined which were similar to each other. However, in all but two noncultivated bovine isolates, the tryptic cleavage sites were invariant.[77d] A monoclonal antibody against VP8* was detected which prevented cleavage of VP3 in NCDV and SA11 when added to these rotaviruses prior to trypsin treatment, suggesting that the cleavage region, or a nearby site, was immunogenic.[78] Therefore, some of the amino acid changes near the cleavage sites may have occurred due to antigenic drift under immune selective pressure, as observed for influenza virus.[79] Another monoclonal antibody against VP5* could discriminate between two similar viruses isolated 3 months apart in the same hospital ward, also suggesting variation had occurred in an epitope.[80] Part of the SA11 VP3 gene, including that encoding the cleavage region, has been expressed in *E. coli* as an MS2 RNA polymerase/VP3 fusion protein. This hybrid polypeptide induced antibodies in mice which inhibited hemagglutination and neutralized SA11 infectivity.[80a]

Monoclonal antibodies raised against VP3 fall into six neutralization groups, and these were used to select antigenic variants which escaped neutralization. The amino acid changes in these variants mapped predominantly to VP8*. However, one group of three antibodies which showed heterotypic neutralization selected variants that mapped to a hydrophobic region in VP5*. This region shares homology with putative fusion sequences of Sindbis and Semliki Forest viruses and represents a logical target for future vaccine studies.[77c] Perhaps this homology also explains why the N terminus of VP5* is not hydrophobic, in contrast to Sendai and influenza viruses, for which cleavage produced a hydrophobic N terminus involved in virus penetration.[81,82] Alternatively, rotaviruses may interact with the cell in a different way. Chymotrypsin, which cleaves VP3, but does not enhance rotavirus infectivity, may cleave VP3 between residues 245/246, generating a different N terminus for VP5*.[77] It is conceivable that the failure of host cells to correctly cleave VP3 could limit the spread of rotavirus infection, as proposed for influenza virus.[83]

Until recently, VP7 was considered to be the major neutralizing antigen of rotaviruses, but it is now recognized that VP3 is also capable of inducing neutralizing antibodies.[84-86] The specificity associated with them can be segregated in reassortant viruses and may differ from the serotype specificity elicited by VP7.[84,86] These observations probably account for the paradox whereby some rotavirus strains appeared to possess dual serotype specificity.[5,87] It has also been reported that a monoclonal antibody against VP3 passively protects suckling mice against rotavirus challenge.[88] Clearly, VP3 cannot be ignored as an antigen in the development of molecular vaccines.

C. Segment 5; Protein NS53

Gene segment 5, which codes for the nonstructural protein NS53 (or NCVP2)[2] has not been reported cloned for SA11. However, the sequence was recently determined for a DNA copy cloned from the bovine rotavirus RF.[88a] The gene is 1581 bases in length with 5′ and 3′ noncoding regions of 32 and 73 nucleotides, respectively. The predicted protein of 491 amino acids has a calculated size of 58,654, but an apparent molecular weight of 54 kdaltons. The full-length, nonfused protein has been stably expressed in soluble form in *E. coli* using an inducible expression vector. Purification of the protein will facilitate the production of monospecific, polyclonal antibodies to probe the function of NS53 which remains unknown.

D. Segment 6; Protein VP6

In all rotaviruses studied so far, gene segment 6 codes for VP6, a nonglycosylated, 41-kdalton protein which is the major component of single-shelled particles.[2-5] VP6 may exist as a trimer in its native state in the virus, the subunits associating via hydrophobic and/or charge interactions, rather than via intermolecular disulfide bonding.[89] Perhaps the association of VP6 into oligomers precedes the formation of viral core particles during infection. VP6 may also have, contribute to, or influence by conformation, the virion-associated RNA polymerase activity; $CaCl_2$-treated cores which lack VP6 lack enzymic activity.[90]

Antigenic relationships of rotaviruses have been studied by a variety of assays such as plaque neutralization reduction, immune electron microscopy (IEM), ELISA, complement fixation, and immune adsorption hemagglutination (IAHA).[2-5] This led to confusion as to the number of types of rotaviruses which had been defined until it was shown that the antigen that reacted with neutralizing antibodies in plaque reduction tests was separate from the antigen detected by IEM, IAHA, ELISA, and complement fixation tests.[5] It is now clear that the latter tests detected VP6, while neutralizing antibodies react with VP7 and VP3. There are at least two rotavirus subgroups and possibly three[4] which can be distinguished. VP6 also has two nonoverlapping antigenic domains which can be resolved using monoclonal antibodies; one of these is recognized as common to all rotaviruses studied so far.[5,91] This common antigen is not shared with viruses in other genera of the Reoviridae.

The segment encoding VP6 has now been cloned and sequenced for two subgroup 1 viruses (SA11,[33,92] Bovine RF strain[93]) and one subgroup 2 virus (Wa).[92] In all cases, the gene is 1356 nucleotides long, with 5′ and 3′ noncoding regions of 23 and 142 nucleotides, respectively. The single, open reading frame (bases 24 to 1212) codes for a protein of 397 amino acids (molecular weight 44,816, for SA11). The bovine (RF) and SA11 subgroup 1 isolates are 97% homologous at the amino acid level and only 2 of the 12 changes are nonconservative.[93] A greater difference exists between the subgroup 1 (SA11) and 2 (Wa) proteins, however, where there is 90% amino acid homology (34 changes). Most changes (25 out of 34) are clustered between residues 39 to 62, 80 to 122, or 281 to 315. This fairly minimal change in protein sequence presumably reflects the limitations imposed on variation by the structural requirements of the rotavirus core.[92] In the absence of information on the tertiary structure of VP6, it is not possible to identify those regions which contribute to the variable and cross-reactive epitopes in the protein.

VP6 has now been expressed in high yields (20 to 150 μg/ 10^6 cells) in the baculovirus system under the control of the strong polyhedrin promoter.[93a] The protein, isolated from the cells or the medium, reacted with monoclonal antibodies against the native structure. Expressed VP6 was also able to spontaneously assemble into morphologic subunits and was immunogenic in guinea pigs. However, while the antisera detected homologous and heterologous rotaviruses by several tests, no neutralizing activity was detected in plaque reduction assays.[93a]

E. Segment 7

SA11 gene segment 7 codes for the nonstructural protein NS34 (or NCVP4).[94,95] However, for most rotaviruses, genome segments 7, 8, and 9 are not easily resolved by polyacrylamide gel electrophoresis, and their relative order of migration varies depending on the strain. In principle, the identity of a gene in this region can now be determined for any isolate by Northern hybridization using available cloned gene probes,[96] or by comparison of partial nucleotide sequences from the unknown segment with those determined for other rotavirus genes.

SA11 gene segment 7 is 1104 nucleotides long with 5' and 3' noncoding regions of 25 and 134 nucleotides, respectively.[97] The longest open reading frame consists of 315 codons (bases 26 to 970). However, it seems likely that the second AUG codon might be used for initiation since it is of the consensus type.[58,59] Therefore, NS34 is probably 312 amino acids in length (molecular weight 36,072). The gene from the UK bovine strain (segment 9) has also been sequenced[98] and is only 1076 nucleotides long. The open reading frame is 313 codons in length, but protein probably consists of 310 amino acids because, again, the second initiation codon appears likely to be used. The genes and proteins are 75.5 and 76.6% homologous, respectively. Although ten complementation groups have been described for rotavirus *ts* mutants,[99] not all have yet been assigned to gene segments.[5,100] The phenotype of such mutants may eventually provide a clue as to the function of NS34, which at present is unknown. Segment 7 appears sufficiently conserved between strains of rotaviruses that DNA clones may prove useful in identifying clinical specimens of the virus in further epidemiological studies.[101]

F. Segment 8

Despite being slightly smaller than segment 7, SA11 gene segment 8 codes for a larger protein, NS35 (or NCVP3). The gene is 1059 base-pairs long with 5' and 3' noncoding regions of 46 and 59 nucleotides, respectively.[11] The equivalent gene (segment 7) from the UK bovine strain is identical in size and arrangement and is 88% homologous at the nucleotide level.[24] The NS35 proteins are 317 amino acids long (SA11 molecular weight, 36,628) with only 12 amino acid differences between them, all of which are conservative.[24] Presumably, this reflects a constraint on variation imposed by function. Viruses carrying *ts* mutations in this gene[100] have an RNA negative phenotype[99] and produce a large proportion of empty particles,[102] suggesting a role for NS35 in assembly of subviral particles or replication of RNA. Without either process, ssRNA cannot be amplified by transcription. The high positive charge of NS35 at neutral pH[11,24] would be consistent with a role in binding RNA. Immunoelectronmicroscopy also shows that NS35 is located in viroplasms where RNA synthesis and assembly occurs.[67] Comparison of NS35 and reovirus σ NS (for which a ssRNA binding role is suggested,[18,62,64] see above), using a variety of computer programmes shows no significant similarity between them.

G. Segment 9

Genome segment 9 of SA11 codes for the glycoprotein VP7[94,95] which is the major serotype antigen and component of the outer capsid in double-shelled rotavirus particles.[2-5] This gene was one of the first studied for two reasons: (1) the antigenic importance of VP7 and its perceived relevance to the eventual development of a molecular vaccine, and (2) the potential of VP7 to serve as a model for the mechanism involved in directing proteins to the ER, a consequence of the unusual mode of replication of rotaviruses (see above).

1. VP7 Gene Structure and Immunogenicity

The first VP7 gene sequenced was that from SA11,[94] but its features are conserved in the equivalent gene from all other strains which have been studied.[26,31,32,103-106,106a] The gene is 1062 base-pairs in length with 5' and 3' noncoding segments of 48 and 36 nucleotides, respectively, and codes for a protein of 326 residues (molecular weight 37,197). The genes from Wa (serotype 1), S2 (serotype 2), and NCDV (serotype 6) are 76.3, 74.2, and 80.4% homologous with SA11 (serotype 3) at the nucleotide level. Particularly notable is the sequence conservation among serotypes between bases 33 and 72 (spanning the first potential initiation codon), where only one nucleotide varies in all four genes. The sequences flanking this region are also highly conserved. Whether this is due to the requirement to conserve some structural feature in the RNA or the protein is not clear.

At the protein level, these same serotypes show homologies of 82.2 (59 changes), 75.4 (80 changes), and 84.7% (50 changes), respectively, with the sequence predicted for SA11 VP7. Comparison of the protein sequences for these four antigenically distinguishable VP7 proteins reveals that the Cys residues and several extensive regions of amino acids are conserved, probably reflecting an overall similar architecture for the different VP7 molecules. In other regions, amino acid differences are clustered, and some of these must account for the antigenic differences between serotypes. Many of these changes also fall in hydrophilic regions of VP7 and may have surface locations. Six peptides corresponding to these hydrophilic, variable domains were made. However, none were able to induce virus-neutralizing antibodies in rabbits.[31] New information has identified residues within two of these peptides as antigenically important. Monoclonal antibodies against VP7 were used to select antigenic variants.[107] The amino acid changes detected mapped to three of the variable regions defined by sequence comparison. Specifically, changes in residues 94, 96 (A region), 147 (B region), 211, and 223 (C region) were selected under antibody pressure. Peptides spanning residues 90 to 103 and 208 to 225 failed to elicit neutralizing antibodies,[31] perhaps because, as suggested by competitive antibody binding studies,[107,108] these residues contribute to a conformational antigenic site which the peptides alone cannot mimic. The immunogenicity of this site is apparently maintained in a nonreduced, 14-kdalton fragment of VP7, whose location within the molecule is unknown.[109] One amino acid change at residue 211 (Asp to Asn)[107] affected the ability of polyclonal antiserum to bind virus, possibly due to the presence of carbohydrate on a newly created glycosylation site.[107,110] Other mutants for which the antigenicity of an immunodominant epitope of SA11 was changed have also been selected using polyclonal antisera; these have not been characterized.[111] The potential for changing immunogenicity by masking antigenic regions with carbohydrate has been observed for influenza viruses[112] and, on this basis, suggested as a possible reason for antigenic differences between VP7 proteins.[31] SA11 VP7 has only one glycosylation site, while other strains in different serotypes have as many as three,[32] two of which are used.[113]

2. Immunogenicity of Expressed VP7

Recently, the crystallographic structures of polio[114] and rhinoviruses[115] have been determined. This work elegantly shows how subunit proteins in a viral capsid can be intimately associated, and helps to explain the poor immunogenic character of purified polio capsid protein VP1. By analogy, the immunogenicity of VP7 (and VP3) as part of a rotavirus capsid may differ from the immunogenicity of the isolated protein, i.e., VP7 expressed from a cloned gene in a heterologous host may not assume the same conformation as it does in mature virus.

Several host/vector systems are being used to assess the immunogenicity of VP7. Residues 15 to 151 of UK bovine VP7 and SA11 VP7, minus its N-terminal hydrophobic regions (see below), have been expressed as β-galactosidase fusion proteins in *E. coli*. Both proteins induced a low level of neutralizing antibodies in mice.[115a,b] Other partial VP7 products expressed in *E. coli* are also capable of inducing in rabbits antibodies which will recognize whole SA11 virus by ELISA (Bellamy, A. R., personal communication).

An alternative system for expressing rotavirus antigens involves the construction of recombinant vaccinia viruses. Using this system, heterologous antigens have been successfully expressed both in tissue culture and in animals.[116-118] Because of the wide host range of

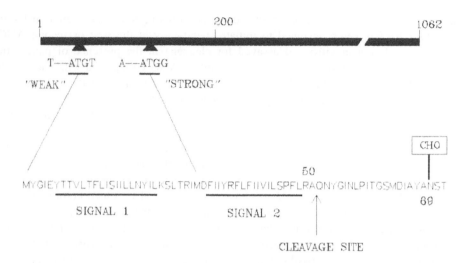

FIGURE 2. Features of the SA11 VP7 gene and protein. The initiation codons, signal peptide regions 1 and 2, major cleavage site (arrow), and single glycosylation site (CHO) are indicated.

vaccinia virus, this system is presently the one of choice for presenting native antigens in vivo where immunogenicity can be assessed. The wild-type SA11 VP7 gene and a mutant encoding a secreted variant[69] (see below) were inserted into plasmids which facilitated the construction of recombinant vaccinia viruses.[119] These viruses, which appropriately expressed the two forms of VP7 in tissue culture were used to inoculate rabbits and sera were taken. The presence of antibodies directed against both wild-type and secreted VP7 was confirmed by immunoblotting and by ELISA using double-shelled rotavirus particles. The latter assay indicated that the antibodies could recognize whole SA11 virus and were serotype specific. These antibodies also neutralized SA11 in a plaque-neutralization reduction test, but much less efficiently than polyclonal antiviral serum.[119a] Thus, some appropriate immunogenicity was present in VP7 expressed by vaccinia, but whether it is sufficient to induce protection in vivo, remains to be determined. The availability of a mouse model system for which EDIM virus, SA11, and a human rotavirus strain have been shown to induce diarrheal disease should prove useful for such protection studies.[120-123]

3. Expression and Processing of VP7

One prominent feature to emerge from analysis of the VP7 gene and protein sequence was the presence of two potential initiation codons, each of which preceded a hydrophobic domain with the characteristics of a signal peptide.[94] The first AUG codon appeared in a context unfavorable for initiation, while the second, 30 codons downstream, appeared favorable (Figure 2).[58,59] This configuration is conserved in all VP7 genes examined, suggesting its functional importance.[31,32,106a] However, a mutant called *(1-14)*, carrying a deletion for the first hydrophobic region, still coded for apparently normal VP7 which was glycosylated and directed to the ER in transfected COS 7 cells,[69] raising doubts as to which AUG codon(s) were used for VP7 expression. Furthermore, processing of a VP7 precursor by cleavage of a signal peptide was observed in vivo and in vitro,[124] but due to the blocked N terminus of the viral protein,[26,77] the cleavage site was uncharacterized.

Using the techniques of gene mutagenesis and expression, it has now been shown that either hydrophobic domain alone can direct VP7 to the ER in transfected COS cells. A protein lacking both hydrophobic regions was not transported.[126] However, it is still unclear whether both initiation codons are used to initiate translation in the infected cell, and if so, in what proportion. Recent work shows, however, that, in contrast with an earlier report,[125] cleavage of the VP7 signal peptide occurred at the same location, no matter where translation began. Mutant VP7 proteins designed to initiate specifically at the first or second AUG codons were translation in vitro and assayed for cleavage in the presence of pancreatic microsomes. The cleaved proteins produced from either AUG comigrated with each other and with a VP7 species present in SA11 rotavirus-infected cells. Examination of the VP7 amino acid sequence for cleavage sites by an improved method[127] revealed a likely cleavage site between residues 50 and 51, C terminal to the second signal peptide domain (Figure 2).[128] When this site was mutated, VP7 cleavage was inhibited, indirectly confirming that the site was used. Cleavage at that location also predicted an N terminal glutamine residue for mature VP7. This suggested a possible basis for the blocked N terminus reported for viral VP7[26,27] i.e., the presence of pyroglutamic acid. The partial sequence of radiolabeled VP7 synthesized in vitro or purified from virus was therefore determined before and after digestion with pyroglutamate amino peptidase, and enzyme which removes the putative blocking group. The results confirmed glutamine 51 as the major N terminal residue.[128] VP7 in SA11 rotavirus is clearly heterogeneous, even after removal of its carbohydrate with endoglycosidase H.[125,128] It is possible that the minor species are the result of cleavage at adjacent residues just upstream of glutamine 51, but the mechanism for this is unknown.[128] Removal of both hydrophobic domains in mature VP7 has implications for the way in which the protein is retained in the ER.

4. Cellular Location of VP7

The targeting of VP7 to the ER as part of the unusual maturation process of rotaviruses (see Section IV.A.) warrants further discussion. VP7, a capsid protein of a nonenveloped virus, behaves quite distinctly from other glycoprotein antigens such as influenza hemagglutinin and neuraminidase, or the G protein of vesicular stomatitis virus. These proteins are synthesized and translocated across the ER membrane, then transported via the Golgi apparatus to the cell surface. In polarized cells, sorting to either the apical or the basolateral plasma membranes also occurs.[129-131] What feature of VP7 allows it to be retained in the ER? The loss of both hydrophobic domains during processing implies that VP7 is not simply anchored via an uncleaved signal peptide. Although there is no precedent, perhaps the VP7 signal peptide(s) have the capacity to direct the protein to the ER while diverting it from the main transport pathway. If this were so, the signal peptide would need to exert its effect prior to cleavage. More likely, perhaps, is that VP7 contains a retention signal within the mature protein. Consistent with this, disruption of the amino acid sequence in the region of the cleavage site resulted in the secretion of mutated proteins from the cell.[69] Furthermore, when rotavirus signal peptide 2 was replaced by the signal sequence from influenza hemagglutinin, cleavage unexpectedly occurred four residues downstream of glutamine 51, even though cleavage at the authentic site was favorable (S.C. Stirzaker and G.W. Both, unpublished results). The loss of four N terminal residues from mature VP7 resulted in rapid secretion of this protein from COS cells, implying that these amino acids have an important role in retention. Whether other parts of VP7 are also involved is under investigation.

H. Segment 10

The RNA genome segments of SA11 display a "long" electrophoretic pattern, while other rotaviruses, e.g., DS-1 and S2 (serotype 2) show a "short" pattern.[3] These differ primarily with respect to the migration of segments 10 and 11, the two smallest. For SA11, the 10th segment codes for NS29 (NCVP5),[48,95] while for viruses with the "short" pattern, NS29 is coded for by segment 11.[3]

Segment 10 has been cloned and sequenced for SA11,[48] bovine rotaviruses,[132,133] and Wa.[134] For the former, the segment is 751 base-pairs in length with 5' and 3' noncoding regions of 41 and 185 bases, respectively. The open reading frame (bases 42 to 566) codes for a protein of 175 amino acids (molecular weight 20,309 for SA11). The Wa gene has a one base deletion at position 684. The SA11 gene and protein are more closely related to their bovine counterparts (92% nucleotide sequence homology and 97% amino acid homology) than to those for Wa (83 and 84% homology, respectively). Nevertheless, there are important features conserved between all three, namely, the two glycosylation sites first observed for SA11 at residues 8 and 18 and three hydrophobic domains between amino acids 7 to 21, 28 to 46, and 70 to 80. The utilization of both glycosylation sites is reflected in the observed molecular weight shift of NS29 partially digested with endo H in vitro.[124] The NS29 carbohydrate is comprised mostly of $Man_9GlcNac_2$, with a small amount of the Man_8 species also present,[48,68] and thus is processed even less than that attached to VP7. The position of these carbohydrate sites also implies that an internal, uncleaved signal peptide is used to translocate NS29 across the membrane.[48,132] Direct analysis of the protein supports this.[124] No size difference could be demonstrated between the protein synthesized in vivo in the presence or absence of tunicamycin, or for the endo H-treated glycoprotein and its nonglycosylated precursor. NS29 also remained trypsin sensitive, suggesting that while it was a transmembrane protein, most of it remained on the cytoplasmic side of the ER membrane. Observations from immunoelectron microscopy support this.[67] From its hydrophobicity profile,[133] it would seem most likely that NS29 associates with membranes via residues 28 to 46 and/or 70 to 80, leaving the hydrophilic C terminal region exposed. Site-directed mutagenesis of the gene followed by transcription and translation of RNA in vitro is being used to define the function of these regions of the protein.

The role of NS29 during infection is not yet elucidated. However, when glycosylation in infected cells was prevented by the addition of tunicamycin, mature virus particles failed to appear and transiently enveloped intermediates accumulated. This was not due to an effect on the glycosylation of VP7, since a carbohydrate-negative isolate of SA11 was completely viable.[66] Thus, NS29 is somehow involved in the removal of the lipid envelope during virus maturation and has been suggested as a scaffolding protein for the assembly of the outer capsid.[135] Conceivably then, mutations in gene 10 could interfere with outer capsid assembly and thereby indirectly affect virus-cell interactions mediated by VP7.[136]

I. Segment 11

Segment 11 of SA11, the smallest, codes for VP9,[48,95] a protein generally considered to be a minor structural component of the outer capsid,[2,4,5] although some consider it a nonstructural protein.[137] For viruses with the "short" electropherotype, the equivalent protein is encoded by segment 10.[3] Segment 11 has been cloned and sequenced for the human rotavirus Wa[22] and the UK bovine strain.[133] These genes are 663 and 667 nucleotides long, respectively, carrying various short deletions and insertions relative to each other. The most significant of these occurs in Wa at nucleotide 389 where a deletion of one base shifts the reading frame for 11 amino acids. A single insertion restores the reading frame by base 418. There are numerous other differences between the genes, yet they retain 86 and 83% homology at the base and amino acid levels, respectively. Each gene has a single, open reading frame containing 197 (Wa) and 198 (UK) codons (molecular weights of 21,560 and 21,700, respectively). Therefore, segment 11, although shorter than segment 10, encodes a slightly larger protein. However, the primary translation product of gene 11, analysed by gel electrophoresis, migrates with an apparent size of 28 kdaltons.[48] Perhaps this is due to the very hydrophilic nature of VP9;[133] the protein contains about 20% serine in addition to clusters of basic or acidic amino acids in the C terminal region.[22,133] It has been suggested that the gene 11 product is a minor neutralizing antigen.[138]

The sequence of SA11 gene 11 has also been determined (D. Mitchell and G.W. Both, unpublished results). Its features are similar to those for the genes from Wa and UK bovine. However, all three segments possess an alternative open reading frame beginning at base

80, which could potentially code for a protein of 82 (SA11 and Wa) or 90 residues (UK). This is the longest alternative reading frame so far found in the rotavirus genome, but whether it is used is not yet known.

V. OTHER REOVIRIDAE GENES

A. Bluetongue Virus Genes and Proteins

1. Segment L2; Protein VP2

VP2, one of two outer capsid proteins of BTV, is the serotype-specific antigen which induces neutralizing antibodies in rabbits and sheep. The protein is also responsible for hemagglutinating sheep erythrocytes and for cellular adsorption.[139] In vitro translation studies[140] and intertypic reassortant viruses[141] have shown that VP2 is coded for by segment L2 which has now been cloned and sequenced for BTV serotype 10.[142] The gene is 2926 base-pairs long, coding for a protein of 956 amino acids (molecular weight 111,122). Expression studies are being done with the gene to produce antigen which might be assessed for vaccine potential.

2. Segment L3; Protein VP3

Segment L3 of BTV codes for the core polypeptide VP3.[140] Full-length clones of this gene from BTV-1, BTV-10, and BTV-17 were obtained[27,143,143a] as judged by the identity of the terminal sequences compared with those of segment 3 dsRNA.[144,145] The genes are 2772 base-pairs long, with a single, open reading frame (bases 18 to 2720) coding for a protein of 901 amino acids (molecular weight 103,412 for BTV-17).[27] The BTV-10 and BTV-17 genes differ by 126 point mutations, only 9 of which resulted in amino acid changes, most of which are conservative. Similarly, the BTV-1 and BTV-17 proteins differ by 19 amino acid changes, 11 of which are conservative.[143a] This probably reflects the structural role of VP3 in the virus core; no other function has so far been assigned to it. The BTV-17 gene shows homology with the equivalent gene from other BTV serotypes under stringent hybridization conditions (consistent with the above homology), but does not hybridize to RNA from a related orbivirus, epizootic hemorrhagic disease virus of deer. Biotinylated L3 DNA has been used to detect the presence of viral RNA in infected cells or in blood samples from infected sheep and may therefore be useful as a diagnostic probe.[146] Further sequence comparisons of BTV isolates have shown that subtle features of the nucleotide sequences can be used to identify strains from the same geographical area.[143a]

3. Segment M5; Protein VP5

The VP5 protein, together with VP2, comprises the outer capsid of BTV. However, the role of VP5 is not clear. It appears not to induce neutralizing antibodies by itself, but in combination with VP2, induces a higher level of neutralizing antibodies than VP2 alone.[139] Since the BTV outer capsid (and hence infectivity) is susceptible to disruption by high salt solutions and low pH, VP2 and VP5 may associate via ionic interactions, with VP5 influencing the conformation of VP2 in the capsid. The complete sequence of the M5 segment has now been determined from overlapping clones of the gene.[147] The segment is 1638 base-pairs in length coding for a protein of 526 amino acids (molecular weight 59,163).

B. Gene Segments of Wound Tumor Virus

While nine of the twelve WTV genes have apparently been cloned as full length copies, only the smallest has been sequenced.[29] Segment 12 of WTV codes for a protein of molecular weight 19,171, consistent with the size of the nonstructural polypeptide Pns12, the predicted product of this gene.[16] The segment is 851 base-pairs long with a single, open reading frame of 534 nucleotides (178 codons). Like the rotavirus VP9 protein, this polypeptide is also predicted to have a substantial serine content (24 of 178 residues) and a similar C terminal distribution of basic and acidic amino acids. Despite this superficial similarity, computer programmes comparing this protein with rotavirus VP9 so far do not reveal any significant homology or suggest a possible similar function.

VI. CONCLUSION

This chapter has attempted to draw together genetic and biochemical information available for certain Reoviridae genes and proteins. The exercise emphasizes how little is still known about some viral proteins, e.g., rotavirus NS34 and reo σ_2, for which only the gene structure and the predicted amino acid sequence are described. The cloned gene in these cases provides basic information and a tool for further research. In contrast, the nucleotide sequence of the reovirus S1 segment revealed its bicistronic nature and the amino acid sequence inferred for the σ_1 protein gave an important clue to the partial coiled-coil nature of the polypeptide. Expression of σ_1 as a fusion protein containing functional hemagglutinating and cell-binding domains will facilitate structure/function studies using site-directed mutagenesis, complementing the more traditional biochemical and genetic approaches that have already been used. Similarly, sequences responsible for the cellular location of rotavirus VP7 are also being identified by mutagenesis and expression. Other studies employing in vitro RNA transcription and translation systems to study mutated genes are also in progress. While these are powerful analytical tools, their efficient use relies very much on narrowing the target area for study using biochemical data and information such as that obtained from mutants selected for altered function. Thus, the gene cloning studies of the last few years have made possible new approaches to studying viral protein structure and function. Perhaps the next major obstacle to this line of research will be to find a way to incorporate a desirable mutation into the genome of an infectious dsRNA virus.

ACKNOWLEDGMENT

I thank Drs. A. R. Bellamy and P. L. Whitfeld for their comments on the manuscript.

REFERENCES

1. **Joklik, W. K.,** The members of the family reoviridae, in *The Reoviridae,* Joklik, W. K., Ed., Plenum Press, New York, 1983, chap. 8.
2. **Estes, M. K., Palmer, E. L., and Obijeski, J. F.,** Rotaviruses: a review, *Curr. Top. Microbiol. Immunol.,* 105, 123, 1983.
3. **Holmes, I. H.,** Rotaviruses, in *The Reoviridae,* Joklik, W. K., Ed., Plenum Press, New York, 1983, 359.
4. **Cukor, G. and Blacklow, N. R.,** Human viral gastroenteritis, *Microbiol. Rev.,* 48, 157, 1984.
5. **Kapikian, A. Z. and Chanock, R. M.,** Rotaviruses, in *Virology,* Fields, B. N., Ed., Raven Press, New York, 1985, chap. 37.
6. **Gorman, B. M., Taylor, J., and Walker, P. J.,** Orbiviruses, in *The Reoviridae,* Joklik, W. K., Ed., Plenum Press, New York, 1983, chap. 7.
7. **Joklik, W. K., Ed.,** *The Reoviridae,* Plenum Press, New York, 1985.
8. **Lamb, R. A.. and Choppin, P. W.,** The gene structure and replication of influenza virus, *Annu. Rev. Biochem.,* 52, 467, 1983.
9. **McCauley, J. W. and Mahy, B. W.,** Structure and function of the influenza virus genome, *Biochem. J.,* 211, 281, 1983.
10. **Cashdollar, L. W., Chmelo, R., Esparza, J., Hudson, G. R., and Joklik, W. K.,** Molecular cloning of the complete genome of reovirus serotype 3, *Virology,* 133, 191, 1984.
11. **Both, G. W., Bellamy, A. R., Street, J. E., and Siegman, L. J.,** A general strategy for cloning double-stranded RNA: nucleotide sequence of the Simian-11 rotavirus gene 8, *Nucleic Acids Res.,* 10, 7075, 1982.

12. **Rixon, F., Taylor, P., and Desselberger, U.**, Rotavirus RNA segments sized by electron microscopy. *J. Gen. Virol.*, 65, 233, 1984.

13. **Payne, C. C. and Mertens, P. P. C.**, Cytoplasmic polyhedrosis viruses, in *The Reoviridae*, Joklik, W. K., Ed., Plenum Press, New York, 1983, chap. 9.

14. **Shatkin, A. J. and Kozak, M.**, Biochemical aspects of reovirus transcription and translation, in *The Reoviridae*, Joklik, W. K., Ed., Plenum Press, New York, 1983, chap. 3.

15. **Zarbl, H. and Millward, S.**, The reovirus multiplication cycle, in *The Reoviridae*, Joklik, W. K., Ed., Plenum Press, New York, 1983, chap. 4.

16. **Nuss, D. L.**, Molecular biology of wound tumor virus, *Adv. Virus Res.*, 29, 57, 1984.

17. **Spencer, E. and Garcia, B. I.**, Effect of S-adenosylmethionine on human rotavirus RNA synthesis, *J. Virol.*, 52, 188, 1984.

18. **Joklik, W. K.**, The reovirus particle, in *The Reoviridae*, Joklik, W. K., Ed., Plenum Press, New York, 1983, chap. 2.

19. **Imai, M., Akatani, K., Ikegami, N., and Furuichi, Y.**, Capped and conserved terminal structures in human rotavirus genome double-stranded RNA segments, *J. Virol.*, 47, 125, 1983.

20. **Joklik, W. K.**, Structure and function of the reovirus genome, *Microbiol. Rev.*, 45, 483, 1981.

21. **Cashdollar, L. W., Esparza, J., Hudson, G. R., Chmelo, R., Lee, P. W., and Joklik, W. K.**, Cloning the double-stranded RNA genes of reovirus: sequence of the cloned S2 gene, *Proc. Natl. Acad. Sci. U.S.A.*, 79, 7644, 1982.

22. **Imai, M., Richardson, M. A., Ikegami, N., Shatkin, A. J., and Furuichi, Y.**, Molecular cloning of double-stranded RNA virus genomes, *Proc. Natl. Acad. Sci. U.S.A.*, 80, 373, 1983.

23. **McCrae, M. A. and McCorquodale, J. G.**, Molecular biology of rotaviruses. IV. Molecular cloning of the bovine rotavirus genome, *J. Virol.*, 44, 1076, 1982.

24. **Dyall-Smith, M. L., Elleman, T. C., Hoyne, P. A., Holmes, I. H., and Azad, A. A.**, Cloning and sequence of UK bovine rotavirus gene segment 7: marked sequence homology with simian rotavirus gene segment 8, *Nucleic Acids Res.*, 11, 3351, 1983.

25. **Gorziglia, M., Cashdollar, L. W., Hudson, G. R., and Esparza, J.**, Molecular cloning of a human rotavirus genome, *J. Gen. Virol.*, 64, 2585, 1983.

26. **Arias, C. F., Lopez, S., Bell, J. R., and Strauss, J. H.**, Primary structure of the neutralization antigen of Simian rotavirus SA11 as deduced from cDNA sequence, *J. Virol.*, 50, 657, 1984.

27. **Purdy, M., Petre, J., and Roy, P.**, Cloning of the bluetongue virus L3 gene, *J. Virol.*, 51, 754, 1984.

28. **Bassel-Duby, R., Jayasuriya, A., Chatterjee, D., Sonenberg, N., Maizel, J. V., Jr., and Fields, B. N.**, Sequence of reovirus haemagglutinin predicts a coiled-coil structure, *Nature (London)*, 315, 421, 1985.

28a. **Bassel-Duby, R., Nibert, M. L., Homcy, C. J., Fields, B. N., and Sawutz, D. G.**, Evidence that the sigma 1 protein of reovirus serotype 3 is a multimer, *J. Virol.*, 61, 1834, 1987.

29. **Asamizu, T., Summers, D., Motika, M. B., Anzola, J. V., and Nuss, D. L.**, Molecular cloning of the genome of wound tumor virus; a tumor-inducing plant reovirus, *Virology*, 144, 398, 1985.

30. **Nagata, L., Masri, S. A., Mah, D. C., and Lee, P. W.**, Molecular cloning and sequencing of the reovirus (serotype 3) S1 gene which encodes the viral cell attachment protein sigma 1, *Nucleic Acids Res.*, 12, 8699, 1984.

31. **Gunn, P. R., Sato, F., Powell, K. F., Bellamy, A. R., Napier, J. R., Harding, D. R., Hancock, W. S., Siegman, L. J., and Both, G. W.**, Rotavirus neutralizing protein VP7: antigenic determinants investigated by sequence analysis and peptide synthesis, *J. Virol.*, 54, 791, 1985.

32. **Glass, R. I., Keith, J., Nakagomi, O., Nakagomi, T., Askua, J., Kapikian, A. Z., Chanock, R. M., and Flores, J.**, Nucleotide sequence of the structural glycoprotein VP7 gene of Nebraska Calf Diarrhea Virus: Comparison with homologous genes from four strains of human and animal rotaviruses, *Virology*, 141, 292, 1985.

33. **Estes, M. K., Mason, B. B., Crawford, S., and Cohen, J.**, Cloning and nucleotide sequence of the simian rotavirus gene 6 that codes for the major inner capsid protein, *Nucleic Acids Res.*, 12, 1875, 1984.

34. **Antczak, J. B., Chmelo, R., Pickup, D. J., and Joklik, W. K.**, Sequence at both termini of the 10 genes of reovirus serotype 3 (strain Dearing), *Virology*, 121, 307, 1982.

35. **Sanger, F., Nicklen, S., and Coulson, A. R.**, DNA sequencing with chain-terminating inhibitors, *Proc. Natl. Acad. Sci., U.S.A.*, 74, 5463, 1977.

36. **Sharpe, A. H. and Fields, B. N.**, Pathogenesis of reovirus infection, in *The Reoviridae*, Joklik, W. K., Ed., Plenum Press, New York, 1983, chap. 6.

37. **Fields, B. N.**, Genetics of reovirus virulence, *Trends Genet.*, 1, 284, 1985.

38. **Kauffman, R. S., Wolf, J. L., Finberg, R., Trier, J. S., and Fields, B. N.**, The sigma 1 protein determines the extent of spread of reovirus from the gastrointestinal tract of mice, *Virology*, 124, 403, 1983.

39. **Kauffman, R. S., Ahmed, R., and Fields, B. N.**, Selection of a mutant S1 gene during reovirus persistent infection of L cells: role in maintenance of the persistent state, *Virology*, 131, 79, 1983.

40. **Armstrong, G. D., Paul, R. W., and Lee, P. W.,** Studies on reovirus receptors of L cells: virus binding characteristics and comparison with reovirus receptors of erythrocytes, *Virology,* 138, 37, 1984.

41. **Noseworthy, J. H., Fields, B. N., Dichter, M. A., Sobotka, C., Pizer, E., Perry, L. L., Nepom, J. T., and Greene, M. I.,** Cell receptors for the mammalian reovirus. I. Syngeneic monoclonal anti-idiotypic antibody identifies a cell surface receptor for reovirus, *J. Immunol.,* 131, 2533, 1983.

42. **Co, M. S., Gaulton, G. N., Fields, B. N., and Greene, M. I.,** Isolation and biochemical characterization of the mammalian reovirus type 3 cell-surface receptor, *Proc. Natl. Acad. Sci. U.S.A.,* 82, 1494, 1985.

42a. **Paul, R. W. and Lee, P. W. K.,** Glycophorin is the reovirus receptor on human erythrocytes, *Virology,* 159, 94, 1987.

43. **Epstein, R. L., Powers, M. L., Rogart, R. B., and Weiner, H. L.,** Binding of 125I-labeled reovirus to cell surface receptors, *Virology,* 133, 46, 1984.

44. **Cashdollar, L. W., Chmelo, R. A., Wiener, J. R., and Joklik, W. K.,** Sequences of the S1 genes of the three serotypes of reovirus, *Proc. Natl. Acad. Sci. U.S.A.,* 82, 24, 1985.

45. **Lee, P. W.,** Glycosylation of reovirus proteins and the effect of 2-deoxy-D-glucose on viral replication and assembly, in *Double-Stranded RNA Viruses,* Compans, R. W. and Bishop, D. H. L., Eds., Elsevier, New York, 1983, 193.

46. **von Heijne, G.,** Patterns of amino acids near signal-sequence cleavage sites, *Eur. J. Biochem.,* 133, 17, 1983.

47. **Both, G. W., Mattick, J. S., Siegman, L. J., Atkinson, P. H., Weiss, S., Bellamy, A. R., Street, J., and Metcalf, P.,** Cloning of SA11 rotavirus genes: gene structure and polypeptide assignment for the type-specific glycoprotein, in *The Double-Stranded RNA Viruses,* Compans, R. W. and Bishop, D. H. L., Eds., Elsevier, New York, 1983, 73.

48. **Both, G. W., Siegman, L. J., Bellamy, A. R., and Atkinson, P. H.,** Coding assignment and nucleotide sequence of simian rotavirus SA11 gene segment 10: location of glycosylation sites suggests that the signal peptide is not cleaved, *J. Virol.,* 48, 335, 1983.

49. **McPhillips, T. H. and Ramig, R. F.,** Extragenic suppression of temperature-sensitive phenotype in reovirus: mapping suppressor mutations, *Virology,* 135, 428, 1984.

50. **Spriggs, D. R., Bronson, R. T., and Fields, B. N.,** Hemagglutinin variants of reovirus type 3 have altered central nervous system tropism, *Science,* 220, 505, 1983.

51. **Masri, S. A., Nagata, L., Mah, D. C. W., and Lee, P. W. K.,** Functional expression in E. coli of cloned reovirus S1 gene encoding the viral cell attachment protein sigma 1, *Virology,* 149, 83, 1986.

52. **Kozak, M.,** Analysis of ribosome binding sites from the S1 message of reovirus: initiation at the first and second AUG codons, *J. Mol. Biol.,* 156, 807, 1982.

53. **Cenatiempo, Y., Twardowski, T., Shoeman, R., Ernst, H., Brot, N., Weissbach, H., and Shatkin, A. J.,** Two initiation sites detected in the small S1 species of reovirus mRNA by dipeptide synthesis in vitro *Proc. Natl. Acad. Sci. U.S.A.,* 81, 1084, 1984.

54. **Ernst, H. and Shatkin, A. J.,** Reovirus hemagglutinin mRNA codes for two polypeptides in overlapping reading frames, *Proc. Natl. Acad. Sci. U.S.A.,* 82, 48, 1985.

55. **Jacobs, B. L. and Samuel, C. E.,** Biosynthesis of reovirus-specified polypeptides: the reovirus S1 mRNA encodes two primary translation products, *Virology,* 143, 63, 1985.

56. **Jacobs, B. L., Atwater, J. A., Munemitsu, S. M., and Samuel, C. E.,** Biosynthesis of reovirus polypeptides: the S1 mRNA synthesized in vivo is structurally and functionally indistinguishable from in vitro-synthesized S1 mRNA and encodes two polypeptides sigma 1a and sigma 1bNS, *Virology,* 147, 9, 1985.

57. **Sarkar, G., Pelletier, J., Bassel-Duby, R., Jayasuriya, A., Fields, B. N., and Sonenberg, N.,** Identification of a new polypeptide coded by reovirus gene S1, *J. Virol.,* 54, 720, 1985.

58. **Kozak, M.,** Compilation and analysis of sequences upstream from the translational start site in eukaryotic mRNAs, *Nucleic Acids Res.,* 12, 857, 1984.

59. **Kozak, M.,** Point mutations that define a sequence flanking the AUG initiator codon that modulates translation by eukaryotic ribosomes, *Cell,* 44, 283, 1986.

60. **Ceruzzi, M. and Shatkin, A. J.,** Expression of reovirus p14 in bacteria and identification in the cytoplasm of infected mouse L cells, *Virology,* 153, 35, 1986.

61. **Roner, M. R. and Cox, D. C.,** Cellular integrity is required for inhibition of cellular DNA synthesis by reovirus type 3, *J. Virol.,* 53, 350, 1985.

62. **Richardson, M. A. and Furuichi, Y.,** Nucleotide sequence of reovirus genome segment S3, encoding non-structural protein sigma NS, *Nucleic Acids Res.,* 11, 6399, 1983.

63. **Kozak, M. and Shatkin, A. J.,** Characterization of ribosome-protected fragments from reovirus messenger RNA, *J. Biol. Chem.,* 251, 4259, 1976.

64. **Richardson, M. A. and Furuichi, Y.,** Synthesis in E. coli of the reovirus non-structural protein sigma NS, *J. Virol.,* 56, 527, 1985.

64a. **Lemieux, R., Lemay, G., Millward, S.,** The viral protein sigma 3 participates in translation of late viral mRNA in reovirus-infected L cells, *J. Virol.,* 61, 2472, 1987.

65. **Giantini, M., Seliger, L. S., Furuichi, Y., and Shatkin, A. J.,** Reovirus type 3 genome segment S4: nucleotide sequence of the gene encoding a major virion surface protein, *J. Virol.,* 52, 984, 1984.

66. **Dubois-Dalcq, M., Holmes, K. V., and Rentier, B., Eds.,** *Assembly of Enveloped Viruses,* Springer Verlag, New York, 1984, chap. 10.

67. **Petrie, B. L., Greenberg, H. B., Graham, D. Y., and Estes, M. K.,** Ultrastructural localization of rotavirus antigens using colloidal gold, *Virus Res.,* 1, 133, 1984.

68. **Kabcenell, A. K. and Atkinson, P. H.,** Processing of the rough endoplasmic reticulum membrane glycoproteins of rotavirus SA11, *J. Cell Biol.,* 101, 1270, 1985.

69. **Poruchynsky, M. S., Tyndall, C., Both, G. W., Sato, F., Bellamy, A. R., and Atkinson, P. H.,** Deletions into an NH2-terminal hydrophobic domain result in secretion of rotavirus VP7, a resident endoplasmic reticulum membrane glycoprotein, *J. Cell Biol.,* 101, 2199, 1985.

70. **Shahrabadi, M. S. and Lee, P. W. K.,** Bovine rotavirus maturation is a calcium-dependent process, *Virology,* 152, 298, 1986.

70a. **Shahrabadi, M. S., Babiuk, L. A., and Lee, P. W. K.,** Further analysis of the role of calcium in rotavirus morphogenesis, *Virology,* 158, 103, 1987.

71. **Musalem, C. and Espejo, R. T.,** Release of progeny virus from cells infected with simian rotavirus SA11, *J. Gen. Virol.,* 66, 2715, 1985.

72. **Kouvelos, K., Petric, M., and Middleton, P. J.,** Oligosaccharide composition of calf rotavirus, *J. Gen. Virol.,* 65, 1159, 1984.

73. **Hubbard, S. C. and Ivatt, R. J.,** Synthesis and processing of asparagine-linked oligosaccharides, *Annu. Rev. Biochem.,* 50, 555, 1981.

74. **Greenberg, H. B., Flores, J., Kalica, A. R., Wyatt, R. G., and Jones, R.,** Gene coding assignments for growth restriction, neutralization and subgroup specificities of the W and DS-1 strains of human rotavirus, *J. Gen. Virol.,* 64, 313, 1983.

75. **Offit, P. A., Blavat, G., Greenberg, H. B., and Clark, H. F.,** Molecular basis of rotavirus virulence: role of gene segment 4, *J. Virol.,* 57, 46, 1986.

76. **Kalica, A. R., Flores, J., and Greenberg, H. B.,** Identification of the rotaviral gene that codes for hemagglutination and protease-enhanced plaque formation, *Virology,* 125, 194, 1983.

77. **Lopez, S., Arias, C. F., Bell, J. R., Strauss, J. H., and Espejo, R. T.,** Primary structure of the cleavage site associated with trypsin enhancement of rotavirus SA11 infectivity, *Virology,* 144, 11, 1985.

77a. **Lopez, S. and Arias, C. F.,** The nucleotide sequence of the 5' and 3' ends of rotavirus SA11 gene 4, *Nucleic Acids Res.,* 15, 4691, 1987.

77b. **Kantharidis, P., Dyall-Smith, M. L., and Holmes, I. H.,** Marked sequence variation between segment 4 genes of human RV-5 and simian SA11 rotaviruses, *Arch. Virol.,* 93, 111, 1987.

77c. **Mackow, E. R., Shaw, R. D., Matsui, S. M., Vo, P. T., Dang, M.-N., and Greenberg, H. B.,** Characterization of the rhesus rotavirus VP3 gene: location of amino acids involved in homologous and heterologous rotavirus neutralization and identification of a putative fusion region, *Proc. Natl. Acad. Sci. U.S.A.,* 1987, in press.

77d. **Lopez, S., Arias, C. F., Mendez, E., and Espejo, R. T.,** Conservation in rotaviruses of the protein region containing the two sites associated with trypsin enhancement of infectivity, *Virology,* 154, 224, 1986.

78. **Ikegami, N. and Akatani, K.,** Characterization of outer capsid proteins of rotaviruses using monoclonal antibodies, *Abstr. Int. Cong. Virol.,* p. 209, 1984.

79. **Both, G. W., Sleigh, M. J., Cox, N., and Kendal, A. P.,** Antigenic drift in influenza virus H3 hemagglutinin from 1968 to 1980: multiple evolutionary pathways and sequential amino acid changes at key antigenic sites, *J. Virol.,* 48, 52, 1983.

80. **Coulson, B. S., Fowler, K. J., Bishop, R. F., and Cotton, R. G. H.,** Neutralizing monoclonal antibodies to human rotavirus and indications of antigenic drift among strains from neonates, *J. Virol.,* 54, 14, 1985.

80a. **Arias, C. F., Lizano, M., and Lopez, S.,** Synthesis in *Escherichia coli* and immunological characterization of a polypeptide containing cleavage sites associated with trypsin enhancement of rotavirus SA11 infectivity, *J. Gen. Virol.,* 68, 633, 1987.

81. **Gething, M. J., White, J. M., and Waterfield, M.,** Purification of the fusion protein of Sendai virus: analysis of the NH2-terminal sequence generated during precursor activation *Proc. Natl. Acad. Sci., U.S.A.,* 75, 2737, 1976.

82. **Daniels, R. S., Downie, J. C., Hay, A. J., Knossow, M., Skehel, J. J., Wang, M. L., and Wiley, D. C.,** Fusion mutants of the influenza virus hemagglutinin glycoprotein, *Cell,* 40, 431, 1985.

83. **Rott, R. and Klenk, H. D.,** Virus determined differences in the pathogenicity of avian influenza viruses, in *Veterinary Viral Diseases,* Della-Porta, A. J., Ed., Academic Press, Sydney, 1985, 35.

84. **Hoshino, Y., Sereno, M. M., Midthun, K., Flores, J., Kapikian, A. Z., and Chanock, R. M.,** Independent segregation of two antigenic specificities (VP3 and VP7) involved in neutralization of rotavirus infectivity, *Proc.Natl. Acad. Sci. U.S.A.,* 82, 8701, 1985.

85. **Taniguchi, K., Urasawa, S., and Urasawa, T.,** Preparation and characterization of neutralizing monoclonal antibodies with different reactivity patterns to human rotaviruses, *J. Gen. Virol.,* 66, 1045, 1985.

86. **Offit, P. A. and Blavat, G.,** Identification of the two rotavirus genes determining neutralization specificities, *J. Virol.,* 57, 376, 1986.

87. **Bridger, J. C. and Brown, J. F.,** Antigenic and pathogenic relationships of three bovine rotaviruses and a porcine rotavirus, *J. Gen. Virol.,* 65, 1151, 1984.

88. **Offit, P. A., Shaw, R., and Greenberg, H. B.,** Passive protection against rotavirus-induced diarrhea by monoclonal antibodies to surface proteins VP3 and VP7, *J. Virol.,* 58, 700, 1986.

88a. **Bremont, M., Charpilienne, A., Chabanne, D., and Cohen, J.,** Nucleotide sequence and expression in *Escherichia coli* of the gene encoding the non-structural protein NCVP2 of bovine rotavirus, *Virology,* 160, 1987, in press.

89. **Gorziglia, M., Larrea, C., Liprandi, F., and Esparza, J.,** Biochemical evidence for the oligomeric (possibly trimeric) structure of the major inner capsid polypeptide (45K) of rotaviruses, *J. Gen. Virol.,* 66, 1889, 1985.

90. **Bican, P., Cohen, J., Charpilienne, A., and Scherrer, R.,** Purification and characterization of bovine rotavirus cores, *J. Virol.,* 43, 1113, 1982.

91. **Greenberg, H. B., McAuliffe, V., Valdesuso, J., Wyatt, R., Flores, J., Kalica, A., Hoshino, Y., and Singh, N.,** Serological analysis of the subgroup protein of rotavirus, using monoclonal antibodies, *Infect. Immun.,* 39, 91, 1983.

92. **Both, G. W., Siegman, L. J., Bellamy, A. R., Ikegami, N., Shatkin, A. J., and Furuichi, Y.,** Comparative sequence analysis of rotavirus genomic segment 6 — the gene specifying viral subgroups 1 and 2, *J. Virol.,* 51, 97, 1984.

93. **Cohen, J., Lefevre, F., Estes, M. K., and Bremont, M.,** Cloning of bovine rotavirus (RF strain): nucleotide sequence of the gene coding for the major capsid protein, *Virology,* 138, 178, 1984.

93a. **Estes, M. K., Crawford, S. E., Penaranda, M. E., Petrie, B. L., Burns, J. W., Chan, W.-K., Ericson, B., Smith, G. E., and Summers, M. D.,** Synthesis and immunogenicity of the rotavirus major capsid antigen using a baculovirus expression system, *J. Virol.,* 61, 1488, 1987.

94. **Both, G. W., Mattick, J. S., and Bellamy, A. R.,** Serotype-specific glycoprotein of simian 11 rotavirus: coding assignment and gene sequence, *Proc. Natl. Acad. Sci. U.S.A.,* 80, 3091, 1983.

95. **Mason, B. B., Graham, D. Y., and Estes, M. K.,** Biochemical mapping of the simian rotavirus SA11 genome, *J. Virol.,* 46, 413, 1983.

96. **Dyall-Smith, M. L., Azad, A. A., and Holmes, I. H.,** Gene mapping of rotavirus double-stranded RNA segments by northern blot hybridization: application to segments 7, 8, and 9, *J. Virol.,* 46, 317, 1983.

97. **Both, G. W., Bellamy, A. R., and Siegman, L. J.,** Nucleotide sequence of the dsRNA genomic segment 7 of Simian 11 rotavirus, *Nucleic Acids Res.,* 12, 1621, 1984.

98. **Ward, C. W., Elleman, T. C., Azad, A. A., and Dyall-Smith, M. L.,** Nucleotide sequence of gene segment 9 encoding a nonstructural protein of UK bovine rotavirus, *Virology,* 134, 249, 1984.

99. **Ramig, R. F.,** Genetic studies with simian rotavirus SA11, in *The Double-Stranded RNA Viruses,* Compans, R. W. and Bishop, D. H. L., Eds., Elsevier, New York, 1983, 321.

100. **Gombold, J. L., Estes, M. K., and Ramig, R. F.,** Assignment of simian rotavirus SA11 temperature sensitive mutants groups B and E to genome segments, *Virology,* 143, 309, 1985.

101. **Lin, M., Imai, M., Bellamy, A. R., Ikegami, N., Furuichi, Y., Summers, D., Nuss, D. L., and Diebel, R.,** Diagnosis of rotavirus infection with cloned cDNA copies of viral genome segments, *J. Virol.,* 55, 509, 1985.

102. **Ramig, R. F. and Petrie, B. L.,** Characterization of temperature-sensitive mutants of simian rotavirus SA11: protein synthesis and morphogenesis, *J. Virol.,* 49, 665, 1984.

103. **Elleman, T. C., Hoyne, P. A., Dyall-Smith, M. L., Holmes, I. H., and Azad, A. A.,** Nucleotide sequence of the gene encoding the serotype-specific glycoprotein of UK bovine rotavirus, *Nucleic Acids Res.,* 11, 4689, 1983.

104. **Dyall-Smith, M. L. and Holmes, I. H.,** Sequence homology between human and animal rotavirus serotype-specific glycoproteins, *Nucleic Acids Res.,* 12, 3973, 1984.

105. **Richardson, M. A., Iwamoto, A., Ikegami, N., Nomoto, A., and Furuichi, Y.,** Nucleotide sequence of the gene encoding the serotype-specific antigen of human (Wa) rotavirus: comparison with the homologous genes from simian SA11 and UK bovine rotaviruses, *J. Virol.,* 51, 860, 1984.

106. **Mason, B. B., Dheer, S. K., Hsiao, C. L., Zandle, G., Kostek, B., Rosanoff, E. I., Hung, P. P., and Davis, A. R.,** Sequence of the serotype-specific glycoprotein of the human rotavirus Wa strain and comparison with other human rotavirus serotypes, *Virus Res.,* 2, 291, 1985.

106a. **Gorziglia, M., Aguirre, Y., Hoshino, Y., Esparza, J., Blumentals, I., Askaa, J., Thompson, L. M., Glass, R. I., Kapikian, A. Z., and Chanock, R. M.,** VP7 serotypespecific glycoprotein of OSU porcine rotavirus: coding assignment and gene sequence, *J. Gen. Virol.,* 67, 2445, 1986.

107. **Dyall-Smith, M. L., Lazdins, I., Tregear, G. W., and Holmes, I.,** Location of the major sites involved in rotavirus serotype-specific neutralization, *Proc. Natl. Acad. Sci. U.S.A.,* 83, 3465, 1986.

108. **Sonza, S., Breschkin, A. M., and Holmes, I. H.,** The major surface glycoprotein of simian rotavirus (SA11) contains distinct epitopes, *Virology,* 134, 318, 1984.

109. **Sabara, M., Gilchrist, J. E., Hudson, G. R., and Babiuk, L. A.,** Preliminary characterization of an epitope involved in neutralization and cell attachment that is located on the major bovine rotavirus glycoprotein, *J. Virol.,* 53, 58, 1985.

110. **Lazdins, I., Sonza, S., Dyall-Smith, M. L., Coulson, B. S., and Holmes, I. H.** Demonstration of an immunodominant neutralization site by analysis of antigenic variants of SA11 rotavirus, *J. Virol.,* 56, 317, 1985.

111. **Knowlton, D. R. and Ward, R. L.,** Effect of mutation in immunodominant neutralization epitopes on antigenicity of rotavirus SA11, *J. Gen. Virol.,* 66, 2375, 1985.

112. **Caton, A. J., Brownlee, G. G., Yewdell, J. W., and Gerhard, W.,** The antigenic structure of the influenza virus A/PR/8/34 hemagglutinin (H1 subtype), *Cell,* 31, 417, 1982.

113. **Kouvelos, K., Petric, M., and Middleton, P. J.,** Comparison of bovine, simian and human rotavirus structural glycoproteins, *J. Gen. Virol.,* 65, 1211, 1984.

114. **Hogle, J. M., Chow, M., and Filman, D. J.,** Three-dimensional structure of poliovirus at 2.9 angstroms resolution, *Science,* 229, 1358, 1985.

115. **Rossmann, M. G., Arnold, E., Erikson, J. W., Frankenberger, E. A., Griffith, J. P., Hecht, H.-J., Johnson, J. E., Kamer, G., Luo, M., Mosser, A. G., Rueckert, R. R., herry, B., and Vriend, G.,** Structure of a human common cold virus and functional relationship to other picornaviruses, *Nature (London),* 317, 145, 1985.

115a. **Arias, C. F., Ballado, T., and Plebanski, M.,** Synthesis of the outer-capsid glycoprotein of the simian rotavirus SA11 in *Escherichia coli, Gene,* 47, 211, 1986.

115b. **McCrae, M. A., and McCorquodale, J. G.,** Expression of a major bovine rotavirus neutralizing antigen (VP7c) in *Escherichia coli, Gene,* 55, 9, 1987.

116. **Kieny, M. P., Lathe, R., Drillien, R., Spehner, D., Skory, S., Schmitt, D., Wiktor, T., Koprowski, H., and Lecocq, J. P.,** Expression of rabies virus glycoprotein from a recombinant vaccinia virus, *Nature (London),* 312, 163, 1984.

117. **Paoletti, E., Lipinskas, B. R., Samsonoff, C., Mercer, S., and Panicali, D.,** Construction of live vaccines using genetically engineered poxviruses: biological activity of vaccinia virus recombinants expressing the hepatitis B virus surface antigen and the herpes simplex virus glycoprotein D, *Proc. Natl. Acad. Sci. U.S.A.,* 81, 193, 1984.

118. **Moss, B., Smith, G. L., Gerin, J. L., and Purcell, R. H.,** Live recombinant vaccinia virus protects chimpanzees against hepatitis B, *Nature (London),* 311, 67, 1984.

119. **Boyle, D. B., Coupar, B. E., and Both, G. W.,** Multiple-cloning-site plasmids for the rapid construction of recombinant poxviruses, *Gene,* 35, 169, 1985.

119a. **Andrew, M. E., Boyle, D. B., Coupar, B. E. H., Whitfeld, P. L., Both, G. W., and Bellamy, A. R.,** Vaccinia virus recombinants expressing the SA11 rotavirus VP7 glycoprotein gene induce serotype-specific neutralizing antibodies, *J. Virol.,* 61, 1054, 1987.

120. **Sheridan, J. F., Smith, C. C., Manak, M. M., and Aurelian, L.,** Prevention of rotavirus-induced diarrhea in neonatal mice born to dams immunized with empty capsids of simian rotavirus SA11, *J. Infect. Dis.,* 149, 434, 1984.

121. **Offit, P. A., Clark, H. F., Kornstein, M. J., and Plotkin, S. A.,** A murine model for oral infection with a primate rotavirus (Simian SA11), *J. Virol.,* 51, 233, 1984.

122. **Offit, P. A. and Clark, H. F.,** Protection against rotavirus-induced gastroenteritis in a murine model by passively acquired gastrointestinal but not circulating antibodies, *J. Virol.,* 54, 58, 1985.

123. **Gouvea, V. S., Alencar, A. A., Barth, O. M., De Castro, L., Fiahlo, A. M., Araujo, H. P., Majerowicz, S., and Pereira, H. G.,** Diarrhea in mice infected with a human rotavirus, *J. Gen. Virol.,* 67, 577, 1986.

124. **Ericson, B. L., Graham, D. Y., Mason, B. B., Hanssen, H. H., and Estes, M. K.,** Two types of glycoprotein precursors are produced by the simian rotavirus SA11, *Virology,* 127, 320, 1983.

125. **Chan, W. K., Penaranda, M. E., Crawford, S. E., and Estes, M. K.,** Two glycoproteins are produced from the rotavirus neutralization gene, *Virology,* 151, 243, 1986.

126. **Whitfeld, P. L., Tyndall, C., Stirzaker, S. C., Bellamy, A. R., and Both, G. W.,** Location of sequences within rotavirus SA11 glycoprotein VP7 which direct it to the endoplasmic reticulum, *Mol. Cell. Biol.,* 7, 2491, 1987.

127. **von Heijne, G.,** A new method for predicting signal sequence cleavage sites, *Nucleic Acids Res.,* 11, 4683, 1986.

128. **Stirzaker, S. C., Whitfeld, P. L., Christie, D. L., Bellamy, A. R., and Both, G. W.,** Processing of the rotavirus glycoprotein VP7: implications for the retention of the protein in the endoplasmic reticulum, *J. Cell Biol.,* 1987, in press.

129. **Rose, J. K. and Bergmann, J. E.,** Altered cytoplasmic domains affect intracellular transport of the vesicular stomatitis virus glycoprotein, *Cell,* 34, 513, 1983.

130. **Roth, M. G., Compans, R. W., Giusti, L., Davis, A. R., Nayak, D. P., Gething, M. J., and Sambrook, J.,** Influenza virus hemagglutinin expression is polarized in cells infected with recombinant SV40 viruses carrying cloned hemagglutinin DNA, *Cell,* 33, 435, 1983.

131. **Jones, L. V., Compans, R. W., Davis, A. R., Bos, T. J., and Nayak, D. P.,** Surface expression of influenza virus neuraminidase, an amino-terminally anchored viral membrane glycoprotein, in polarized epithelial cells, *Mol. Cell. Biol.,* 5, 2181, 1985.

132. **Baybutt, H. N. and McCrae, M. A.,** The molecular biology of rotaviruses. VII. Detailed structural analysis of gene 10 of bovine rotavirus, *Virus Res.,* 1, 533, 1984.

133. **Ward, C. W., Azad, A. A., and Dyall-Smith, M. L.,** Structural homologies between RNA segments 10 and 11 from UK bovine, simian SA11 and human Wa rotaviruses, *Virology,* 144, 328, 1985.

134. **Okada, Y., Richardson, M. A., Ikegami, N., Nomoto, A., and Furuichi, Y.,** Nucleotide sequence of human rotavirus genome segment 10, an RNA encoding a glycosylated virus protein, *J. Virol.,* 51, 856, 1984.

135. **Petrie, B. L., Estes, M. K., and Graham, D. Y.,** Effects of tunicamycin on rotavirus morphogenesis and infectivity, *J. Virol.,* 46, 270, 1983.

136. **Sabara, M. and Babiuk, L. A.,** Identification of a bovine rotavirus gene and gene product influencing cellular attachment, *J. Virol.,* 51, 489, 1984.

137. **Arias, C. F., Lopez, S., and Espejo, R. T.,** Gene products of SA11 simian rotavirus genes, *J. Virol.,* 41, 42, 1982.

138. **Matsuno, S., Hasegawa, A., Kalica, A. R., and Kono, R.,** Isolation of a recombinant between simian and bovine rotaviruses, *J. Gen. Virol.,* 48, 253, 1980.

139. **Huismans, H., van der Watt, N. T., Cloete, M., and Erasmus, B. J.,** The biochemical and immunological characterization of bluetongue virus outer capsid polypeptides, in *Double-Stranded RNA Viruses,* Compans, R. W. and Bishop, D. H. L., Eds., Elsevier, New York, 1983, 165.

140. **Sangar, D. V. and Mertens, P. P. C.,** Comparison of type 1 bluetongue virus protein synthesis in vivo and in vitro, in *Double-Stranded RNA Viruses,* Compans, R. W., Bishop, D. H. L., Eds., Elsevier, New York, 1983, 183.

141. **Kahlon, J., Sugiyama, K., and Roy, P.,** Molecular basis of bluetongue virus neutralization, *J. Virol.,* 48, 627, 1983.

142. **Purdy, M. A., Ghiasi, H., Rao, C. D., and Roy, P.,** Complete sequence of bluetongue virus L2 RNA that codes for the antigen recognized by neutralizing antibodies, *J. Virol.,* 55, 826, 1985.

143. **Ghiasi, H., Purdy, M. A., and Roy, P.,** The complete sequence of bluetongue virus serotype 10 segment 3 and its predicted VP3 polypeptide compared with those of BTV serotype 17, *Virus Res.,* 3, 181, 1985.

143a. **Gould, A. R.,** The complete nucleotide sequence of bluetongue virus serotype 1 RNA 3 and a comparison with other geographic serotypes from Australia, South Africa and the United States of America and with other orbivirus isolates, *Virus Res.,* 7, 169, 1987.

144. **Rao, C. D., Kuichi, A., and Roy, P.,** Homologous terminal sequences of the genome double-stranded RNAs of bluetongue virus, *J. Virol.,* 46, 378, 1983.

145. **Mertens, P. P. C. and Sangar, D. V.,** Analysis of the terminal sequences of the genome segments of four orbiviruses, *Virology,* 140, 55, 1985.

146. **Roy, P., Ritter, G. D., Jr., Akashi, H., Collisson, E., and Inaba, Y.,** A genetic probe for identifying bluetongue infections in vivo and in vitro, *J. Gen. Virol.,* 66, 1613, 1985.

147. **Purdy, M. A., Ritter, G. D., Jr., and Roy, P.,** Nucleotide sequence of cDNA clones encoding the outer capsid protein, VP5, of bluetongue virus serotype 10, *J. Gen. Virol.,* 67, 957, 1986.

Chapter 10

REPLICATION OF dsRNA MYCOVIRUSES

Jeremy Bruenn

TABLE OF CONTENTS

I.   Introduction 196

II.   dsRNA Mycovirus Capsid Structure 196

III.   RNA Polymerases of dsRNA Mycoviruses 197
        A.   Mode of Transcription and Replication 198
                1.   AfV-S Polymerase 199
                2.   ScV Transcriptase and Replicase 199
                3.   PsV-S Replication 200

IV.   Host Genes in Replication 201

V.   Viral dsRNA Sequences Required in *Cis* 203

References 205

I. INTRODUCTION

Double-stranded RNA (dsRNA) mycoviruses have been demonstrated in at least 16 species of fungi.[66] Although most of these viruses are propagated from cell to cell only by mating, meiosis, and mitosis, their properties are sufficiently similar to those of infectious dsRNA viruses of plant and animal cells that they are nevertheless considered viruses.[66] In some cases, it has been possible to demonstrate infectivity of viral particles to protoplasts.[37,93] Replication of the dsRNA mycoviruses was previously reviewed in 1979.[29]

The mycoviruses are sufficiently well integrated into host cell metabolism that they usually cause no symptoms in infected cells. In fact, they are so widespread that they may be present in at least a majority of fungal species isolated from the wild. For instance, more than 60% of *Ustilago maydis* strains isolated from the wild have viral dsRNAs.[34] In two fungi, *U. maydis* and *Saccharomyces cerevisiae*, there are viral dsRNAs that encode extracellular toxins lethal to cells without these viral dsRNAs (see reviews in References 16, 20, 98). In these cases, the endogenous viruses may even confer a selective advantage on their host cells. We might expect the fungal viruses to utilize many host functions, other than those normally used by RNA viruses, in their successful attempts to persist in their host cells without killing them. In the best studied cases, the *S. cerevisiae* viruses, this is clearly true.[112]

Although most dramatic in the fungi, this phenomenon of permanent persistent infection with dsRNA viruses is not unique to the fungi. There are such dsRNA viruses in plants,[47,64,65] insects,[87] and protozoans.[101,102] Persistent infections can also occur with mammalian dsRNA viruses.[94] The ubiquitous nature of the fungal viruses is not the result of integration of provirus, at least in the most well-studied cases.[51,105,106] This also appears to be the case among other lower eucaryotes.[101,102]

There are two major reasons why replication of the dsRNA mycoviruses is of general interest to virologists. First, their replication must be intimately controlled by the host, a situation atypical of RNA viruses, but perhaps illustrative of mechanisms of persistence. Second, the mycoviruses are fundamentally simpler than the reoviruses, so that understanding the enzyme activities intrinsic to their virions should prove much simpler than understanding those activities proven among the reoviruses. Since the enzymatic activities responsible for replication of the dsRNA viruses have been intractible to purification and characterization, the investigation of replication of the dsRNA mycoviruses may provide useful generalizations.

II. dsRNA MYCOVIRUS CAPSID STRUCTURE

The capsid structure of the dsRNA mycoviruses is simple in three respects. First, every well-studied dsRNA mycovirus has a single major capsid polypeptide, rather than the 7 characteristic of the reoviruses.[56] Second, each virus particle contains only a single type of dsRNA segment (although it may have more than one copy of this segment), rather than the 10 or 12 different segments characteristic of reoviruses or the three of phi6.[63,66] The viral dsRNAs, which range from around 0.7 to 5 kilobases in length, are therefore packaged independently into similar particles. The replication and transcription of any segment can then be studied independently of any other segment, and in some cases, a single segment is sufficient for all viral functions.[39,54,66,82] Third, preliminary evidence indicates that the best-studied dsRNA mycoviruses have simple T=1 icosahedral capsids, the simplest type of icosahedral particle.

The capsid polypeptides of the best-characterized dsRNA mycoviruses are summarized in Table 1. All of these have a single major capsid polypeptide, a situation more characteristic of plant viruses than of animal viruses and completely different from that of the reoviruses or of phi6. With a few exceptions, these capsid polypeptides are large, greater than 50

Table 1
MAJOR CAPSID POLYPEPTIDES OF dsRNA MYCOVIRUSES

Virus species	Abbr.	Molecular weight (kdaltons)	Number/ virion	Ref.
Allomyces arbuscala		28—38		59
Aspergillus foetidus virus	AfV-S	87	120	25
Aspergillus foetidus virus	AfV-F	83	120	25
Colletotrichum lindemuthianum virus		53		81
Gaeumannomyces graminis virus				
Group I		54—60		27
Group II		68—72		27,80
Group III		78—87		27
Helminthosporium maydis		121		11
Penicillium crysogenum virus	PcV	87	60	22,70
		83		76
Penicillium stoliniferum virus S	PsV-S	42	120	26
Penicillium stoliniferum virus F	PsV-F	47		26
Saccharomyces cerevisiae virus 1	ScV-L$_1$	88	60	54,82
		81		92
		75		75
Saccharomyces cerevisiae virus 2	ScV-L$_2$	88		82
		86		36
Saccharomyces cerevisiae virus a	ScV-L$_a$	76—80		82
		73—77		92
		82		36
Ustilago maydis virus	UmV	75		60

kdaltons, and in 10 of the 15 cases, larger than 70 kdaltons. There have been reports of minor capsid polypeptides in some of these viruses,[26,75,81,92] but these may be major capsid polypeptides of secondary viruses in the same species (it is common for a single cell to harbor several different dsRNA viruses), undissociated multimers or degradation products of the major capsid polypeptide,[82] or host proteins contaminating the virus particle preparation. In those cases in which there is more than one virus in the same cell (e.g., PsV and ScV), there is no mixing of components; each viral dsRNA is separately encapsidated in its own capsid polypeptide and never in that of the co-infecting virus.[25,36,39,82,92] There is as yet no well-documented case of more than one capsid polypeptide in a dsRNA fungal virion. However, in the ScV system, in which only one viral dsRNA (L$_1$) is sufficient for its own replication and packaging, the sequence predicts one viral polypeptide in addition to the major capsid polypeptide.[19,19a]

Various physical techniques, including scanning transmission electron microscopy,[82] have been used to measure the molecular weight of mycovirus virions and thus to estimate the number of capsid polypeptide molecules per virion. In two cases, the result has been a number close to 60, consistent with an icosahedral capsid with three molecules per face (T=1), the simplest possible icosahedral structure. In several cases, the proposed capsid structure is a T=1 icosahedron with dimers of the capsid polypeptide as subunits, so that there would be 120 monomers per capsid.[25,26] Electron microscopy of all of the well-characterized dsRNA mycoviruses demonstrates particles with apparent icosahedral symmetry, but in no case has detailed electron microscopy or X-ray diffraction been performed.

III. RNA POLYMERASES OF dsRNA MYCOVIRUSES

Like viruses whose genomes are single-stranded RNAs complementary to the viral mRNAs (negative strand RNA), the dsRNA viruses are obliged to have capsid-associated RNA

Table 2
dsRNA MYCOVIRUSES FOR WHICH RNA POLYMERASES ARE KNOWN

Virus species	Abbreviation	Ref.
Allomyces arbuscula virus		58
Aspergillus foetidus virus S	AfV-S	77,78,79
Aspergillus foetidus virus F	AfV-F	77
Gaeumannomyces graminis virus (groups I and II)	GgV	24,27
Helminthosporium victoriae virus	HvV-A	45
Penicillium stoloniferum virus S	PsV-S	21,23
Penicillium stoloniferum virus F	PsV-F	31
Penicillium chrysogenum virus	PcV	70
Penicillium cyaneo-fulvum virus		67
Saccharomyces cerevisiae virus	ScV	19,53,103,104
Ustilago maydis virus	UmV	5

polymerases (transcriptases) to synthesize the viral mRNAs. Unlike the negative strand RNA viruses, the reoviruses synthesize genomic dsRNAs on plus strand templates within nascent viral particles.[56] Synthesis of reovirus mRNAs and dsRNAs is conservative: that is, the parental RNA strands remain associated and do not appear among the progeny molecules.[56] Less is known about replication among the fungal viruses. The dsRNA viruses of 11 species of fungi have been shown to have particle-associated RNA polymerase activities (Table 2). Typically, the predominant products of these reactions are single-stranded RNAs, and the reactions are not significantly stimulated by detergent or protease treatments (as opposed to activition of the reovirus transcriptase by chymotrypsin), not inhibited by actinomycin D, not stimulated by *S*-adenosylmethionine (SAM), and dependent on magnesium ion (see Reference 24 for example). Typically, maximal activities are seen at a magnesium ion concentration of about 5 mM.[24,104] *S*-adenosylhomocysteine, alpha-amanitin, and rifampicin are not inhibitory,[104] but ethidium bromide[77,104] and sodium pyrophosphate[104] are. Radioactive label in the methyl group of SAM is not incorporated into RNA.[18] This is as expected, since at least in the case of ScV, neither the viral dsRNAs[14] nor their transcripts[13,73] are capped or methylated.

A. Mode of Transcription and Replication

In most cases, the fungal virus RNA polymerases have been shown to synthesize single-stranded RNA which is extruded from the virion. In only one case (ScV) has this been shown to be the viral plus strand (the messenger strand). In one case, the RNA synthesized appears predominantly in dsRNA (PsV-S). In only one case (ScV) has incorporation into both dsRNA and ssRNA been sufficiently well characterized that transcriptase (ssRNA-producing) and replicase (dsRNA-producing) activities have been demonstrated.

In two cases the mode of synthesis of single-stranded RNA has been determined. In AfV-S it is semiconservative,[78] while in ScV it is conservative.[72] Again, in only two cases has the mode of synthesis of dsRNA been examined. In ScV it is conservative,[43,72,74,86,113] while in PsV-S it may be semi-conservative.[21,23]

These conclusions about the modes of transcription and replication of dsRNA fungal viral genomes are based on examinations in vitro (with AfV-S, PsV-S, and ScV) and in vivo (with ScV). Three methods have been used to determine whether viral RNA synthesis is conservative (analogous to transcription of dsDNA genomes) or semiconservative (resulting in displacement of a parental strand). One method is analogous to the Meselson-Stahl experiment: newly synthesized RNA is density labeled with nucleotides containing isotopes of either higher or lower atomic weight than those of the parental RNA, or with modified nucleotides of higher molecular weight. Progeny dsRNA should be of hybrid density after one cycle of replication, if replication is semiconservative. This method has been used in vitro with PsV-S[23] and AfV-S[78] and in vivo with ScV.[74,86] A second method is to uniformly label viral dsRNAs in vivo, purify particles, allow RNA synthesis to occur in vitro in the presence of differentially labeled nucleotides, and look for the appearance of the first label in the ssRNA produced in vitro. A positive result indicates semiconservative transcription. This method has been used for ScV.[19,72,113] A final method is to characterize the strandedness of the label incorporated into dsRNA in vitro during polymerase reactions. Semiconservative transcription results in synthesis of labeled plus strands in the dsRNA. Conservative replication results in synthesis of labeled minus strands. This method has been used for AfV-S[78] and ScV.[43,72,113]

1. AfV-S Polymerase

The virion RNA polymerase of AfV-S is typical in some respects of the fungal virus RNA polymerases. The activity is maximal at about 3 mM Mg^{++}, is inhibited by ethidium bromide, but not by actinomycin D, is not significantly increased by heat shock or detergent treatment, and results primarily in the synthesis of single-stranded RNA.[77] The molecular weights of the single-stranded products are half those of their parental RNAs,[79] and as many as 8 mol of product RNA are produced per mole of parental RNA.[79] A significant proportion of the label incorporated, about 35% early in the reaction, appears in dsRNA rather than in single-stranded RNA.[77,79] The label incorporated into dsRNA can be displaced from the dsRNA by hybridization of denatured dsRNA to unlabeled ssRNA synthesized in vitro.[78] Although this is a negative experiment, it can be interpreted to mean that the ssRNA synthesized in vitro is only of one strand and that synthesis is semiconservative.[78] It does not signify which strand is synthesized, however, although for obvious teleological reasons it is almost certainly the plus strand. Consequently, we cannot yet say if the AfV-S polymerase is a transcriptase. A density labeling experiment with bromodeoxyuridine (Budr) also demonstrates a density shift of an appropriate portion of the dsRNA to a value consistent with a hybrid density.[78] In this case, as with similar experiments on ScV in vivo, no hybrid density controls were used.[74,78,86] Finally, the label incorporated into dsRNA by AfV-S during the reaction can be chased out of dsRNA into ssRNA by further incubation with unlabeled nucleotides.[78] These three experiments make a fairly conclusive case for semiconservative synthesis of one strand (presumably the plus strand) by AfV-S.

2. ScV Transcriptase and Replicase

Despite the early demonstration of an RNA polymerase in ScV particles[53] in 1977, it was not until recently that definitive evidence for conservative replication and transcription was obtained.[72] This was primarily because it is inherently more difficult to demonstrate conservative synthesis than to demonstrate semiconservative synthesis. Of the three methods discussed, only one, the characterization of labeled dsRNA, is capable of giving positive, rather than negative, results. The ScV polymerase synthesizes primarily ssRNA,[19,53,103,104] but not, as described by some authors,[86] solely ssRNA. The polymerase exhibits properties similar to those of other fungal virus polymerases, and its ssRNA product is of only one strand.[19,103] The ssRNA product was first shown to be of the plus strand by in vitro translation[18] and later by hybridization to cDNA clones of known sequence coding for a viral product of known amino acid sequence[72] and by direct sequence analysis.[48,72,96] The ssRNA polymerase activity of ScV is thus a transcriptase, like that of reovirus. The transcript is a complete and accurate copy of its parental minus strand, with the addition of a terminal A$_{OH}$ or G$_{OH}$.[73,96] Initiation occurs in vitro without the addition of any primers or host factors.[73]

Typically, about 3% of the label incorporated in vitro appears in the viral dsRNAs.[19,53]

However, if particles labeled in vivo with ^{32}P-phosphate are used for in vitro transcription, less than 2% of the parental ^{32}P label is incorporated into ssRNA.[19] More recent results limited this estimate to 0.7%.[72] Similarly, in vivo experiments have shown less than 2% of the dsRNA after a single round of replication is of hybrid density in density-shift experiments.[74,86] Unfortunately, these experiments can only be interpreted to mean that, at most, 2% (or 0.7%) of the particles are engaged in semiconservative synthesis. Since no estimate of the number of viral particles engaged in transcription (or replication) is yet available, these results are difficult to interpret. More definitive experiments have been performed using the dsRNA radiolabeled in vitro. These experiments were performed with particles containing S14, a defective-interfering dsRNA derived from the viral toxin encoding dsRNA (M_1) by internal deletion.[61] The label incorporated into dsRNA is predominantly in the viral minus strand, as demonstrated both by hybridization to appropriate M13 cDNA clones and by strand separation.[72] Less than 0.004% of the particles incorporate label into (+) strands of dsRNA. Thus, if transcription were semiconservative, only 0.004% of the particles would have to be responsible for all the transcript synthesized.[72] Since transcription was far more active than replication in these experiments and replication (synthesis of minus strand) occurred in 0.05% of the particles, this is very unlikely. ScV transcription and replication are thus conservative, like those processes in reovirus. Similar experiments have now shown that ScV-L transcription and replication are conservative,[43,113] although results for ScV-M particles are more equivocal (see below).[38,113]

It is clear that at least a small proportion of the ScV particles present in vivo are capable of replication. ScV thus has a full spectrum of particles analogous to those present in reovirus infection, if the particles undergoing transcription and those undergoing replication are different. ScV-L replicating particles, like reovirus subviral particles, are encapsidated plus strands undergoing synthesis of minus strands.[6,38] A fraction of the ScV particles containing L_1 (the capsid-encoding dsRNA) have a lower molecular weight than mature L dsRNA particles and have partially completed L dsRNA with a full-length ssRNA and a variable dsRNA portion. About 90% of the RNA synthesized in vitro by these particles is dsRNA.[6] These have the characteristics expected of replicating particles. As expected, they are synthesizing the L_1 minus strand.[43]

Replicating M_1 particles may have either one or two dsRNAs after replication.[38,113] There should therefore be two classes of M_1 (and S) viral particles undergoing minus strand synthesis: those with only an encapsidated plus strand, and those with a plus strand and a dsRNA. This second class of particles, with a dsRNA and a nascent, encapsidated (not extruded) plus strand exists.[38] Since synthesis of both plus strands and of minus strands destined to become dsRNA occurs in vitro,[38,113] the asymmetry of strand production observed during in vitro synthesis of dsRNA[72] reflects the proportion of particles in which the plus strand was completed prior to in vitro synthesis in the particle preparation used. There are differences in the life-cycle of ScV-L and ScV-M particles, but it is unclear whether these affect the actual process of replication (see Section IV.).

3. PsV-S Replication

In *Penicillium stoliniferum* virus S particles, the major RNA polymerase activity is not synthesis of single-stranded RNA as in ScV, but rather synthesis of dsRNA.[21] The particles with this polymerase activity (H particles) contain one dsRNA copy of either of two genomic RNA species of about 1.6 kilobases, and a single molecule of ssRNA, whose length varies up to at most 1.6 kilobases, the size of the individual dsRNA strands.[28] Equilibrium cesium chloride density gradient centrifugation shows that the particles that result from this RNA polymerase activity are denser than their percursor particles. The precursor particles have a density of about 1.36 and the product particles about 1.39 g/cm^3.[21,28] The amount of label incorporated is consistent with the synthesis of varying amounts of RNA, reaching a maximum of 1.6 kilobase-pairs (or two strands of 1.6 kilobases). These data have been interpreted to mean that a particle of PsV-S containing one molecule of genomic dsRNA synthesizes a complete new copy of this dsRNA, which remains encapsidated in the same particle.[21] This is similar to the process of replication by ScV-M$_1$, also a smaller dsRNA of about 1.8 kilobases.[38,113]

Density-label experiments were performed with those H particles that have the least single-stranded RNA, and thus should synthesize, according to this model, essentially two new strands per particle. The progeny particles labeled in vitro in the presence of 5-bromo-UTP have dsRNA of higher average density than that of the precursor H particles.[23] This RNA has a density thought to be consistent with hybrid-density molecules in which one strand is an unlabeled parental strand and the other is labeled in vitro in the presence of the density label. This was interpreted as supportive of semiconservative replication.[23] However, the density-label experiments do not necessarily show semiconservative replication, since the H particles used must have been heterogeneous, some containing preformed, complete single-stranded RNAs. The resultant dsRNAs have heterogeneous densities. The greatest density might actually represent RNA labeled in both strands. There were no controls to establish the actual densities of hybrid molecules and of molecules with 5-Br-U in both strands. In addition, the strandedness of the product RNA was never established. If replication of PsV-S is as similar to that of ScV-M$_1$ as it appears, there should be a class of PsV-S particles that have only the viral plus strand engaged in synthesis of minus strand. There is an appropriate class of particles (M) with single-stranded RNA of unknown polarity, but no polymerase activity has yet been demonstrated in them.[21]

It seems well established that dsRNA fungal virus transcription (synthesis of viral plus strands) may be either conservative (ScV) or semiconservative (AfV-S). As in reovirus, replication appears to consist of synthesis of viral minus strands on the plus strand templates that result from transcription.

IV. HOST GENES IN REPLICATION

S. cerevisiae is the only fungus whose dsRNA viruses have been studied extensively for which genetic analysis is routine. There is extensive literature on host genes affecting the maintenance and expression of the ScV dsRNAs (see References 20, 98, 111, and 112 for reviews). Most laboratory and commercial strains of yeast have the ScV-L$_1$ virus and the unrelated ScV-L$_a$ virus. These appear under various names in the literature. L$_1$ is synonymous with L$_A$ or L$_{1A}$. L$_a$ is synonymous with L$_{BC}$. L$_1$ encodes a capsid polypeptide[54] that also encapsidates M$_1$, a viral dsRNA encoding a killer toxin.[9] Within this viral family is another viral dsRNA, L$_2$ (L$_{2A}$) with some sequence homology to L$_1$[7,39] encoding a related viral capsid polypeptide[8] that can also encapsidate a toxin-encoding dsRNA (M$_2$). The L$_1$ and L$_2$ capsid polypeptides are in some circumstances interchangeable.[50] There are thus two families of dsRNA viruses in *S. cerevisiae:* the L$_1$ family and the L$_a$ family. These are generally affected by different host functions, as shown in Table 3.

Most of the host genes necessary for maintenance of ScV or modulating its numbers have been defined by recessive mutations resulting in the loss of M$_1$ from haploids. These are the *mak* genes. Table 3 summarizes what is known about the viruses affected by each host gene. Note that in many cases not all the viral subtypes have been combined with each mutation, so that those viruses listed as unaffected or affected are only those that have been tested. In general, a given mutation affects only one family of viruses (the L$_1$ or the L$_a$ family). Yet, in many cases, ScV-L$_1$ is unaffected while ScV-M$_1$ is affected by a host gene (for instance, by *mak4, 5, 6, 7, 14, 15,* etc.). In some cases, the host function even discriminates between viruses whose genomes differ by probably only a few base-pairs (see *spe10* in Table 3, Reference 100). One possible explanation for the multiplicity of host gene

Table 3
HOST GENES AFFECTING ScV FUNCTIONS

Gene	Affected	Not affected	Host function	Ref.
mak1	ScV-M$_1$, M$_2$	ScV-L$_1$, L$_a$	DNA topoisomerase I	3,97
spe2	ScV-M$_1$, M$_2$	ScV-L$_1$, L$_a$	SAM decarboxylase	32,33,100
spe10[a]	ScV-M$_1$, M$_2$, L$_{1E}$	ScV-L$_a$, L$_{1HN}$	Polyamine synth.	100
mak8	ScV-M$_1$, M$_2$	ScV-L$_1$	Ribosomal protein L3	110
mak3,10	ScV-L$_1$, M$_1$, M$_2$	ScV-L$_a$		39,91,92,109
pet18	ScV-L$_1$, M$_1$, M$_2$	ScV-L$_a$		62,91,92,109
mak16	ScV-M$_1$, M$_2$	ScV-L$_1$		112
mak4—7	ScV-M$_1$	ScV-L$_1$, L$_a$[b]		3,91,92,109
mak14—15	ScV-M$_1$	ScV-L$_1$		3,91,92,109
mak17,19	ScV-M$_1$	ScV-L$_1$, L$_a$[b]		3,91,92,109
mak24,26	ScV-M$_1$	ScV-L$_1$, L$_a$[b]		3,91,92,109
mak27	ScV-M$_1$	ScV-L$_1$		3,91,92,109
mak2,9	ScV-M$_1$			112
mak11—13	ScV-M$_1$			112
mak18	ScV-M$_1$			112
mak20—23	ScV-M$_1$			112
mak25	ScV-M$_1$			112
mkt1,2	ScV-M$_2$	ScV-L$_1$, M$_1$		108
ski2—8	ScV-M$_1$, M$_2$, L$_1$, L$_a$			3,84,99
clo	ScV-L$_a$	ScV-L$_1$, M$_1$, M$_2$		105

[a]   L$_{1HN}$ and L$_{1E}$ are alleles of L$_1$

[b]   *mak4,7,17,* and *24* do not affect ScV-L$_a$.[3]

functions necessary for M$_1$, but not for L$_1$, is that packaging of L$_1$ plus strands may take place coincident with their translation, so that plus strands and capsid polypeptide are already in physical proximity. M$_1$ plus strands must be brought to the sites of capsid formation, and this may require many host gene functions. This model is consistent with the observation that defective L$_1$ genomes, which no longer encode the capsid polypeptide, but retain the terminal sequences of L$_1$ require many of the same host gene products required by M$_1$ for replication.[38a]

Many of the mutations that affect maintenance of ScV are pleiotropic. For instance, *pet18* also results in respiratory deficiency due to loss of mitochondria and temperature sensitivty of growth;[62] *spe2* mutants are defective in sporulation and grow very slowly; and *mak1, 13, 15, 16, 17, 20, 22,* and *27* confer temperature sensitivity of growth.[107] In a few cases, the host functions encoded by these genes are known. These host functions are DNA topoisomerase I (*mak1*), SAM decarboxylase (*spe2*), polyamine synthetase *(spe10)*, and ribosomal protein L3 (*mak8*); see Table 3. The viral functions of none of these host gene products are known, although a role for the *spe2* gene product can be deduced (see below). The role of topoisomerase I may be very indirect.[97] It has been proposed that L3 may play a role similar to that of *E. coli* ribosomal protein S1 in Qβ bacteriophage infection (see chapter by Biebricher and Eigen),[110] although there is no evidence for such a role.

Host gene products could conceivably be involved in any process necessary for maintenance of the virus: for instance in translation, transcription, replication, packaging, or segregation during mitosis. One of the host functions identified, the *spe2* gene product SAM decarboxylase, is required for synthesis of the polyamines spermidine and spermine,[32,33,100] and *spe10* is required for synthesis of all polyamines.[100] Since there are numerous cases of both DNA[1,46] and RNA[4,55] viruses that use polyamines to neutralize the charges of their genomes for packaging, this is probably the role of polyamines implicated by *spe2* and *spe10*. This would imply that the packaging process is different for ScV-L$_1$ and ScV-M$_1$, even though both are packaged in virions with the same capsid polypeptide. Recent evidence on the function of *pet18* supports this view.

The *pet18* mutation confers on ScV-L$_1$, but not on ScV-M$_1$, particles a thermolabile RNA polymerase activity.[42] The thermolability correlates with sensitivity of particles to disruption, monitored by release of dsRNA. These data could be interpreted in several ways. The thermolability could be the result of a thermolabile host protein present in ScV-L$_1$ particles, but absent in ScV-M$_1$ particles. The *pet18* protein could be required for thermostable packaging of L$_1$, but not present in viral particles and not required for packaging of M$_1$. The *pet18* protein could be a component of ScV-L$_1$ particles that confers thermostability to otherwise thermolabile particles. One of the latter two explanations seems to be correct, since all the *pet18* mutants used in these experiments have complete deletions of the single, wild-type gene.[42] The *pet18* protein could play a number of roles in packaging of L$_1$. One possibility is that it is a scaffolding protein, analogous to the lambda Nu3 protein, the phiX174 D gene product, or the P22 gp8.[44,52] These proteins are necessary for proper maturation of viral particles, but are removed from the particle in the process of maturation. The known examples of scaffolding proteins, however, are all viral gene products, not host gene products. Another possibility is that *pet18* is a small basic protein serving to neutralize the charges of L$_1$, analogous to the polyamines required for packaging of M$_1$. The *mak10* gene product similarly affects L$_1$ particles.[43]

There are other indications that the enzymatic activities of ScV-L$_1$ and ScV-M$_1$ particles are not identical. Like the reoviruses,[114] ScV-L$_1$ particles make a series of oligonucleotides corresponding to the 5' end of the transcript, varying in size from several to 25 nucleotides in length.[13] There do not appear to be corresponding "pause products" made by ScV-S particles (see below).[73] Finally, the viral dsRNA sequences required in *cis* appear to be different in M and L (see below).

Host proteins have also been implicated in RNA polymerase activities in other viral systems. One of the cellular RNA pol II subunits also appears to be a subunit in the rabbit poxvirus DNA dependent RNA polymerase.[68] Since actin has been implicated as a positive regulatory protein for RNA polymerase II,[35,85] it might also be required for viral RNA polymerases. These experiments stimulated an examination of the dependence of RNA virus polymerases on actin and, since it interacts with actin, tubulin. The VSV and Sendai transcriptases appear to be dependent on tubulin molecules packaged in the virions, but not on actin.[69] There is only one gene for tubulin in *S. cerevisiae*,[71] but it has not yet been implicated in the ScV polymerase activities. Immunological evidence indicates that none of the yeast RNA polymerase II subunits are found in ScV particles.[88]

There is relatively little information on whether ScV replication is limited by coordination with cell division. Experiments with cell cycle arrest by temperature shifts of temperature-sensitive mutants and by alpha factor[86,115] suggest that ScV replication does not take place during S phase. Experiments with normally growing cells separated by size,[74] imply that ScV replication takes place throughout the cell cycle. This question remains to be resolved, since cell cycle arrest may result in conditions not normally encountered by replicating ScV particles; for instance, reduction of the rNTP pool size.

V. VIRAL dsRNA SEQUENCES REQUIRED IN *CIS*

There is strong evidence that all the dsRNA viruses require specific sequences at both ends of their genomic dsRNAs. For instance, all 10 genomic dsRNAs of reovirus preserve 4 base-pairs at both ends.[2] Since the viral transcriptase must initiate at one end and the viral replicase at the other, we might expect these sequences to play a role in transcription and replication. They might also be necessary for packaging.

There does not appear to be any great degree of conservation of ends among the fungal virus dsRNAs. For instance, within the ScV-L$_1$ family, the L$_1$ plus strand has the 5' sequence pppGAAAAAUUUUUAAA, while the M$_1$ plus strand has the 5' sequence ppp-GAAAAAUAAAUGAC,[10,12,17,61,89,96] but there is less similarity at the 3' end of the plus strand, which is GAAACACCCAUCA$_{OH}$ in M$_1$ and UUACCCAUAUGCA$_{OH}$ in L$_1$.[10,12,17,61,89,95,96] When the sequences of L$_2$ and M$_2$, which belong to the same family, are included, the sequence similarities are reduced even further.[12] Nevertheless, there is good evidence that sequences at the ends of the ScV dsRNAs are necessary. There are defective interfering dsRNAs (S) derived from M$_1$, which, when introduced into the same cells as M$_1$, replace it by faster replication.[83,90] These are all derived from M$_1$ by internal deletion[15,17,41,57,61,96] and, therefore, preserve the end sequences of the dsRNAs. Just how much of the end sequences are necessary has not been determined, but the three defective dsRNAs best characterized retain several hundred bases from the 3' end of the plus strand and at least 44 to 82 bases from the 5' end of the plus strand.[61] The large region at the 3' end of the plus strand has no extensive open reading frames and is probably required in *cis*. It has 5 repeats of a 11-base-pair consensus sequence with dyad symmetry, a sequence which also appears in M$_2$,[49,61] but not in the sequenced regions of L$_1$. It has been proposed that this sequence is recognized by some of the many host proteins required for maintenance of ScV-M$_1$, but not for maintenance of ScV-L$_1$.[61] At the 5' end of the plus strand, only a small region (less than 82 bases) is necessary. Since the only M$_1$ open reading frame, which encodes the toxin, begins at position 14,[10,89] it may be that only a very small noncoding region is necessary at this end for viral dsRNA replication.

A similar defective derivative of L$_1$ has also been isolated.[38a] As with the defective dsRNAs derived from M$_1$, this dsRNA retains several hundred bases from the 3' end of the L$_1$ plus strand and a small region from the 5' region of the plus strand. Like the S dsRNAs, but unlike its parental dsRNA L$_1$, it is dependent on many of the *Mak* gene products for its maintenance. This may be the result of differences in packaging between capsid polypeptide encoding and noncoding dsRNAs (see Section IV).

Some information about the viral polymerases may also be derived by determination of the break points of defective interfering dsRNAs. The break points of the defective interfering dsRNAs from ScV have been determined in two cases. These are shown in Figure 1. The only obvious generalization is that the deleted portion is flanked by 5' and 3' C residues in both cases. In the transcriptase copy choice mechanism which has been proposed for deletion generation,[61] this would correspond to premature transcription termination and reinitiation with a C in each case. This is also true of termination of the wild-type plus strands, since their penultimate residue is invariably C; the terminal A$_{OH}$ or G$_{OH}$ is not template encoded.[14,73] In both cases, the termination event occurred within the sequence CCU.[61]

The only other fungal virus dsRNAs for which any sequences have been determined are those of UmV.[30,40] Even within one viral subtype of UmV there is little sequence homology at the ends of the viral dsRNAs, no more than 3 nucleotides. Many (but not all) of the UmV dsRNAs have one (deduced) 5' end with the sequence pppGAAAAA, similar to the 5' ends of the ScV plus strands. It was proposed that this was the 5' end of the UmV plus strands, and there is now evidence that, in at least one case, it is.[30] One of the UmV dsRNAs is only 357 base-pairs long, has no extensive open reading frames, and apparently arises by packaging and replication of a specific 3' fragment of the plus strand of a larger dsRNA of 1.5 kilobases.[30,40] This is not a defective-interfering dsRNA, and since it does not preserve the putative transcriptase initiation site, it may not be transcribed. Evidence on this point is equivocal.[5] These data do demonstrate, however, that packaging does not require sequences at the 5' ends of the viral plus strands.

```
           220          230          240          250              -550          -540

            *            *            *            *                *             *

S14     CAAGTCGATCACCTGGGGTTCATTCGTAGCGAGC                         CTCACCTTGAG

S3      CAAGTCGATCACC                                   CACAAGCACACTCACCTTGAG
```

FIGURE 1. Break points of yeast defective interfering dsRNAs. The sequences of two defective interfering ScV dsRNAs derived from their parental dsRNA, M_1, by internal deletion are shown around their break points, as deduced from complete sequences[61] or partial sequences.[61,96] The plus strands are shown. Numbers on the left refer to distance in nucleotides from the 5' end of the plus strand of M_1. Numbers on the right refer to distances in nucleotides from the 3' end of the plus strand of M_1. The sequences shown are those of the cDNAs.

Further information about *cis*-acting sequences in viral dsRNAs will probably await in vitro mutagenesis of cDNA clones.

REFERENCES

1. **Ames, B. N., Dubin, D. T., and Rosenthal, S. M.,** Presence of polyamines in certain bacterial viruses, *Science,* 127, 814, 1958.
2. **Antczak, J. B., Chmelo, R., Pickup, D. J., and Joklik, W. K.,** Sequences at both termini of the 10 genes of reovirus serotype 3 (Strain Dearing), *Virology,* 121, 307, 1982.
3. **Ball, S., Tirtiaux, C., and Wickner, R.,** Genetic control of L-A and L-(BC) dsRNA copy number in killer systems of *Saccharomyces cerevisiae, Genetics,* 107, 199, 1984.
4. **Beer, S. V. and Kosuge, T.,** Spermidine and spermine-polyamine components of turnip yellow mosaic virus, *Virology,* 40, 930, 1970.
5. **Ben-Tzvi, B., Koltin, Y., Mevarech, M., and Tamarkin, A.,** RNA polymerase activity in virions from *Ustilago maydis, Mol. Cell. Biol.,* 4, 188, 1984.
6. **Bevan, E. A. and Herring, A. J.,** The killer character in yeast: preliminary studies of virus-like particle replication, in *Genetics, Biogenesis and Bioenergetics of Mitochondria,* EURATOM Symp., Bandelow, W., Scheweyen, R. J., Thomas, D. Y., Wolf, K., and Kaudewitz, F., Eds., W. de Gruyter, Berlin, 1976, 153.
7. **Bobek, L. A., Bruenn, J. A., Field, L. J., and Gross, K. W.,** Cloning of cDNA to a yeast viral double-stranded RNA and comparison of three viral RNAs, *Gene,* 19, 225, 1982.
8. **Bobek, L. A., Bruenn, J. A., Field, L. J., Reilly, J. D., and Gross, K. W.,** Cloning of cDNAs to a yeast viral dsRNA, in *Double-Stranded RNA Viruses,* Compans, D. W. and Bishop, D. H. L., Eds., Elsevier, New York, 1983, 451.
9. **Bostian, K. A., Hopper, J. E., Rogers, D. T., and Tipper, D. J.,** Translational analysis of the killer-associated virus-like particle dsRNA genome of *S. cerevisiae:* M dsRNA encodes toxin, *Cell,* 19, 403, 1980.
10. **Bostian, K. A., Elliott, Q., Bussey, H., Burn, V., Smith, A., and Tipper, D. J.,** Sequence of the preprotoxin dsRNA gene of type 1 killer yeast: multiple processing events produce a two-component toxin, *Cell,* 36, 741, 1984.
11. **Bozarth, R. F.,** Biophysical and biochemical characterization of virus-like particles containing a high molecular weight dsRNA from *Helminthosporium maydis, J. Virol.,* 80, 149, 1977.
12. **Brennan, V. E., Field, L., Cizdziel, P., and Bruenn, J. A.,** Sequences at the 3' ends of yeast viral dsRNAs: proposed transcriptase and replicase initiation sites, *Nucleic Acids Res.,* 9, 4007, 1981.
13. **Brennan, V. E., Bobek, L. A., and Bruenn, J. A.,** Yeast dsRNA viral transcriptase pause products: identification of the transcript strand, *Nucleic Acids Res.,* 9, 5049, 1981.
14. **Bruenn, J. and Keitz, B.,** The 5' ends of yeast killer factor RNAs are pppGp, *Nucleic Acids Res.,* 3, 2427, 1976.
15. **Bruenn, J. and Kane, W.,** Relatedness of the double-stranded RNAs present in yeast virus-like particles, *J. Virol.,* 26, 762, 1978.
16. **Bruenn, J.,** Virus-like particles of yeast, *Annu. Rev. Mirobiol.,* 34, 49, 1980.
17. **Bruenn, J. and Brennan, V.,** Yeast viral double-stranded RNAs have heterogeneous 3' termini, *Cell,* 19, 923, 1980.

18. **Bruenn, J. Bobek, L., Brennan, V., and Held, W.,** Yeast viral RNA polymerase is a transcriptase, *Nucleic Acids Res.,* 8, 2985, 1980.
19. **Bruenn, J., Madura, K., Siegel, A., Miner, Z., and Lee, M.,** Long internal inverted repeat in a yeast viral dsRNA, *Nucleic Acids Res.,* 13, 1575, 1985.
19a. **Bruenn, J. A.,** unpublished data.
20. **Bruenn, J.,** The killer systems of *Saccharomyces cerevisiae* and other yeasts, in *Fungal Virology,* Buck, K. W., Ed., CRC Press, Boca Raton, Fla., 1986, 85.
21. **Buck, K.,** Replication of double-stranded RNA in particles of *Penicillium stoloniferum* virus S, *Nucleic Acids Res.,* 2, 1889, 1975.
22. **Buck, K. W. and Girvan, R. F.,** Comparison of the biophysical and biochemical properties of *Penicillium cyano-fulvum* virus and *Penicillium chrysogenum* virus, *J. Gen. Virol.,* 34, 145, 1977.
23. **Buck, K. W.,** Semi-conservative replication of double-stranded RNA by a virion-associated RNA polymerase, *Biochem. Biophys. Res. Commun.,* 84, 639, 1978.
24. **Buck, K. W., Romanos, M. A., McFadden, J. J. P., and Rawlinson, C. J.,** In vitro transcription of double-stranded RNA by virion-associated RNA polymerases of viruses from *Gaeumannomyces graminis,* *J. Gen. Virol.,* 57, 157, 1981.
25. **Buck, K. W. and Ratti, G.,** Biophysical and biochemical properties of two viruses isolated from *Aspergillus foetidus, J. Gen. Virol.,* 27, 211, 1975.
26. **Buck, K. W. and Kempson-Jones, G. F.,** Capsid polypeptides of two viruses isolated from *Penicillium stoloniferum, J. Gen. Virol.,* 22, 441, 1974.
27. **Buck, K. W., Almond, M. R., McFadden, J. J. P., Romanos, M. A., and Rawlinson, C. J.,** Properties of thirteen viruses and virus variants obtained from eight isolates of the wheat take-all fungus, *Gaeumannomyces graminis* var. *tritici. J. Gen. Virol.,* 53, 235, 1981.
28. **Buck, K. W. and Kempson-Jones, G. F.,** Biophysical properties of *Penicillium stoloniferum* virus S, *J. Gen. Virol.,* 18, 223, 1973.
29. **Buck, K. W.,** Replication of double-stranded RNA mycoviruses, in *Viruses and Plasmids in Fungi,* Lemke, P. A., Ed., Marcel Dekker, New York, 1979, 94.
30. **Chang, T. S., Bruenn, J. A., Held, W., Peery, T., and Koltin, Y.,** A very small viral dsRNA, submitted.
31. **Chater, K. and Morgan, D.,** Ribonucleic acid synthesis by isolated viruses of *Penicillium stoloniferum, J. Gen. Virol.,* 23, 307, 1974.
32. **Cohn, M. S., Taber, C. W., Taber, H., and Wickner, R. B.,** Spermidine or spermine requirement for killer dsRNA plasmid replication in yeast, *J. Biol. Chem.,* 253, 5225, 1978.
33. **Cohn, M. S., Tabor, C. W., and Tabor, H.,** Isolation and characterization of Saccharomyces cerevisiae mutants deficient in *S*-adenosylmethionine decarboxylase, spermidine, and spermine, *J. Bacteriol.,* 134, 208, 1978.
34. **Day, P. R.,** Fungal virus populations in corn smut from Connecticut, *Mycologia,* 73, 379, 1981
35. **Egly, J. M., Miyamoto, N. G., Monocollin, V., and Chambon, P.,** Is actin a transcription initiation factor for RNA polymerase B?, *EMBO J.,* 3, 2363, 1984.
36. **El-Sherbeini, M., Tipper, D. J., Mitchell, D. J., and Bostian, K. A.,** Virus-like particle capsid proteins encoded by different L-dsRNAs of *S. cerevisiae:* their roles in maintenance of M-dsRNA killer plasmids, *Mol. Cell. Biol.,* 4, 2818, 1984.
37. **El-Sherbeini, M. and Bostian, K. A.,** Viruses in fungi: infection of yeast with the K1 and K2 killer viruses, *Proc. Natl. Acad. Sci. U.S.A.,* 84, 4293, 1987.
38. **Esteban, R. and Wickner, R. B.,** Three different M_1 RNA-containing virus-like particle types in *Saccharomyces cerevisiae:* in vitro M_1 double-stranded RNA synthesis, *Mol. Cell. Biol.,* 6, 1552, 1986.
38a. **Esteban, R. and Wickner, R. B.,** A deletion mutant of L-A double-stranded RNA replicates like M_1 double-stranded RNA, submitted.
39. **Field, L. J., Bobek, L., Brennan, V., Reilly, J. D., and Bruenn, J.,** There are at least two yeast viral dsRNAs of the same size: an explanation for viral exclusion, *Cell,* 31, 193, 1982.
40. **Field, L. J., Bruenn, J. A., Chang, T. H., Pinhasi, O., and Koltin, Y.,** Two *Ustilago maydis* viral dsRNAs of different size code for the same product, *Nucleic Acids Res.,* 11, 2765, 1983.
41. **Fried, H. M. and Fink, G. R.,** Electron microscopic heteroduplex analysis of "killer" double-stranded RNA species from yeast, *Proc. Natl. Acad. Sci. U. S.A.,* 75, 4224, 1978.
42. **Fujimura, T. and Wickner, R. B.,** Thermolabile L-A virus-like particles from pet18 mutants of *Saccharomyces cerevisiae, Mol. Cell. Biol.,* 6, 404, 1986.
43. **Fujimura, T., Esteban, R., and Wickner, R. B.,** In vitro L-A double-stranded RNA synthesis in virus-like particles from *Saccharomyces cerevisiae, Proc. Natl. Acad. Sci. U.S.A.,* p. 4433, 1986.
44. **Georgopoulos, C., Tilly, K., and Casjens, S.,** Lambdoid phage head assembly, in *Lambda II,* Hendrix, R. W., Roberts, J. W., Stahl, F. W., and Weisberg, R. H., Eds., Cold Spring Harbor Laboratory, Cold Spring Harbor, New York, 1983, 279.
45. **Ghabrial, S. A. and Havens, W. M.,** Transcription of double-stranded RNA in virions of *Helminthosporium victoriae* virus, *Am. Phytopathol. Soc. Monogr.,* 76, 1090, 1986.

46. **Gibson, W. and Roizman, B.**, Compartmentalization of spermine and spermidine in the herpes simplex virion, *Proc. Natl. Acad. Sci. U.S.A.*, 68, 2818, 1971.

47. **Grill, L. K. and Garger, J.**, Identification and characterization of double-stranded RNA associated with cytoplasmic male sterility in *Vicia faba*, *Proc. Natl. Acad. Sci. U.S.A.*, 7043, 1981.

48. **Hannig, E. M., Thiele, D., and Leibowitz, M. J.**, *Saccharomyces cerevisiae* killer virus transcripts contain template-coded polyadenylate tracts, *Mol. Cell. Biol.*, 4, 92, 1984.

49. **Hannig, E. M. and Leibowitz, M. J.**, Structure and expression of the M_2 genomic segment of a type 2 killer virus of yeast, *Nucleic Acids Res.*, 13, 4379, 1985.

50. **Hannig, E. M., Leibowitz, M. J., and Wickner, R. B.**, On the mechanism of exclusion of M_2 double-stranded RNA by L-A-E double stranded RNA in *Saccharomyces cerevisiae*, *Yeast*, 1, 57, 1985.

51. **Hastie, N., Brennan, V., and Bruenn, J.**, No homology between double-stranded RNA and nuclear DNA of yeast, *J. Virol.*, 28, 1002, 1978.

52. **Hayashi, M.**, Morphogenesis of the isometric phages, in *The Single-Stranded DNA Phages*, Denhardt, D. T., Dressler, D., and Ray, D. S., Eds., Cold Spring Harbor Laboratory, Cold Spring Harbor, New York, 1978, 531.

53. **Herring, A. J. and Bevan, E. A.**, Yeast virus-like particles possess a capsid-associated single-stranded RNA polymerase, *Nature (London)*, 268, 464, 1977.

54. **Hopper, J. E., Bostian, K. A., Rowe, L. B., and Tipper, D. J.**, Translation of the L-species dsRNA genome of the killer-associated virus-like particles of *Saccharomyces cerevisiae*, *J. Biol. Chem.* 252, 9010, 1977.

55. **Johnson, M. W. and Markham, R.**, Nature of polyamine in plant viruses, *Virology*, 17, 276, 1962.

56. **Joklik, W.**, Reproduction of reoviridae, in *Comprehensive Virology*, Vol. 2, Plenum Press, New York, 1974, 231.

57. **Kane, W. P., Pietras, D., and Bruenn, J.**, Evolution of defective interfering double-stranded RNAs in yeast, *J. Virol.*, 32, 692, 1979.

58. **Khadijian, E. W. and Turian, G.** In vitro RNA synthesis by double-stranded mycovirus from *Allomyces arbuscula*, *FEBS Lett.*, 2, 121, 1977.

59. **Khandijian, E. W., Turian, G., and Eisen, H.**, Characterization of the RNA mycovirus infecting *Allomyces arbuscula*, *J. Gen. Virol.*, 35, 415, 1977.

60. **Koltin, Y., Mayer, I., and Steinlauf, R.**, Killer phenomenon in *Ustilago maydis*: mapping viral functions, *Mol. Gen. Genet.*, 166, 181, 1978.

61. **Lee, M., Pietras, D. F., Nemeroff, M., Corstanje, B., Field, L., and Bruenn, J. A.**, Conserved regions in defective interfering viral double-stranded RNAs from a yeast virus, *J. Virol.*, 58, 402, 1986.

62. **Leibowitz, M. J. and Wickner, R. B.**, pet18: a chromosomal gene required for cell growth and the maintenance of m DNA and the killer plasmid of yeast, *Mol. Gen. Genet.*, 165, 115, 1978.

63. **Lemke, P. A. and Nash, C. H.**, Fungal viruses, *Bacteriol. Rev.*, 38, 29, 1974.

64. **Lisa, V., Luisoni, E., and Milne, R. G.**, A possible virus cryptic in carnation, *Ann. Appl. Biol.*, 98, 431, 1981.

65. **Lisa, V., Boccardo, G., and Milne, R. G.**, Double-stranded RNA from carnation cryptic virus, *Virology*, 115, 410, 1981.

66. **Matthews, R. E. F., Ed.**, Classification and nomenclature of viruses, *Fourth Rep. Int. Comm. Taxonomy of Viruses*, S. Karger, Basel, 1982; as cited in *Intervirology*, 7, No. 1 to 3, 199.

67. **McGinty, R. M., Buck, K. W., and Rawlinson, C. J.**, Transcriptase activity associated with a type 2 double-stranded RNA mycovirus, *Biochem. Biophys. Res. Commun.*, 98, 501, 1981.

68. **Morrison, D. K. and Moyer, R. W.**, Detection of a subunit of cellular pol II within highly purified preparations of RNA polymerase isolated from rabbit poxivrus virions, *Cell*, 44, 587, 1986.

69. **Moyer, S. A., Baker, S. C., and Lessard, J. L.**, Tubulin: a factor necessary for the synthesis of both Sendai virus and vesicular stomatitis virus RNAs, *Proc. Natl. Acad. Sci. U.S.A.*, 83, 5405, 1986.

70. **Nash, C., Douthart, R., Ellis, L., VanFranke, R., Burnett, J., and Lemke, P.**, On the mycophage of *Penicillium chrysogenum*, *Can. J. Microbiol.*, 19, 97, 1973.

71. **Neff, N. F., Thomas, J. H., Grisafi, P., and Botstein, D.**, Isolation of the beta-tubulin gene from yeast and demonstration of its essential function in vivo, *Cell*, 33, 211, 1983.

72. **Nemeroff, M. E. and Bruenn, J. A.**, Conservative replication and transcription of yeast viral double-stranded RNAs in vitro, *J. Virol.*, 57, 754, 1986.

73. **Nemeroff, M. E. and Bruenn, J. A.**, Initiation by the yeast viral transcriptase in vitro, *J. Biol. Chem.*, 262, 6785, 1987.

74. **Newman, A. M., Elliott, S. G., McLaughlin, C. S., Sutherland, P. A., Warner, R. C.**, Replication of double-stranded RNA of the virus-like particles in *Saccharomyces cerevisiae*, *J. Virol.*, 38, 263, 1981.

75. **Oliver, S. E., McCready, S. J., Holm, C., Sutherland, P. A., McLaughlin, C., and Cox, B. S.**, Biochemical and physiological studies of the yeast VLP, *J. Bacteriol.*, 130, 1303, 1977.

76. **Petrovic, S. L., Sumonja, B. D., and Vasiljevic, R. B.,** Fractionation of nucleic acids from *Penicillium chrysogenum* and associated ribonucleic acid viruses by selective exclusion and retention in agarose gels, *Biochem. J.*, 139, 157, 1974.

77. **Ratti, G. and Buck, K. W.,** RNA polymerase activity in double-stranded ribonucleic acid virus particles from *Asperigillus foetidus*, *Biochem. Biophys. Res. Commun.*, 66, 706, 1975.

78. **Ratti, G. and Buck, K. W.,** Semi-conservative transcription in particles of a double-stranded RNA mycovirus, *Nucleic Acids Res.*, 5, 3843, 1978.

79. **Ratti, G. R. and Buck, K. W.,** Transcription of double-stranded RNA in virions of *Aspergillus foetidus* virus S, *J. Gen. Virol.*, 42, 59, 1979.

80. **Rawlinson, C. J., Hornby, D., Pearson, V., and Carpenter, J. M.,** Virus-like particles in the take-all fungus *Gaeumannomyces graminis*, *Ann. Appl. Biol.*, 74, 197, 1973.

81. **Rawlinson, C. J., Carpenter, J. M., and Muthyalu, G.,** Double-stranded RNA virus in *Colletotrichum lindemuthianum*, *Trans. Br. Mycol. Soc.*, 65, 305, 1975.

82. **Reilly, J. D., Bruenn, J., and Held, W.,** The capsid polypeptides of the yeast viruses, *Biochem. Biophys. Res. Commun.*, 121, 619, 1984.

83. **Ridley, S. P. and Wickner, R. B.,** Defective interference in the killer system of *Saccharomyces cerevisiae*, *J. Virol.*, 45, 800, 1983.

84. **Ridley, S., Sommer, S., and Wickner, R.,** Killer yeast: superkiller mutations suppress exclusion of M_2 dsRNA by L-A-HN and confer cold sensitivity in the presence of M and L-A-HN, *Mol. Cell. Biol.*, 4, 761, 1984.

85. **Scheer, U., Minssen, J., Franke, W. W., and Jockusch, B. M.,** Microinjection of actin-binding proteins and actin antibodies demonstrates involvement of nuclear actin in transcription of lampbrush chromosomes, *Cell*, 39, 111, 1984.

86. **Sclafani, R. A. and Fangman, W. L.,** Conservative replication of yeast double-stranded RNA by displacement of progeny single strands, *Mol. Cell. Biol.*, 4, 1618, 1984.

87. **Scott, M. P., Fostel, J. M., and Pardue, M. L.,** A new type of virus from cultured Drosophila cells: characterization and use in studies of heat-shock response, *Cell*, 22, 929, 1980.

88. **Sentenac, A.,** personal communication, 1984.

89. **Skipper, N., Thomas, D. Y., and Lau, P.,** Cloning and sequencing of the preprotoxin-coding region of the yeast M1 double-stranded RNA, *EMBO J.*, 3, 107, 1984.

90. **Somers, J. M.,** Isolation of suppressive sensitive mutants from killer and neutral strains of *Saccharomyces cerevisiae*, *Genetics*, 74, 571, 1973.

91. **Sommer, S. S. and Wickner, R. B.,** Co-curing of plasmids affecting killer double-stranded RNAs of *Saccharomyces cerevisiae*: [HOK], [NEX], and the abundance of L are related and further evidence that M_1 requires L, *J. Bacteriol.*, 150, 545, 1982.

92. **Sommer, S. S. and Wickner, R. B.,** Yeast L dsRNA consists of at least three distinct RNAs; evidence that non-Mendelian genes [HOK], [NEX], and [EXL] are on one of these dsRNAs, *Cell*, 31, 429, 1982.

93. **Stanway, C. A. and Buck, K. W.,** Infection of protoplasts of the wheat take-all fungus, *Gaeumannomyces graminis* var. *tritici*, with double-stranded RNA viruses, *J. Gen. Virol.*, 65, 2061, 1984.

94. **Taber, R., Alexander, V., and Whitford, W.,** Persistent reovirus infection of CHO cells resulting in virus resistance, *J. Virol.*, 17, 513, 1976.

95. **Thiele, D. J. and Leibowitz, M. J.,** Structural and functional analysis of separated strands of killer double-stranded RNA of yeast, *Nucleic Acids Res.*, 10, 6903, 1982.

96. **Thiele, D. J., Hannig, E. M., and Leibowitz, M. J.,** Genome structure and expression of a defective interfering mutant of the killer virus of yeast, *Virology*, 137, 20, 1984.

97. **Thrash, C., Voelkel, K. A., DiNardo, S., Sternglanz, R.,** Identification of *Saccharomyces cerevisiae* DNA topoisomerase I mutants, *J. Biol. Chem.*, 259, 1375, 1984.

98. **Tipper, D. J. and Bostian, K. A.,** Double-stranded ribonucleic acid killer systems in yeasts, *Microbiol. Rev.*, 48, 125, 1984.

99. **Toh-e, A., Guerry, P., and Wickner, R. B.,** Chromosomal superkiller mutants of *S. cerevisiae*, *J. Bacteriol.*, 136, 1002, 1978.

100. **Tyagi, A., Wickner, R. B., Tabor, C. W., and Tabor, H.,** Specificity of polyamine requirements for the replication and maintenance of different double-stranded RNA plasmids in *Saccharomyces cerevisiae*, *Proc. Natl. Acad. Sci. U.S.A.*, 81, 1149, 1984.

101. **Wang, A. L. and Wang, C. C.,** The discovery of a specific double-stranded RNA virus in *Giardia lamblia*, *Mol. Biochem. Parasitol.*, 21, 269, 1986.

102. **Wang, A. L. and Wang, C. C.,** The double-stranded RNA in *Trichomonas vaginalis* may originate from virus-like particles, *Proc. Natl. Acad. Sci. U.S.A.*, 83, 7956, 1986.

103. **Welsh, J. D. and Leibowitz, M. J.,** Transcription of killer virion double-stranded RNA in vitro, *Nucleic Acids Res.*, 8, 2365, 1980.

104. **Welsh, J. D., Leibowitz, M. J., and Wickner, R. B.,** Virion DNA-independent RNA polymerase from *Saccharomyces cerevisiae*, *Nucleic Acids Res.*, 8, 2349, 1980.

105. **Wesolowski, M. and Wickner, R.,** Two new double-stranded RNA molecules showing non-Mendelian inheritance and heat inducibility in *Saccharomyces cerevisiae, Mol. Cell. Biol.,* 4, 181, 1984.

106. **Wickner, R. B. and Leibowitz, M. J.,** Dominant chromosomal mutant bypassing chromosomal genes needed for killer RNA plasmid replication in yeast, *Genetics,* 87, 453, 1977.

107. **Wickner, R. B.,** mak mutants of yeast: mapping and characterization, *J. Bacteriol.,* 140, 154, 1979.

108. **Wickner, R. B.,** Plasmids controlling exclusion of the k_2 killer double-stranded RNA plasmid of yeast, *Cell,* 21, 217, 1980.

109. **Wickner, R. B., and Toh-e, A.,** [HOK], a new yeast non-Mendelian trait, enables a replication-defective killer plasmid to be maintained, *Genetics,* 100, 159, 1982.

110. **Wickner, R. B., Ridley, S. P., Fried, H. M., and Ball, S. G.,** Ribosomal protein L3 is involved in replication or maintenance of the killer double-stranded RNA genome of *Saccharomyces cerevisiae, Proc. Natl. Acad. Sci. U.S.A.,* 79, 4706, 1982.

111. **Wickner, R. B.,** Killer systems in *Saccharomyces cerevisiae:* three distinct modes of exculsion of M_2 double-stranded RNA by three species of double-stranded RNA, M_1, L-A-E and L-A-HN, *Mol. Cell. Biol.,* 3, 654, 1983.

112. **Wickner, R. B.,** Genetic control of replication of the double-stranded RNA segments of the killer systems in *Saccharomyces cerevisiae, Arch. Biochem. Biophys.,* 222, 1, 1983.

113. **Williams, T. L. and Leibowitz, M. J.,** Conservative mechanism of the in vitro transcription of killer virus of yeast, *Virology,* 158, 231, 1987.

114. **Yamakawa, M., Furuichi, Y., Nakashima, K., LaFiandra, A. J., and Shatkin, A. J.,** Excess synthesis of viral mRNA 5'-terminal oligonucleotides by reovirus transcriptase, *J. Biol. Chem.,* 256, 6507, 1981.

115. **Zakian, V. A., Wagner, D. W., and Fangman, W. L.,** Yeast L double-stranded RNA is synthesized during the G1 phase but not the S phase of the cell cycle, *Mol. Cell Biol.,* 1, 673, 1981.

Index

INDEX

A

Actinomycin D, 57, 93, 100, 118
AfV-F, 197—198
AfV-S, 197—199, 201
Alfalfa mosaic virus (AlMV), 50—52, 81, 92—93, 101
Alphaviruses, 71—82, 87—90, 98, 105
  evolution, 78—81
  genome organization, 72—75
  nonstructural proteins of, 98
  relationship to certain plant viruses, 81—82
  RNA replication, 75—78
Amino acid homologies, 98
Aminoacylatability, 96, 103
Aminoacylation, 103
Aminoacyl tRNA synthetase, 96
Animal viruses, see Alphaviruses; Flaviviruses
Antibodies, 92
Anticodon, 104
Antigenic site, 182
Antigenic variants, 182
Antitermination, 152, 163—164, 167
Aphthoviruses, 29
Arthropod vectors, 72
ATP, 35
Autocatalytic cleavage, 29
Autoproteolytic cleavage, 74

B

Bacteriophage Qβ, 34
Baculovirus system, 162, 180
Barley protoplasts, 96, 99—100, 103, 105
Barley stripe mosaic virus, 93
Bipartite genome viruses, 51, 53
Blocked N terminus, 183
Bluetongue virus, 172, 186
Brome mosaic virus (BMV), 50, 52, 81
  genome organization, 95—98
  proteins, functions and roles of, 98—99
  replicase, 99
    endogenous activity, 100
    (-) sense RNA templates, 104—105
    (-) strands synthesis, 101—104
    products, double-stranded, 100
    products, full-length complementary, 100
    RNA specific, 100
    specificity, 100—101
    template dependency, 100—101
    template recognition, 103
  RNA plant viruses related to, 93
  RNA replication of, 91—106, 108—109
    aminoacylation, 96
    cDNA clones, 103
    composition of active replication complex, 108
    cycle of, 99
    form of actively replicating intermediates, 108

  gene 1, 98—99
  gene 2, 98
  gene 3, 99
  genomic organization, 95—98
  (-) strand synthesis, 99
  perspective on research, 92—93
  (+) strand synthesis, 99
  polypeptide components, 107—108
  preparation of BMV replicase from infected
    barley leaves, 99—100
  preparation of TYMV replicase from infected
    plants, 106—107
  subgenomic RNA4 synthesis, 105—106
  synthesizing activities, 93—95
  template dependency and specificity, 100—101, 104—105
  tRNA-like activities, 103—104
  tRNA-like properties, 96—98
  tyrosylation, 96
  tRNA-like structures, 108—109
Bunyaviruses, 144

C

Canine distemper virus, 139—140
Cap-binding protein complex, 28
Capped (m⁷GpppNm-containing) RNA fragments, 160
Capped RNA-primed viral mRNA synthesis, 167
Capping, 98, 105
Capsids, 96, 196—197
Carbohydrate, 178, 182, 185
Cardioviruses, 29
CCMV, see Cowpea chlorotic mottle virus
cDNA clones, 41
Cell cycle, 203
Cell-to-cell spread of virus, 99
Cellular proteases, 74, 82
Cellular proteins, 144—146
Chandipura virus, 139
Chattering, 147
Chinese cabbage, 106
Chloroplast, 94, 107
Chymotrypsin, 179
Cleavage, 27, 183
CMV, see Cucumber mosaic virus
Coat proteins, 96, 99
Codon usage randomization, 84—85
Coliphages, 3—4
Comovirus, 49—69, see also Cowpea mosaic virus
  cowpea mosaic virus, 52—53
  expression of RNAs, 53—55
  polyprotein processing, 53
  RNA replication in, 49—65
    characterization of purified CPMV RNA
      complex, 58—60
    comovirus expression, 52—55
    different from CPMV, TMV, BMV, and AMV,

61—62
distinctive features, 55—57
initiation, 62
model, 63—65
role of protein processing, 63
similarities with picornaviruses, 60—61
Competition among RNA species, 13—16
Complementary strands, interaction of, 11, 13
Conserved sequence elements, 75, 86
Cordycepin, 102
Core polymerases, 98, 107
Core RNA polymerase 3D^{pol}, 28
Coronaviruses, 140, 144
  defective-interfering RNA, 128, 131
  lipid, 117
  mRNA structure, 120—122
  negative strand RNA, 119—120
  open reading frames, 120—121
  replicative forms, 125—126
  RNA-dependent RNA polymerases, 128—130
  RNA genome, 117
  RNA recombination, 127—128, 131
  RNA regulatory proteins, 130
  RNA replication, 115—132
    DI RNA, 131
    general pathway, 117—119
    genome, 117
    lipid, 117
    mechanism of subgenomic RNA transcription, 122—125
    mRNA structure and coding functions, 120—122
    negative strand RNA, 119—120
    perspectives, 131—132
    polymerases and regulatory proteins, 128—130
    recombination, 131
    RI and RF, 125—126
    structure and organization, 116—117
    temperature-sensitive mutants, 130
    structural proteins, 116—117
    subgenomic mRNA transcription, 122—125
    temperature-sensitive mutants, 129—130
    unique features of, 132
Covalent linkage, 35, 39
Cowpea chlorotic mottle virus (CCMV), 100—101
Cowpea mosaic virus (CPMV), 50, 52—55, 94
  picornaviruses compared with, 60—61
  poliovirus compared, 60—61
  protein processing, 63
  RNA replication, 57—59
    differing from other comoviruses, 61—62
    initiation of, 62
    model for, 63—65
    structure of RNA, 55—57
  terminal uridylyl transferase, 65
Coxsackie virus, 29
CPMV, see Cowpea mosaic virus
Crude replication complex, 36—38
CTP, ATP:tRNA nucleotidyl-transferase, 96, 109
Cucumber mosaic virus (CMV), 93
Cypoviruses, 172
Cytopathic structures, 57

Cytoplasmic polyhedrosis virus, 172

D

De novo synthesis, 5, 14, 34, 57, 101, 105, 128
Defective-interfering (DI) RNA, 41, 75, 128, 131
Defective interfering dsRNAs, 204—205
Deletion mutants, 41
Dengue viruses, 82, 84—85
Density-shift experiments, 200
DI, see Defective-interfering RNA
Diagnostic probe, 186
DNA-dependent RNA polymerase, 95
DNA genome, 2
DNA topoisomerase I, 202
Dodecyl-β-ᴅ-maltoside, 92, 100
Double-length RNA molecules, 33
Double-strand RNA, 11, 13—15, 33, 35, 108, 195—209
  capsid structure, 196—197
  host genes in replication, 201—203
  major capsid polypeptides of, 197
  RNA polymerases of, 197—201, see also RNA polymerases
  viral dsRNA sequences required in *cis*, 203—205

E

Ebola virus, 139
Elongation factor, 31, 36, 39, 41, 96, 107
Encephalomyocarditis (EMC) virus, 27, 29—30, 37
Encoded proteins, 24—27
Endogenous RNA, 31, 100
Endonucleases, 36
Endonucleolytic cleavage, 33
Endoplasmic reticulum, 178
Enterovirus, 24
Error propagation, 2
Escherichia coli, 4, 182
Evolution, 13, 78—81
Exogenous viral RNA, 31

F

Fijiviruses, 172
Fingerprints, 36
Flaviviruses, 71—72, 82—87
  comparison of sequences, 83—86
  genome organization, 82—83
  RNA replication, 86
  yellow fever strains 17D and Asibi, 86—87
Foot-and-mouth disease virus, 29—30
14-kdalton protein, 176

G

Gene clothing, 172—173
Gene remnants, 142—143
Genetic organization, 140—143
Genetic recombination, 31
Genome-linked viral protein (VPg), 24—25, 27—29,

32—33, 50
antibodies, 35
cowpea mosaic virus, 56—57
host factor, 39
initiation of viral RNA replication, role in, 62
initiation of viral RNA synthesis, role in, 65
primer for 3D^{pol}, 37
RNA in vitro, 36
Genome organization alphaviruses, 72—75, 95—98
Genomic RNA, 33
GgV, 198
Glycoprotein, 116, 138
Glycoprotein E, 84
Glycoprotein VP7, 181—184, see also Segment 9
Glycosylation sites, 185
Golgi apparatus, 74, 178
gp65, 117
Guanidine, 31—32, 61
Guanylyl transferase, 98

H

Hairpin, 33, 37, 39—40, 62, 127
Heat-shock proteins, 95
Hemagglutinin, 178
Hepatitis A virus, 29
Host cell primers, 160
Host-cell proteins, 32—33
Host-encoded proteins, 92
Host-encoded RNA polymerase, 58
Host factor, 32, 34—36, 38—40, 63, 81, 199
Host genes in replication, 201—203
H particles, 200—201
Human parainfluenza virus, 138—140
Human rhinovirus, 29
HvV-A, 198
Hydrophobic domain, 32, 36, 84, 183

I

IBV, see Infectious bronchitis virus
Identification, 172—173
Ilarviruses, 52, 92
Immunogenicity, 181—183
Immunoprecipitation, 32
In vitro translation, 28, 35, 163
Incorporation profile, 15—16
Infection cycle, 3—4, 6, 27—29
Infectious bronchitis virus (IBV), 116
Infectious hematopoietic necrosis virus, 139, 141
Infectious poliovirus RNA, 39
Influenza virus, 159—167, 179
  negative strand RNA virus, 160
  replication, 160
  segmented genome, 160
  template RNA synthesis, 160, 163—164
  transcription, 160
  viral gene expression, regulation of, 165—167
  viral mRNA synthesis, 160—163
  virion RNA synthesis, 160, 165—167
Initiation codons, 183

Interferon, 162—163, 167
Intergenic junctions, 146—148, 151
Intramolecular base-pairing, 3
Intramolecular cleavage, 27

J

Japanese encephalitis virus, 82

L

L_A, 201
L_{BC}, 201
L_1, 202, 204
L_2, 201, 204
Leader polypeptides, 29
Leader-primed transcription, 122
Leader RNA, 122, 144, 148, 151—152
Leviviruses, 3, 4, 6
Lipid, 117
Lubrol, 107

M

M_1, 200, 202, 204
M_2, 201, 204
Major capsid polypeptides, 197
mak genes, 201
Marburg virus, 139
Measles virus, 139—140
Membrane-associated replication systems, 36—37
Membrane-bound polypeptide, 38
Membrane bound viral RNA replication complex,
    32, 58
Membraneous replication complexes, 34
Membrane vesicles, 94
Methyl transferase, 98
MHV, see Mouse hepatitis virus
Micrococcal nuclease, 100
Minus strand, 33—36, 41, 99, 103—104
Monoclonal antibodies, 146, 179—180
Monopartite viruses, 51, 93, 96
Morphogenetic cleavage, 29
Mouse hepatitis virus (MHV), 116, 122—123, 130
mRNA, 33, 96, 99, 120—122, 143—144
Mumps virus, 139
Murray Valley encephalitis virus, 82, 84—86
Mutant RNAs, 103
Mutant spectra, 11
Mutation, 13, 16
Mx protein, 162—163, 167
Mycoviruses, see Double-stranded RNA my-
    coviruses

N

Nascent strands, 38, 41
Nascent viral RNAs, 32
Negative strand RNA, 119—120, 150, 160, 167
Neutralizing antibodies, 182—183, 186
Neutralizing antigen, 179

Newcastle disease virus, 139—140
Nicking mechanism, 40
Nonsegmented negative strand RNA viruses, 137—
      152
   defined, 138
   genetic organization, 140—143
   intergenic junctions, 146—148
   morphology, 138—140
   protein composition, 138—140
   transcription, 143—150
      basic phenomena of, 143—148
      models of, 148—150
      roles of viral and cellular proteins in, 144—146
      transcription reaction, 143—144
      viral replication, 150—152
Nonstructural proteins, 98, 118
Northern blot analysis, 119
NSl protein, 163—165, 167
NS29, 184—185
Nucleic acids, 2
Nucleocapsid protein, 117, 130, 139, 145—146, 160,
      162—165
Nucleolytic cleavage, 62
Nucleotide incorporation, 7—11
Nucleotide sequences, 78, 86
Nucleus, 160

O

Oligonucleotide fingerprinting analysis, 128
Oligonucleotide primer, 36, 39
Open reading frames, 120—121, 181
Orthomyxoviruses, 144
Overlapping reading frames, 142

P

PA protein, 160—162
Paramyxoviruses, 138—140, 142, 147
Pathogenesis-related proteins, 95
PBl protein, 160—162
PB2 protein, 160—162
PcV, 197—198
Peptides, 182
Periodate oxidation, 36
pe tl8, 203
Phage-encoded polymerase, 34
Phosphorylation, 33
Phytoreoviruses, 172
Picornaviruses, 24, 29—30, 34, 60—61, 83, 98, 140,
      144
Plants, 93—95
Plant viruses, see Comoviruses; Cowpea mosaic
      virus
Plus strands, 33—36, 41, 99
Polio mRNA, 25
Poliovirus, 29, 60—61
Poliovirus genome
   encoded proteins, 24—27
   functional virus polypeptides, generation of, 24—
      30

infection cycle, survey of, 27—29
models for RNA synthesis and initiation, 37—41,
      see also Poliovirus RNA initiation
replication of, 23—42
   elongation, 41
   experimental observations, in vivo, 30—34
   generation of functional virus polypeptides, 24—
      30
   identification of virus-specific complexes, 30
   initiation of RNA synthesis by protein priming,
      37—39
   in vitro replication, 34—37
   membrane-associated systems, 36—37
   poliovirus proteins associated with, 30—32
   problem of template recognition/initiation at
      different termini, 41
   selected differences, 29—30
   soluble, purified systems, 34—36
   survey of poliovirus infection cycle, 27—29
   viral RNA structures found in infected cells, 33—
      34
   structure, 24—27
Poliovirus RNA
   difference between genomic RNAs of other
      picornaviruses and, 29—30
   elongation, 41
   membrane-associated replication systems, 36—37
   soluble, purified replication systems, 34—36
   viral RNA structures found in infected cells, 33—
      34
Poliovirus RNA initiation, 37—41
Poliovirus RNA polymerase, 33, 37
Poliovirus RNA replication
   host-cell proteins involved in, 32—33
   identification of virus-specific replication
      complexes, 30
   in vitro, 34—37
   in vivo, 30—34
   proteins associated with, 30—32
Polyadenylate, 160
Polyadenylation, 145, 147—148
Polyamine synthetase, 202
Poly(A)tail, 50, 56, 60, 62
Polycistronic mRNAs, 141—142
Polyclonal antibodies, 151
Poly(C) tract, 29
Polypeptides, 34, 85
   leader, 29
   membrane-bound, 38
   poliovirus genome, 24—30
   precursor, 32
   Qβ replication apparatus, 7
   VPg-related, 35
Polyprotein precursors, 72
Polyprotein processing, 24—27, 50, 52—53, 60, 74
Polyribosomes, 33
Polysome-associated virus-specific mRNA, 74, 120,
      127
Positive strand animal viruses, 72, see also
      Alphaviruses; Flaviviruses
Positive-strand plant RNA viruses, 2, 50—52

Posttranscriptional processing model, 122—123
Potato virus Y, 50, 52
Precursor polypeptides, 32
Preinitiated nascent strands, 106
Preinitiated replication complexes, 94
Primer-dependent polio RNA polymerase, 31, 34, 37
Primer for initiation of RNA synthesis, 32—33, 63, 199
Primer-independent, 34
Processing events, 27
Progeny RNAs, 28—29
Promoter, 103—105, 108
Protease protection, 36
Protein composition, 138—140
Protein kinase, 32, 35
Protein NS34, 180—181
Protein NS35, 181
Protein NS53, 180
Protein priming, 37—39
Protein processing, 63, 65
Protein σ_1, 174—176
Protein σ_2, 176
Protein σ_3, 177
Protein σ NS, 177
Protein synthesis requirement for RNA replication, 150
Protein VP2, 186
Protein VP3, 178—179, 186
Protein VP5, 186
Protein VP6, 180
Protein VP7, see Segment 9
Protein VP9, 185—186
Proteolytic cleavage, 178
Pseudoknot, 96
PsV, 197
PsV-F, 197—198
PsV-S, 197—201
Pyrophosphate, 93

Q

Qβ replicase, 1—18, 93, 107—108
    competition among RNA species, 13—16
    complementary strands, interaction of, 11, 13
    infection cycle of RNA coliphages, 3—4
    noninstructed RNA synthesis, 14, 16
    nucleotide incorporation, kinetics of, 7—11
    replication apparatus, 4—5, 7
    replication of viral RNA, 5, 7
    RNA replication mechanism, 7—9, 16
    template dependent, 107
    templates for, 10
Quasispecies, 13

R

Rabies virus, 139
Recombinant DNA, 39
Recombinant virus, 124
Reoviridae, 171—187, see also Segments
Reovirus genes and proteins, 173—177

Replicase, 28, 81, 92—94, 99
Replicase template activity, 104
Replica strand, 5
Replication complexes, 2, 24, 26, 28, 30, 41, 94—95, 98, 108, 126
Replication nucleocapsids, 152
Replication rates, 8, 10—11, 14
Replicative form, 5—6, 33, 86, 125—126
Replicative intermediate, 5—6, 33, 108, 122, 125—126
Respiratory syncytial virus, 138—139, 141
Retention times, 14
Reverse transcriptases, 98
Rhabdoviruses, 138—140, 142, 147
Rhinovirus, 29
Ribosomal protein L3, 202
Ribosome-binding signals, 25
RNA
    amplification, 92—94
    chain elongation, 34, 37
    coliphages, 3—4
    dependent RNA polymerase, 92—95, 98, 106—107, 118, 128—130
    engineering, 92
    genome, 2, 117
    initiation, 34—35, see also Poliovirus RNA initiation
    plant viruses, 92—93
    polymerase, 24, 140—143, 145, 197—201
    polymerases II, 95, 144, 160, 203
    recombination, 127—128, 131
    regulatory proteins, 130
RNA replication complex, 30—32, 58—59
    RNA:RNA base-pairing, 151
    synthesis, 2, 12, 14, 16, 28, 32—33, 37
    viruses, 2, see also specific types
Rotaviruses, 172, 177—187
    cellular location of VP7, 184
    gene structure, VP7, 181—182
    expression and processing of VP7, 183—184
    immunogenicity of expressed VP7, 182—183
    morphogenesis, 178
    other genes, 186—187
    segment 4, 178—179
    segment 5, 180
    segment 6, 180
    segment 7, 180—181
    segment 8, 180
    segment 9, 180
    segment 10, 184—185
    segment 11, 185—186

S

Saccharomyces cerevisiae, 196
SA11, 180—181, 184—185
SAM decarboxylase, 202
Scaffolding proteins, 203
ScV, 197—202, 204
ScV-L$_a$, 197, 201
ScV-L$_1$, 197, 200—201, 203—204

ScV-L$_2$, 197
ScV-M, 200
ScV-M$_1$, 201, 203
ScV transcriptase and replicase, 199—200
Segment L2, 186
Segment L3, 186
Segment M5, 186
Segment S1, 174—176
Segment S2, 176
Segment S3, 177
Segment S4, 177
Segment 4, 178—179
Segment 5, 180
Segment 6, 180
Segment 7, 180—181
Segment 8, 181
Segment 9, 181—184
Segment 10, 184—185
Segment 11, 185—186
Segmented genome, 160
Selection, 13, 16
Selective transport, 167
Self-priming, 37, 40
Self-replicating RNA, 5
Semliki Forest virus, 78—80, 82, 98, 179
Sendai virus, 139—140
Serotype antigen, 181
Serotype-specific antibodies, 183
S14, 205
Shut-off, 24, 28, 86, 138
Signal peptide, 183
Simian virus, 5, 139
Sindbis virus, 72—76, 78—82, 179
Single-stranded RNA, 33, 122
Smooth membrane fraction, 30
Snap-back molecule, 33, 35, 37, 39—41
Soluble, purified replication systems, 34—36, 107
Soluble, template-dependent poly(U) polymerase, 34
SP6 RNA polymerase, 103
spe10, 201, 203
Spring viremia of carp virus, 139
S3, 205
St. Louis encephalitis virus, 82
Strand displacement, 41
Structural proteins, 116—117
Subgenomic mRNA transcription coronavirus, 122—125
Subgenomic RNA synthesis, 99, 105
Subgenomic RNA4, 104—106
Subgroup antigen, 180

T

Temperature-sensitive mutants, 129—130
Template RNA, 40—41, 63, 94—95, 160, 163—164
Template specificity, 32
Terminal transferase, 92, 95, 101
Terminal uridylyl transferase, 32, 33, 35, 39, 65
Thermolabile host protein, 203
Thermostability, 86
Three P proteins, 160—162

Tobacco mosaic virus (TMV), 50, 52, 81, 92, 93, 96—98
Tobacco streak virus, 93
Tomato black ring virus, 50
Transcriptases, 198—199, 204
Transcription, 63, 93, 99, 118, 126, 127, 143—150, 160, 204
Transcriptional mapping, 119
Transcription nucleocapsids, 152
Translation, 55
Transmembrane protein, 185
Tripartite genome viruses, 51—52, 93, 96
tRNA-like properties, 96—98, 105, 108—109
Tryptic cleavage sites, 179
Tubulin, 146
Turnip yellow mosaic virus (TYMV), 50, 52, 91—96, 106—113
12-M, see Dodecyl-β-D-maltoside
Tyrosine, 96
Tyrosylation, 96, 103—104

U

Ultraviolet (UV) mapping, 141—142
UmV, 197—198, 204
Unlinking enzyme, 27
Unprimed initiation, 163, 167
Ustilago maydis, 196

V

Vaccinia virus, 99, 182—183
Variant, 5
Vesicular stomatitis virus, 99, 130, 138—140, 148—152, 164, 167
Vesicular stomatitis virus transcriptase, 143—144
Viral cap-dependent endonuclease, 160
Viral dsRNA sequences required in *cis*, 203—205
Viral gene expression, regulation of influenza virus, 165—167
Viral gene products, 98—99, 203
Viral mRNA synthesis, 160—163
Viral nucleocapsid, 139, 160—161
Viral particles, 203
Viral phosphoprotein, 145—146
Viral protease, 29
Viral proteinases, 25, 27
Viral proteins, 144—146
Viral protein 3CD, 31
Viral protein 3D^{pol}, 31
Viral replicase complexes, 28
Viral replication, 118
Viral RNA, 3, 5, 7, 13, 33—34, 93, 94, 96, 98, 99
Viral RNA-dependent RNA polymerase, 2
Viral RNA-specific replicase, 58, 95
Viral-specific RNA-dependent RNA polymerase, 99
Viral toxin, 200
Viral transcription, see Transcription
Virion RNA synthesis, 33, 99, 160, 165—167
Viroid RNA templates, 95
Virulent Asibi strain, 86

Virus encoded protease, 54—55, 74, 82—83
Virus-encoded proteins, 92—93
Virus-specific replication complexes, 30
Virus-specific RNA polymerase, 30
VPg, see Genome-linked viral protein
VPg-linked template RNA, 35
VPg-related polypeptides, 35
VPg uridylylation, 37
VP7, see Segment 9

W

West Nile virus, 82—86
Wild-type virus, 152
Wound tumor virus gene segments, 186—187

Y

Yellow fever virus, 82—86

Printed and bound by CPI Group (UK) Ltd, Croydon, CR0 4YY

22/10/2024

01777630-0007